Bionanomaterials for Environmental and Agricultural Applications

Bionanomaterials for Environmental and Agricultural Applications

Edited by
Ravindra Pratap Singh
Department of Biotechnology, Indira Gandhi National Tribal University, Madhya Pradesh, India

Kshitij RB Singh
Department of Chemistry, Govt. V. Y. T. PG. Autonomous College, Chhattisgarh, India

IOP Publishing, Bristol, UK

ISBN 978-0-7503-3863-9 (ebook)
ISBN 978-0-7503-3861-5 (print)
ISBN 978-0-7503-3864-6 (myPrint)
ISBN 978-0-7503-3862-2 (mobi)

DOI 10.1088/978-0-7503-3863-9

Version: 20211101

IOP ebooks

British Library Cataloguing-in-Publication Data: A catalogue record for this book is available from the British Library.

Published by IOP Publishing, wholly owned by The Institute of Physics, London

IOP Publishing, Temple Circus, Temple Way, Bristol, BS1 6HG, UK

US Office: IOP Publishing, Inc., 190 North Independence Mall West, Suite 601, Philadelphia, PA 19106, USA

Dedicated

to those who

care, conserve, protect

but do

not spoil/destroy the beauty

and

unique characteristics

of

Kshit (Earth)

Jal (Water)

Pawak (Fire)

Gagan (Sky) and

Sameera (Air)

Contents

Preface

The capability of developing new materials at a nanoscale level (1–100 nanometers) offers exhilarating advancements in various domains like biomedical, agricultural, environmental, industrial, etc. Nanomaterials exhibit distinctive properties, namely mechanical, optical, etc, compared to microscopic and macroscopic structures. Recent technological advancements like neutron and optical spectroscopy, electron and scanning probe microscopy have enhanced the designing of sophisticated experimental methodologies, which have led to rapid development towards the wide exploration of materials at the nanoscale range. The term nanotechnology is derived from the *Greek* word *nano*, meaning dwarf, and is one billionth of a meter. This term was coined in 1974 by Norio Taniguchi. The nanotechnology field is an emerging and innovative field of science, which deals with varied applications of various materials (chemical, physical, and biological) and devices for building, controlling, and structuring at the nanoscale level. The advancement in the nanotechnological field is at a faster pace, and to promote nanotechnology, in 2001, the US Federal Government introduced the National Nanotechnology Initiative (NNI) to boost nanotechnological advancements as such material at a nanoscale level exhibits unique properties and functions.

The major factor in nanotechnology is nano-based materials, which are called nanomaterials. Nanomaterials can be made up of synthetic or biological materials; a nanomaterial made up of molecular material which is collectively composed of partial or non-partial biological molecules, namely antibodies, proteins, enzymes, genetic material (DNA/RNA), lipids, oligosaccharides, viruses, etc are commonly known as bionanomaterial. Furthermore, bionanomaterial shows potential applications as sensors, adhesives, fibers, etc and can also be utilized to fabricate complex devices through self-assembling in eco-friendly and mild conditions. Several research areas like immunology, physics, mathematics, chemistry, biochemistry, materials science, and engineering focus on and play crucial roles in this area. For example, by utilizing recombinant DNA techniques, biochemists can isolate, characterize, and engineer biomolecules (proteins, antibodies, etc), which are vital components in bionanomaterial research and development.

Nanotechnology is identified as one of the six 'Key Enabling Technologies' by the European Commission as it has the potentiality to play an important role in sustainable growth in various domains. Nanotechnology has advanced so much that as a tool, it can handle the most critical issues faced globally related to water, energy, health, agriculture, and biodiversity (WEHAB)-related problems, and these five issues are very critical having been identified in the 2002 Summit for Sustainable Development at United Nations (UN), Johannesburg. Moreover, as per a survey conducted by the UN, it was identified that by utilization of nanotechnology, energy conversion and storage might be attained first; secondly, the most important feature that might be attained is millennium development goals for agricultural yield enhancement in developing countries, and water treatment is the third vital area which nanotechnological applications can achieve. In addition to this, nanotechnology has evolved over the past two decades and has shown incredible progress in the design, innovation, creation, and novel utilization of artificial nanomaterials to meet the major challenges in environmental sustainability; these nanomaterials in various ways are generating numerous substantial applications in the environmental domain. Currently, environmental pollution and shortage of energy are the two major global challenges faced by human civilization. In trying to recover our ailing environment, nanotechnology researchers and developers are approaching various avenues like using silver nanoparticles for creating low pollution during materials manufacturing, silicon-nanowires for producing solar cells, carbon-nanotubes to increase the electricity generated by windmills, etc. Apart from these nanoparticles, the bionanomaterials can be more beneficial than these materials for utility in the environment.

On the other hand, the bionanomaterials have too much potentiality for an agricultural application like in the food industry, pesticide detection, etc. The need for the situation is to excel in agriculture as there are limited water and land resources; and to excel in this domain agricultural growth must be targeted, which can be obtained by increasing per unit natural resources productivity and farm income through judicious use of advanced technologies. As in the era of the green revolution, to increase the potential yields and farm incomes in a short time. Fertilizers, insecticides, and pesticides were intensively utilized, and sustainability and efficient utilization of natural resources were neglected. This was very beneficial at that time, but the intensive use of fertilizers, pesticides, etc, resulted in environmental pollution and end-product yield, which was harmful to human health. Apart from this, the need of the hour is to focus on technologies that can increase agricultural production and product quality, and for overcoming these issues nanotechnological advancements have given so many solutions, among which bionanomaterials have played a vital role in the agricultural sector. Furthermore, advanced physical techniques, like electron/helium ion microscopic techniques and nanofabrication procedures, are very useful in manufacturing and analyzing nano-sized devices. By utilizing these sophisticated technologies, the role of bionanomaterials in the biomedical domain has been extensively studied. But there is very scanty literature data related to the role of bionanomaterials in the agriculture and environment fields compared to biomedical. Therefore, this creates an opportunity for researchers to explore the potentiality of bionanomaterials in the environmental, agricultural, and industrial domains.

This book, 'Bionanomaterials for Environmental and Agricultural Applications,' will be an outstanding collection of reviews based on current research and developments on bionanomaterials utilization in environmental and agricultural applications, along with their prospects. Here in this book, we have provided an extensive concept of bionanomaterials' limitations and applications in the environmental and agricultural context to help students and professionals to understand the importance of bionanomaterials applications in these two domains. This will also help them understand and address problems in the agricultural and environmental sectors by utilizing bionanomaterials. Hence, this book will serve as a textbook for students and will also be of great importance for professionals, scientists, researchers, and academicians in this research field.

Aims and scope

This book will help attract a wide range of readers from all the fields by introducing the bionanomaterials' limitations and highlighting their agricultural and environmental applications; the major points this book will cover are listed in the facets below.

- Bionanomaterials' potentialities for antimicrobial applications.
- Various bionanocomposites for agricultural and environmental applications.
- Various bionanomaterials for environmental and agricultural applications.
- Nanobiosensors' utility for various environmental and agricultural aspects, such as monitoring and analysis of the environment; pesticides, insecticides, and other harmful agricultural components detection, and many other aspects pertaining to these two emerging domains.

Ravindra Pratap Singh
Kshitij RB Singh

Acknowledgements

It gives us immense pleasure to acknowledge Professor Shri Prakash Mani Tripathi, Honorable Vice-Chancellor, Indira Gandhi National Tribal University, Amarkantak, India, Professor Ajaya Kumar Singh, Department of Chemistry, Govt. V. Y. T. PG. Autonomous College, Durg, Chhattisgarh, India, and Dr Jay Singh, Department of Chemistry, Institute of Science, Banaras Hindu University, Varanasi, Uttar Pradesh, India for providing constant assistance in all the possible ways. It is also our great pleasure to acknowledge and express our enormous debt to all the contributors who have provided their quality material to prepare this book. We are grateful to our beloved family members (Dr Babita Singh and Aparajita RB Singh), who joyfully supported and stood with us in many hours of our absence to finish this book project. Thanks, are also due to Caroline Mitchell, Robert Trevelyan, and the entire publishing team for their patience and extra care in publishing this book.

Ravindra Pratap Singh
Kshitij RB Singh

Editors biography

Dr Ravindra Pratap Singh

Dr Singh did his BSc from Allahabad University India and his MSc and PhD in Biochemistry from Lucknow University, India. He is currently working as an Assistant Professor in the Department of Biotechnology, Indira Gandhi National Tribal University, Amarkantak, Madhya Pradesh, India. He has previously worked as a scientist at various esteemed laboratories globally, namely Sogang University, IGR, Paris, etc. His work and research interests include biochemistry, biosensors, nanobiotechnology, electrochemistry, material sciences, and biosensors applications in biomedical, environmental, agricultural and forensics. He has to his credit several reputed national and international honors/awards. Dr Singh has authored over 40 articles in international peer-reviewed journals, three edited books and more than 20 book chapters of international repute, and he serves as a reviewer of many reputed international journals and is also a member of many international societies. He is currently also involved in editing various books, which will be published in internationally reputed publication houses, namely IOP Publishing, CRC Press, Elsevier, Wiley, and Springer Nature. He is also a book series editor of 'Emerging advances in Bionanotechnology,' CRC Press, Taylor and Francis Group. Moreover, he has successfully edited a special issue on 'Smart and Intelligent Nanobiosensors: Multidimensional applications' as a Guest managing editor for *Materials Letters*, Elsevier.

Mr Kshitij RB Singh

Mr Singh is a Postgraduate in Biotechnology from Indira Gandhi National Tribal University, Amarkantak, Madhya Pradesh, India. He is currently working in the laboratory of Dr Ajaya Singh, Department of Chemistry, Government V.Y.T. PG. Autonomous College Durg, Chhattisgarh, India. He has many publications to his credit, has edited three books, and has authored more than ten book chapters published in the internationally reputed press, namely Elsevier, IOP Publishing, Springer Nature, Wiley, and CRC Press. He is currently also involved in the editing of books with international publishing houses, including CRC Press, IOP Publishing, Elsevier, and Springer Nature. His research interest is in biotechnology, biochemistry, epidemiology, nanotechnology, nanobiotechnology, biosensors, and materials sciences.

List of contributors

Wan Nazwanie Wan Abdullah
School of Chemical Science, Universiti Sains Malaysia, 11800 Minden, Pulau Pinang, Malaysia

Charles Oluwaseun Adetunji
Applied Microbiology, Biotechnology and Nanotechnology Laboratory, Department of Microbiology, Edo State University Uzairue, Edo State, Nigeria

Uyiosa Osagie Aigbe
Department of Mathematics and Physics, Faculty of Applied Sciences, Cape Peninsula University of Technology, PO Box 1906, Cape Town, South Africa

Onoyivwe Monday Ama
Department of Chemical Engineering, Vaal University of Technology, PMB X021, Vanderbijlpark 1900, South Africa

Vijay Bhaskar Reddy Ambavaram
Centre of Research in Ionic Liquids, Universiti Teknologi, PETRONAS, Seri Iskandar, Malaysia

Gilberto José de Arruda
Postgraduate Program in Natural Resources, State University of Mato Grosso do Sul, Rodovia Dourados Itahum Km 12, Caixa postal 351, (79804-970), Dourados, MS, Brazil

Samuel Ogochukwu Azi
Department of Physics, Faculty of Physical Sciences, University of Benin, P.M.B. 1154, 300283, Benin City, Edo State, Nigeria

Stephen Rathinaraj Benjamin
Department of Physiology and Pharmacology, Faculty of Medicine, Drug Research and Development Center (NPDM), Federal University of Ceará, Rua Cel. Nunes de Melo 1127, Porangabussu, (60430-270), Fortaleza, Ceará-CE, Brazil

Ayodotun Oluwafemi Bobadoye
Forestry Research Institute of Nigeria (FRIN), P.M.B. 5054, Ibadan, Nigeria

Merve Çalışır
Department of Chemistry, Hacettepe University, Ankara, 06800, Turkey

Adil Denizli
Department of Chemistry, Hacettepe University, Ankara, 06800, Turkey

Joseph Onyeka Emegha
Department of Physics, Faculty of Physical Sciences, University of Benin, P.M.B. 1154, 300283, Benin City, Edo State, Nigeria

Ese Lawrence Esiekpe
Department of Physics and Computer Education, Federal College of Education (Technical), PM.B. 1044, Asaba, Delta State, Nigeria

Ilgım Göktürk
Department of Chemistry, Hacettepe University, Ankara, 06800, Turkey

Abel Inobeme
Laboratory for Reproductive Biology and Developmental Programming, Department of Physiology, Edo State University Uzairue, Edo State, Nigeria

Umesh Kumar
School of Nano Sciences, Central University of Gujarat, Sector-30, Gandhinagar, Gujarat, India

Cansu İlke Kuru
Graduate School of Natural and Applied Sciences, Department of Biotechnology, Ege University, 35100, İzmir, Turkey, and
Buca Municipality Kızılçullu Science and Art Center, 35390, Izmir, Turkey

Heri Septya Kusuma
Department of Chemical Engineering, Faculty of Industrial Technology, Universitas Pembangunan Nasional 'Veteran' Yogyakarta, Indonesia

Fábio de Lima
Postgraduate Program in Natural Resources, State University of Mato Grosso do Sul, Rodovia Dourados Itahum Km 12, Caixa postal 351, (79804-970), Dourados, MS, Brazil

A Manjceevan
Department of Chemistry, University of Jaffna, Jaffna, NP 40000, Sri Lanka

Eneyew Tadesse Melaku
Department of Microbial and Food Technology, Addis Ababa Science and Technology University, Addis Ababa, Ethiopia

Elaine Gabutin Mission
Instituto de Bioeconomia de la Universidad de Valladolid, Valladolid, Spain

Nurul Najidah Mohamed
Unisza Science and Medicine Foundation Centre, Universiti Sultan Zainal Abidin, Gong Badak Campus, 21300 Kuala Nerus, Terengganu, Malaysia

Nur Atiqah Nasir
Faculty of Science and Technology, Universiti Sains Islam Malaysia, 71800, Nilai, Negeri Sembilan, Malaysia

Vanya Nayak
Department of Biotechnology, Indira Gandhi National Tribal University, Amarkantak, M.P., India

Ankur Ojha
Department of Food Science and Technology, National Institute of Food Technology Entrepreneurship and Management, Sonipat, Haryana, India

Benedict Okundaye
School of Art, Design, and Architecture, University of Huddersfield, Queensgate Huddersfield, UK HD1 3DH

Olugbemi T Olaniyan
Department of Chemistry, Edo University Iyamho, PMB 04, Auchi, Edo State University Uzairue, Edo State, Nigeria

Robert Birundu Onyancha
Department of Physics and Space Science, School of Physical Sciences and Technology, Technical University of Kenya, PO Box 52428—00200, Nairobi, Kenya

Otolorin Adelaja Osibote
Department of Mathematics and Physics, Faculty of Applied Sciences, Cape Peninsula University of Technology, PO Box 1906, Cape Town, South Africa

Peter Ogbemudia Osifo
Department of Chemical Engineering, Vaal University of Technology, PMB X021, Vanderbijlpark 1900, South Africa

Pratibha Pandey
Department of Chemistry, Hansraj College, University of Delhi, Delhi, India

Shweta Rathee
Department of Food Science and Technology, National Institute of Food Technology Entrepreneurship and Management, Sonipat, Haryana, India

Eli José Miranda Ribeiro Junior
Faculty of Inhumas (FacMais), Avenida Monte Alegre, n° 100 Residencial - Monte Alegre, Inhumas, (75400-000), Goiás-GO, Brazil

Salmiah Jamal Mat Rosid
Unisza Science and Medicine Foundation Centre, Universiti Sultan Zainal Abidin, Gong Badak Campus, 21300 Kuala Nerus, Terengganu, Malaysia

Sarina Mat Rosid
Advanced Membrane Technology Research Centre (AMTEC), Universiti Teknologi Malaysia, 81310 UTM Skudai, Johor, Malaysia

Israel Uzuazor Siloko
Department of Mathematics, Edo State University Uzairue, P.M.B. 04, Auchi, 312101, Edo State, Nigeria

Jay Singh
Department of Chemistry, Institute of Science, Banaras Hindu University, Varanasi, U.P., India

Kshitij RB Singh
Department of Chemistry, Govt. V. Y. T. PG. Autonomous College, Durg, Chhattisgarh, India

Ravindra Pratap Singh
Department of Biotechnology, Indira Gandhi National Tribal University, Amarkantak, M.P., India

Deniz Türkmen
Department of Chemistry, Hacettepe University, Ankara, 06800, Turkey

Kingsley Eghonghon Ukhurebor
Department of Physics, Edo State University Uzairue, P.M.B. 04, Auchi, 312101, Edo State, Nigeria

Fulden Ulucan-Karnak
Graduate School of Natural and Applied Sciences, Department of Biomedical Technologies, Ege University, 35100, İzmir, Turkey

Tanvi Vats
Department of Applied Chemistry, School of Vocational Studies and Applied Sciences, Gautam Buddha University, Greater Noida, India

K Velauthamurty
Department of Chemistry, University of Jaffna, Jaffna, NP 40000, Sri Lanka

Madhavi Vemula
BVRIT Hyderabad College of Engineering for Women, Hyderabad, India

Gunjan Vyas
School of Nano Sciences, Central University of Gujarat, Sector-30, Gandhinagar, Gujarat, India

Fatma Yılmaz
Vocational School of Gerede, Department of Chemistry Technology, Bolu Abant Izzet Baysal University, Bolu, Turkey

Chapter 1

Introduction: potentialities of bionanomaterials towards the environmental and agricultural domain

Kshitij RB Singh, Vanya Nayak, Jay Singh, Charles Oluwaseun Adetunji and Ravindra Pratap Singh

Bionanotechnology combines biotechnology and nanotechnology, which utilizes bionanomaterials, nanosized materials made up of various biological entities. Bionanomaterials are referred to as state-of-the-art materials as they can be utilized in multiple domains like aeronautics, agriculture, environment, energy, etc. They are also known to cover the gap between molecular approaches and traditional microtechnology. Several bionanomaterials have been utilized for applications in agriculture and environment domains for air, water and soil pollutant detection, remediation of heavy metals, and act as biopesticides, insecticides, etc, to prevent crop and soil damage. Hence, this chapter introduces various potentialities of bionanomaterials in agricultural and environmental domains.

1.1 Introduction

The exciting enhancement in science and technology is the biggest gain to humankind. Its potentiality to maneuver new materials of different dimensions has created a revolution in the science domain, which has eased the lives of human beings, and nanotechnology remains one of the biggest contributors to this revolution. Nanotechnology is an advanced science domain and is considered one of the cleanest technologies that study the properties, synthesis, and applications of nanosized (1–100 nm) materials [1]. The possibility to enhance a material's potentiality at the nano level has grabbed the attention of researchers and has, therefore, successfully attracted global success in almost all domains. The nanosize of the material helps them exhibit unique properties like optical, chemical,

doi:10.1088/978-0-7503-3863-9ch1

biological, physical, structural, mechanical, electrical, and catalytic, that distinguish them from the bulk matter.

Moreover, the exclusive fusion of biotechnology with nanotechnology has given rise to the new multi-disciplinary field known as bionanotechnology, aiming to fill the gap between traditional micro-technology and molecular approaches. This powerful combination of two strong fields of science and technology is expected to uplift material potential and eliminate the gap present between different science fields like physics, chemistry, biology, etc. A nanosized material that is composed of various biological molecules like nucleic acids, metabolites, proteins, antibodies, lipids, etc, is known as a bionanomaterial. However, there are some synthetic bionanomaterials like peptide nucleic acid (PNA), xeno-nucleic acid (XNA), aptamers, etc, that mimic the properties of biological nanomaterials [2]. These bionanomaterials are state-of-the-art materials that have found their profound use in different domains like agriculture, environment, biomedical, veterinary sciences, pharmacology, sensors, etc. Recently, these bionanomaterials have drawn significant attention in the agricultural field as the agriculture domain is considered as one of the most important and stable sectors, as it produces raw materials for food and feed industries and, therefore, development in the agriculture field is necessary and important [3–5]. Thus, these bionanoparticles enhance traditional farming practices like the development of nanopesticides, nanoinsecticides, and nanofertilizers, enhancing crop production and not polluting the environment.

Similarly, various bionanoparticles are nowadays being utilized to develop various types of sensors that can monitor soil and water quality in the agriculture field to maintain the proper growth of the plant. Moreover, these bionanomaterials are also used to detect pollutants from the environment, used for dye removal, wastewater treatment, bioremediation, etc and, therefore, can be utilized in the environment domain. Hence, this chapter of this book entitled 'Bionanomaterials for Environmental and Agricultural Applications' will highlight bionanomaterial's basic utility in the agricultural and environmental domains along with the prospects and recent trends, and the details of this chapter are illustrated in figure 1.1. Further, this book will elaborate on the antimicrobial utility of metal and metal oxide-based bionanomaterials in **chapter 2**, followed by **chapter 3, chapter 4, chapter 5**, and

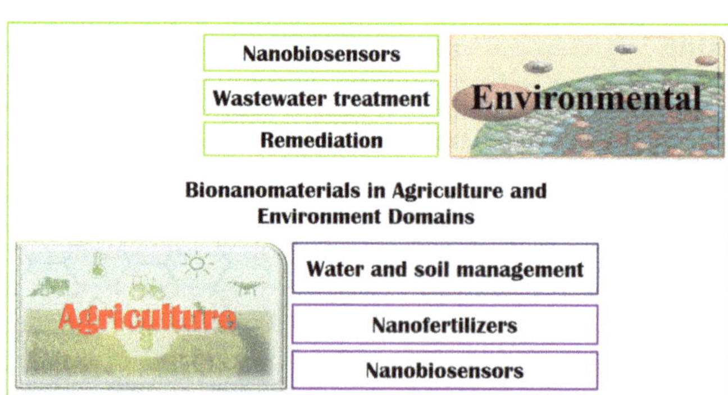

Figure 1.1. Diagrammatic illustration of the details discussed in the chapter.

chapter 6 about bionanocomposites, nanobiosensor, bionanopesticides, and bionanoformulations for various applications in the agricultural sector. Furthermore, the utility of bionanocomposites, metal and metal oxide-based bionanocomposites, and bionanocomposite hydrogels for wastewater treatment is discussed in **chapter 7, chapter 8, and chapter 9**. **Chapter 10** will deal with bionanomaterials' utility in the food sector, followed by **chapter 11**, which deals with polymeric bionanomaterial's potentialities in the agricultural and environmental domains. Further, **chapter 12, chapter 13, and chapter 14** of this book will deal with the role of bionanomaterials and nanobiosensors for environmental remediation, risk assessment, sustainability and safety.

1.2 Utilities of bionanomaterial for agriculture

As the population, industrialization, and globalization are increasing rapidly, the demand for food production is also growing, and it is predicted that the world would face food scarcity by 2050; therefore, the importance of agriculture has now increased. Agriculture is the domain that provides raw materials to the food-based industries and is the most important sector as it also helps increase the country's economy. But due to the increase in the use of chemical-based pesticides, insecticides, fertilizers, and composites, the agricultural land and water have got polluted, which has affected plants, human beings and animals in certain ways. Therefore, researchers have focused on increasing crop production by using minimum chemicals and land areas and developing sustainable resources for agricultural practices. Bionanotechnology has contributed greatly to develop new and advanced technologies in the agriculture domain as it uses eco-friendly, rapid and cost-effective technology [1]. The potentiality of nanotechnology in agriculture aims to decrease the excessive use of chemicals and enhance nutrient intake and productivity. The bionanomaterials are derived from various biopolymers like proteins and carbohydrates, which negatively impact both human health and the environment. Bionanomaterials are known to affect plant growth in many ways and are, therefore, highly known to be utilized in the agricultural field to increase crop yield, as shown in figure 1.2. Some bionanomaterials are known to combat the harmful effects of insects, decrease the impact of toxic herbicides, fungicides and insecticides, deliver the required gene in the cell to develop a good yield plant, and help in the detection of nutrients and pollutants present in the soil [6, 7]. Many metallic bionanoparticles

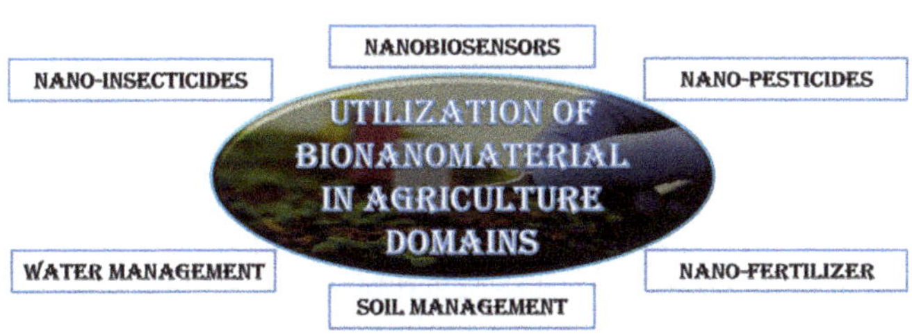

Figure 1.2. Schematic representation of various utilities of bionanomaterials in agriculture domains.

stimulate seed germination and seed growth in many plants like wheat, corn and spinach [8–12].

Further, bionanotechnology has also shown its potentialities in the food industry as it possesses a great potential in ensuring modification and enhancement of flavor, color and nutritional values, increasing the shelf life of food, and monitoring the integrity of food through barcodes such as cold chain, that is, whenever there is a change in the food storage conditions it will be detected [13]. Therefore, bionanotechnology is considered one of the emerging areas in science and technology that carries the potential to develop radically new products and processes in the food sector by using bionanomaterials. It is commonly distinguished between two forms of applications: food additives, also known as nano inside and food packaging or nano outside. [14]. Food processing primarily deals with modifying agricultural and animal products into food and processing the food into edible form. Bionanomaterials help improve food processing through smart delivery of nutrients, nanoencapsulation of nutraceuticals, bioseparation of proteins, and rapid biological and chemical contaminant determination [15]. Similarly, various nanobiosensors have been developed utilizing bionanomaterials to detect traces of pesticides and microorganisms in the food and monitor the food's quality and safety. These sensors showed high sensitivity and excellent response time [16]. For this purpose, carbon nanotubes, semiconductors and metal bionanomaterials have been utilized as they exhibit exceptional electrical or optical characteristics [17]. Moreover, several bionanodelivery vehicles have been developed, like lipid-based nanoencapsulators or bionanocarriers, bionanoemulsions, bionanolaminates, biopolymeric nanoparticles, etc. These bionanodelivery systems can greatly enhance the bioavailability of bioactives through various pathways [18]. Various bionanocomposite materials are utilized in active packaging to avoid oxygen, carbon dioxide, and moisture from reaching the food. Similarly, various antimicrobial agents are also utilized in packaging to prevent the entry and growth of pathogenic microorganisms, which can directly influence the shelf life of products and is considered more beneficial than the direct addition of some antimicrobial agents onto foods through drips and sprays [19]. Furthermore, a detailed overview of bionanomaterial's utility in the food industry is discussed in **chapter 10**.

Since bionanomaterials are known to show antimicrobial activity due to their extraordinary biological properties, they have been utilized in the agricultural domain to reduce the growth of harmful bacteria on plants that hampers their growth resulting in a decrease in the yield of the crop. For this purpose, metal bionanomaterials like green synthesized silver nanoparticles (AgNPs) have been used against the nitrogen-fixing bacteria, which directly affected the cell wall and cytoplasm of the bacterial cell, hindered the respiration system and bacterial DNA replication system, dysfunctioning the cytoplasmic membrane from the cell wall, and obstructed the ATP levels [20–22]. These AgNPs efficiently reduced the chances of developing a tumor caused by the bacteria, making the crop bacteria-free. Moreover, besides bacteria, these metal-based bionanoparticles are known to exhibit antifungal activity against various plant pathogenic fungi like *Bipolaris sorokiniana, Magnaporthe grisea*, etc [23].

Many bionanomaterials like proteins, carbon derivatives, polymers, semiconductor quantum dots (QDs), silicates, ceramics, emulsions, dendrimers, lipids, metal and metal oxides show extreme potentialities to be used as an alternative for chemical insecticides and pesticides for natural based pest control [24–26]. It was observed that the green synthesized silver nanoparticle (AgNPs) could act as ovicides and adulticides when observed in some pest species like the cotton bollworm and *Helicoverpa armigera* (Hübner). Similar observations were observed when green synthesized AgNPs were treated on the larvae of the mosquitoes *Anopheles subpictus* and *Culex quinquefasciatus* [27]. These nano-biopesticides can be utilized to develop a sustainable agriculture system to maintain and preserve the quality of the natural resources and the environment for the next generation. These biopesticides made up of bionanomaterials are safe to use for both living beings and the environment [28]. Sustainable agriculture aims to develop different techniques and technologies that are effective and create a minimum environmental footprint. RNA-based biopesticides are currently gaining much attention as a narrow-spectrum alternative to chemical-based pesticides, with enhanced specificity and accuracy towards targeting pests and pathogens. They are developed by using RNA interference (RNAi), a eukaryotic process in which transcript expression is reduced in a sequence-specific manner, and therefore can be utilized to control and target various plant pests and pathogens. Double-stranded RNA (dsRNA) acts as the key trigger molecule of RNAi, as it provides protection to RNAi without integrating dsRNA-expressing constructs as transgenes. A biopesticide was developed using the RNAi approach to kill western corn rootworm (*Diabrotica virgifera virgifera* LeConte), which was orally-applied through artificial diet or on transgenic maize. A similar approach was used to impair the growth of cotton bollworm (*Helicoverpa armigera*) by feeding plant leaf material that expresses a dsRNA specific to a cytochrome P450 gene. Moreover, RNAi-based biopesticides have also been used to control various plant pests and pathogens like, fungi such as powdery mildew, bacteria such as *Agrobacterium*, and nematodes such as root knot nematodes [29].

The European Commission has recognized green nanotechnology and termed it the 'key enabling technologies,' which has modified and revolutionized the concept of sustainable agriculture and acts as a paradigm shift in precision agriculture. In this context, nanofertilizers have gained significant attention due to the serious drawbacks of conventional fertilizers. It was found that only 20%–50% of nitrogen, 35%–40% potassium, 10%–25% of phosphorous were taken up by the plant, and the rest of the fertilizer was leached. Further, the conventional fertilizers also led to the degradation of porosity of the soil, making it infertile, disturbing the soil mineral balance, generation of greenhouse gases, and eutrophication. Therefore, to overcome these drawbacks, the utilization of nanofertilizers was initiated that developed smart delivery methods of the nutrients which prevented the excess loss of the fertilizers, and is eco-friendly, cost-effective and can be utilized for sustainable agriculture. It was observed that when chitosan and carbon nanotubes were used as nanofertilizers on French beans, they showed improved water absorption, better nutrient uptake and plant growth [30]. Similarly, when zinc oxide and titanium dioxide bionanoparticles were utilized as nanofertilizer, they showed better fruit and dry biomass yield, improved nutrient quality and enhanced fruit development and chlorophyll content in tomatoes [31].

Moreover, these bionanomaterials also help remove toxic pollutants from the soil, which, if left, can lead to soil degradation, increased soil porosity, decreased nutrient content, and cause damage to both plant and other living beings by accumulating in the food chain. Some pollutants like organic solvents, pesticides, pharmaceuticals, personal care products (PPCPs), polychlorinated biphenyls (PCBs), polycyclic aromatic hydrocarbons (PAHs), etc, can be easily and rapidly reduced, photodegraded, or adsorbed from the soil after it has been injected with bionanomaterials. Certain integrated systems are being utilized, like metal ions with *Sphingomonas* sp. to degrade the gamma-hexachlorocyclohexane (γ-HCH) in just six days. Further, certain degradation products like 1,2,3,4,5-pentachlorocyclohexene, 2,4-trichlorobenzene are effectively degraded through anaerobic procedures by bacteria by converting them into benzene [32]. Further, these bionanomaterials help increase the water retention and absorption of the soil, leading to sustainable agriculture, which helps solve the future drought problems and the downfall of the water table. Superabsorbent polymers are well known to hold a large volume of water, and when given a particular condition, they can also desorb moisture. It was observed that the hydrogel based on poly(acrylamide-co-acrylic acid)/$AlZnFe_2O_4$ nanocomposite showed excellent water holding capacity when observed on the sandy loam soil. It was also observed that it also enhanced the seed germination and growth of the wheat plants [33].

The nanosensors are utilized in the agricultural domain to detect soil quality, manage diseases, detect contaminants, proteins, DNA, etc. These nanosensors are rapid, highly sensitive, and with a small response time, but with the addition of bionanomaterials, the sensitivity, specificity and rapidity of a nanosensor are efficiently enhanced [34–36]. Various bionanomaterials have been utilized to develop nanosensors that can detect pesticide residues, such as humic acid-capped Ag NPs were used to detect sulfurazon-ethyl (herbicide) at 100 ppm of detection level [37]. Nowadays, fluorescent nanosensors are utilized over the colorimetric sensors because they are rapid and show high sensitivity, and specificity helps them quickly detect the pesticides present in the soil [38–40]. Moreover, fluorescent liposome-based bionanomaterials are used to detect organophosphorus pesticides, like dichlorvos and paraoxon [41].

Similarly, quantum dot-based nanosensors were also used to detect plant pathogens like *Citrus tristeza* [42]. Moreover, polypyrrole nanoribbon-based nano-sensor covered with antibodies shows the capability to detect various viruses in a broad range of 10 to 100 $\mu g\ mL^{-1}$ [43]. Another innovative use of nanosensors is developing a wireless sensor network (WSN) that can determine the moisture levels in soil along with the water levels present in the field. These can be placed at the root level to get accurate moisture levels at the root system. These sensors are helpful as they help in controlling water usage and build a sustainable water preservation system [44, 45]. It was also observed that when these sensors were integrated with the drip irrigation in the potato, tomato, and plum trees field, they efficiently showed a positive effect and reduced the overflow of water and can be very useful in drought-affected areas [46–49].

The increase in globalization, population, and industrialization has led to the depletion of water bodies and agricultural land; therefore, for a few decades, the reuse of treated domestic wastewater for agricultural irrigation has increased to end water scarcity and improve means for local food production [50]. The bionanomaterial treated wastewater is used to plant various crops, vegetables, orchards, and irrigation. For this purpose, various nanotechnological methods are utilized that enhance the water quality. The wastewater contaminants are removed using various bionanomaterials like ZnO and TiO2 nanoceramics to remove the *Escherichia coli* from the wastewater [51, 52]. Similarly, heavy metals like cadmium and mercury were removed with the help of nanochitosan and nano-hydroxyapatite (nHA) [53, 54]. Another method, named the micro-nano bubble (MNB), is highly used to utilize the nanotreated wastewater in irrigation as it uses the modified dip technologies, which generate small bubbles of size ranging from 200 nm to 50 μm. This technique helps enhance and modify the irrigation system by enhancing the waste water quality with the help of MNB water, which is produced by a bubble generator. It also promotes the increase in soil oxygen levels and improves the supply of oxygen in the root area and hence eases the root respiration, improves photosynthetic activity, enhances crop water absorption and fertilizer consumption, etc. Research has shown the positive impact of utilizing this technique to grow various crops like rice, tomato, and cucumber [55–57]. Moreover, more detailed overviews of bionanocomposites, nanobiosensors, bionanopesticides, and bionano-formulations potentialities for enormous agricultural applications are provided in **chapter 3**, **chapter 4**, **chapter 5**, and **chapter 6**.

1.3 Utilities of bionanomaterial for the environment

Energy shortage, along with environmental pollution, has become a major global challenge, and with the rapid increase in population, globalization and industrialization, it has now become an urgent need to overcome these challenges, or the planet could face a global crisis of natural resources by 2050. Currently, researchers are focusing on developing sustainable materials that can ease human lifestyle and simultaneously do not cause harm to the environment. For these purposes, bionanotechnology has played a commendable role in utilizing nanoscaled materials like biosensors to detect pollutants in the air, water and soil, bioremediation methods, and wastewater treatment. The unique design, synthesis approaches and modifications of the materials help them find their potentialities in the environment domain [58]. With the rapid advancement in catalysis and material sciences, it has become easy to establish a bionanomaterials' structure-activity and controlled synthesis relationship; these bionanomaterials find a considerable number of applications in the field of environment (figure 1.3). Currently, the supply and storage of freshwater are considered of major concern because of the depletion of freshwater bodies due to the increase in the contamination of water reservoirs by different organic and inorganic pollutants [59]. Although most of the inorganic

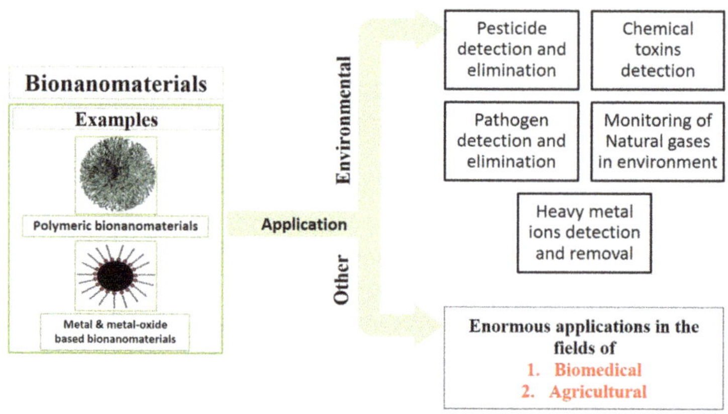

Figure 1.3. Diagrammatic illustration depicting potentialities of bionanomaterials in the environmental domain.

contaminants are naturally present in nature, with the rise in anthropogenic activities, their concentration has increased, and the accumulation of these non-biodegradable inorganic pollutants in the human body can cause many serious diseases like cancer, neurodegenerative diseases, etc [60, 61]. On the other hand, the organic pollutants are significantly caused by surface water runoff, discharge of industrial and domestic pollutants, agriculture activities, improper sewage disposal, etc, and these organic pollutants can also easily accumulate in the biological system, and therefore they have become the most persistent environmental pollutant which threatens human life globally [61]. The treatment of wastewater to convert it into a reusable form is now an urgent requirement to meet future demands.

Nowadays, various biopolymer-based bionanocomposites, including clay hydrogel nanocomposites, gum polysaccharide-based bionanocomposites, chitosan-based polymer bionanocomposites, cellulosic nanocomposites, have been utilized to treat wastewater. Bionanocomposites exhibit excellent mechanical properties, biocompatibility, selectivity, adsorption capacity, stability, and biodegradability, making them the ideal green nanocomposites. They also show better selectivity, adsorption capacity, and stability than the other nanoparticles. Novel stable metal–organic framework (MOF)-based luminescent probes were developed to identify bisulfite anions (HSO_3) in water. The chemically stable MOF, NH_2-MIL-68 (In), was decorated with a particular recognition site through post-synthetic modification. This post-synthetically modified MOF showed a selective turn as a response to a bisulfite anion and simultaneously maintained its structural integrity [62]. Similarly, TiO_2 nanorods (NRs) along with the functionalized gum ghatti (Gg) biopolymer-based bionanocomposite hydrogel (TGB-hydrogel) was synthesized by using a free-radical graft polymerization which was utilized to remove a toxic dye named brilliant green (BG) from the water. The ability of TGB-hydrogel to adsorb the dye was studied in varying conditions like adsorbent dosage, initial dye concentration, pH, and temperature. It was suggested that the presence of various functional groups, like, hydroxyl, carboxylic acid, an amide, along with TiO_2 nanorods in the hydrogel are greatly responsible for this extraordinary high adsorption capacity [63]

An innovative and new water-soluble, cost-effective and sustainable bacterial polysaccharide known as salecan was utilized as a matrix with poly (acrylamide-co-itaconic acid) (PAI) as the synthetic component developing salecan g-PAI hydrogels by the chemical cross-linking method. Salecan was seen on the interaction sites to control the developing hydrogels' water content and modified their morphology. This salecan-PAI hydrogel showed extraordinary adsorption properties toward methylene blue (MB) dye and, therefore, can be used for dye decontamination [64]. Moreover, due to the excess use of drugs and pharmaceutical residues, water contamination is also increasing at alarming levels. These pollutants are also known as new emerging pollutants, and they cause a dangerous effect on the endocrine system, hormones and can also affect the genetic makeup. These new emerging pollutants can also convert into chiral metabolites, which are receptor-specific and cause acute side effects. As the world is suffering from a global pandemic of severe acute respiratory syndrome-coronavirus-2 (SARS-CoV-2), water pollution by these emerging contaminants has now become a major subject of global concern [65–67]. Naproxen is a non-steroidal anti-inflammatory drug (NSAID) that decreases various pain and inflammation-causing hormones [68, 69]. A novel method was developed to eliminate naproxen from the aqueous medium by modifying nanoclay with β-cyclodextrin and then polymerizing it with polyvinylpyrrolidone (PVP) to produce more surfaces for higher adsorption. The results showed that the adsorption was affected by certain factors like the initial concentration of naproxen, the amount of adsorbent, and the pH of the solution. The optimum values of the adsorbent dosage, contact time, initial naproxen concentration, and pH were estimated at 1 g, 120 min, 10 mgL 21 and 6, respectively. Moreover, the efficiency of adsorption of 92.2% was also achieved and based on these results, it can be concluded that this facile adsorption method can be used to efficiently eliminate the naproxen from an aqueous medium by utilizing the modified nanocomposite [69].

However, various processes have been utilized for remediation of these inorganic and organic pollutants from wastewater, but these days nanotechnology has gained significant attention as it uses bionanomaterials that efficiently perform photo-catalysis and adsorption activities. Moreover, carbon-based bionanomaterials have received a lot of attention as they exhibit unique properties, are highly reactive, chemically and thermally stable, easily available, and act as catalysts and adsorbents for the remediation of the pollutants present in the wastewater [33, 61, 70, 71]. Carbon-based bionanomaterials exist in different allotropic forms, namely, diamond, graphite, fullerenes, carbon nanotubes, graphitic carbon nitrate, etc and all these carbon-based bionanomaterials have contributed to pollutant removal and generating renewable and clean forms of energy [72]. The graphene-based bionanomaterials are known to remove organic pollutants like methyl blue, bisphenol A (BPA), phenol, cationic red, etc, through the adsorption process as the oxygen-containing functional groups in graphene can efficiently weaken the hydrophobic effect with non-polar organics [73, 74]. Similarly, carbon nanotubes are also known to remove organic pollutants via the adsorption process by interacting with the organic contaminants through π–π stacking interactions, electrostatic interaction, hydrophobic interactions, hydrogen bonding interactions and van der Waal force [75, 76]. Furthermore, the large surface area, tunable properties, and the presence of surface oxygen-containing functional groups in both carbon nanotubes and graphene-based bionanomaterials help them to remove the inorganic pollutants like heavy metals that are discharged from industries like electroplating, textile, fertilizers, tannery, etc [77–80]. Moreover, bionanocomposites are widely used for wastewater treatment, and to read in more detail, please refer to **chapter 7**, **chapter 8**, and **chapter 9** of this book.

It is well known that the textile industries produce high levels of dyes and other solid materials, and it is estimated that each year approx 5000 tons of toxic dye materials like degreasing solvents, cyanides, alkaline-based cleaning agents, oils, fats, etc, are discharged into the environment [81, 82]. These toxic dye materials are harmful to the environment and are also dangerous for living beings as they makes water bodies toxic, which subsequently kills the marine environment and affects human lives [83]. Although researchers have now focused on removing dye from water bodies, the photocatalytic technique is the best-suited technique to remove the harmful chemicals. For this purpose, bionanomaterials have been utilized as they are small in size and have a high surface-to-volume ratio, which helps them increase the captivation of several micro-organic and organic pollutants [55]. Various metal-based bionanomaterials are known to exhibit photocatalytic activity in colloidal water, such as AgNPs or titanium dioxide nanoparticles (TiO_2NPs) that interact with the help of photons, leading to the excitation of the electrons at the surface. These excited electrons are then accepted by the water-dissolved oxygen, which is further converted into oxygen anion radicals forming unsaturated bonds with the organic molecules of the dye and converting them into simple organic molecules, resulting in the quick degeneration of the dye by color change or disappearance. Similarly, other bionanomaterials like polysaccharide-based hydrogels have also demonstrated their potentiality in removing synthetic dyes as they consist of various functional groups like –OH, –COOH, and $–NH_2$ that help them adsorb both the cationic and anionic dyes [84, 85]. Anionic polysaccharides-based hydrogels such as xanthan gum, karaya gum etc, are used to adsorb the cationic dyes, whereas cationic polysaccharides-based hydrogels like chitosan (CS) are used to adsorb only anionic dyes [86].

Nanotechnology also helps detect various pollutants, toxic intermediates, and heavy metals from water, soil, and air by utilizing nanobiosensors [77, 87–89]. These nanobiosensors are considered a revolution in this field as they are highly accurate, highly sensitive, portable, and minimize the load of traditional laboratory-based techniques [90, 91]. These nanobiosensors are used to detect the organic compounds, biochemical oxygen demand (BOD), soil contaminants like heavy metals, insecticides, etc, and to detect toxins and pollutants and their concentration in soil, water, and air [92–94]. Recently, carbon nanotube-based sensors have gained much attention as they exhibit exceptionally good chemical stability, electrical conductivity, high surface area, and mechanical stiffness. These properties also help them functionalize the conventional carbon electrode sensors, which can help in the efficient detection of the various types of organic pollutants [95, 96]. Similarly, for the gas detection, an ionization sensor was build which had an anode made up of arrays of multiwalled nanotubes (MWNTs) grown on silicon oxide (SiO_2) substrate [97]. These multiwalled carbon nanotubes (MWCNTs) based sensors were portable, battery operated, and generated high electric fields at low voltages. The gas was detected with the help of electric field decomposition of the sample, followed by the cathode registration of a unique fingerprint for each gaseous analyte [98].

Moreover, CNT nanowire sensors have efficiently revolutionized the sensor domain as they have established the correlation between the concentration of the analyte and current fluctuation. It is well known that a high concentration of ammonia (NH_3) in the atmosphere possess a great danger to human health, and the immediate exposure (approx. 500–1000 ppm) of ammonia vapors results in severe irritation in both nose and throat, leading to the generation of pulmonary edema [99]. Therefore, the Occupational Safety and Health Administration (OSHA) have set the total permissible weight-average (TWA) of ammonia exposure limit from 25 to 50 ppm for eight hours [100]. So, various sensors were built that had a coating of various metal nanoparticles like silver, gold, tin, cobalt, etc on single-walled carbon nanotubes (SWCNTs) that helped in the detection of ammonia, in which the metal ions played a role in increasing the sensitivity, selectivity, and recovery rates of the sensor and it was found that the sensor could detect up to 20 ppm of ammonia from the environment. The sensor worked on the principle of oxidation of metal nanoparticles that formed the metal oxides in the composite mixture. The presence of oxygen in the composite mixture helped develop positive charge holes that adsorbed the electron-rich ammonia, resulting in the enhancement sensor's sensitivity [101–105]. Nowadays, aptamer, DNAzyme, NAzymes based nanobiosensors have been developed for real-time and on-site detection of heavy metal ions like mercury, arsenic, cadmium, lead, copper, chromium, etc [106]. DNAzymes based nanobiosensors show the capability to detect Pb^{2+} ions up to 10 nM with high selectivity over other metal ions. Following this, a gold nanomaterials-based biosensor was fabricated, which was affordable and showed enhanced sensing capability [107]. The DNAzyme-based sensors show LOD in the pM range to detect metal ions in the drinking water, which is well below the guideline value assigned by the WHO in 2017 to detect heavy metal ions like lead, arsenic, selenium etc in drinking water. Moreover, a portable sensor named ANDalyze is commercially used to detect various metal ions in a large range of water types [108]. Further, various aptamer-based sensors have been developed to detect microorganisms like bacteria, viruses, fungi, etc, that are present in water [109]. It was observed that when the aptamer was hybridized with complementary DNA, grafted on gold nanoparticles and upconversion nanoparticles, respectively, it efficiently detected the ATCC8739 strain of *Escherichia coli* obtained from pond and tap water with a detection limit of 3 cfu mL^{-1}. However, this sensitivity does not meet the criteria set by the WHO for drinking water but is recommended to check the quality of recreational water [110–112]. Further, more detailed overviews of bionanomaterials, bionanocomposites, and nanobiosensor potentialities for an enormous number of environmental applications are discussed in **chapter 11**, **chapter 12**, **chapter 13**, and **chapter 14**.

1.4 Conclusion and prospects

Bionanomaterials are the new generation nanosized materials composed of biological molecules like nucleic acids, antibodies, proteins, etc. The onset of nanotechnology has efficiently created a revolution in science and technology, and has found its various utilities in many domains like aeronautics, robotics, sensors, biomedicine, the environment, agriculture, etc. The unique properties like optical, physical, catalytic, chemical, electrical, biological, etc, make bionanomaterials different from bulk materials. This chapter discusses various utilities of different bionanomaterials in the agriculture and environment domains.

The agriculture sector is considered as one of the largest sectors that provides raw materials to food and feed industries, and with the increase in the population and globalization, food demand is increasing, but at the same time, agricultural land is decreasing, which makes an urgent need to develop and modernize traditional agricultural practices. With the introduction of nanotechnology, the agriculture field has gained much growth as novel and innovative techniques to increase crop production and nutrient content has been initiated. Nanotechnology utilizes bionanomaterials like carbon-derived bionanomaterials, silica-based bionanomaterials, etc, in the agriculture field. These bionanomaterials are highly used in developing chemical-less nanofertilizer, nanopesticides, and nanoinsecticides. Moreover, bionanomaterials are also used to develop nanosensors to detect pollutants and nutrient levels present in the soil. A new wireless nanosensor has been developed which can easily predict the moisture content of the soil by using bionanomaterials. Although the agriculture sector demands a lot of attention and precision as it directly affects human health, the safety and regulations to develop nanofertilizers are very important and need to be further explored.

Similarly, bionanomaterials show their versatile utilities in the environment domain by helping in the remediation of heavy metals, inorganic and organic pollutants, and the detection of pollutants in water, soil, and air. There are many reviews, books and book chapters that have purely focused on the green synthesis of various metals and their utilities in different domains [113–115]. With the increase in global warming and pollution, novel alternative techniques to reduce pollution have now become a necessity. The production of solar cells to derive solar energy, bioremediation materials that can easily degrade the pollutants present in water and soil, and the development of nanobiosensors to detect the pollutants have been possible due to the introduction of bionanomaterials. These bionanomaterials also play an important role in treating wastewater as it contaminates freshwater bodies. With the modernization and globalization comes certain drawbacks which need to be resolved as quickly as possible. Bionanomaterials show the possibility to combat these drawbacks and hold a promising future.

Acknowledgment

KRBS would like to thank Professor A K Singh for giving constant encouragement throughout the work. JS expresses their gratitude for the DST-INSPIRE faculty Fellowship, BHU (IoE grant), and UGC New Delhi to provide financial assistance. COA is thankful to his respective Universities for helping him to prepare this chapter. RPS and VN are thankful to Vice-Chancellor, IGNTU, Amarkantak, India, for providing financial assistance and for motivating good science.

References

[1] Altindal N and Altindal D 2020 Agriculture and nanoparticles *Biogenic Nano-Particles and their Use in Agro-Ecosystems* (Singapore: Springer) pp 59–66

[2] Singh K R, Sridevi P and Singh R P 2020 Potential applications of peptide nucleic acid in biomedical domain *Eng. Reports.* **2** e12238

[3] Yunlong C and Smit B 1994 Sustainability in agriculture: a general review *Agric. Ecosyst. Environ.* **49** 299–307

[4] Mukhopadhyay S S 2014 Nanotechnology in agriculture: prospects and constraints *Nanotechnol. Sci. Appl.* **7** 63–71

[5] Prasad R, Bhattacharyya A and Nguyen Q D 2017 Nanotechnology in sustainable agriculture: recent developments, challenges, and perspectives *Front. Microbiol.* **8** 1014

[6] Worrall E, Hamid A, Mody K, Mitter N and Pappu H 2018 Nanotechnology for plant disease management *Agronomy* **8** 285

[7] Singh R P 2021 Recent trends, prospects, and challenges of nanobiosensors in agriculture *Biosensors in Agriculture: Recent Trends and Future Perspectives* ed R N Pudake, U Jain and C Kole (Berlin: Springer) pp 3–13

[8] Taran N, Storozhenko V, Svietlova N, Batsmanova L, Shvartau V and Kovalenko M 2017 Effect of zinc and copper nanoparticles on drought resistance of wheat seedlings *Nanoscale Res. Lett.* **12** 60

[9] Li R, He J, Xie H, Wang W, Bose S K and Sun Y *et al* 2019 Effects of chitosan nanoparticles on seed germination and seedling growth of wheat (*Triticum aestivum* L.) *Int. J. Biol. Macromol.* **126** 91–100

[10] Jhanzab H, Razzaq A, Bibi Y, Yasmeen F, Yamaguchi H and Hitachi K *et al* 2019 Proteomic analysis of the effect of inorganic and organic chemicals on silver nanoparticles in wheat *Int. J. Mol. Sci.* **20** 825

[11] Mahakham W, Theerakulpisut P, Maensiri S, Phumying S and Sarmah A K 2016 Environmentally benign synthesis of phytochemicals-capped gold nanoparticles as nanopriming agent for promoting maize seed germination *Sci. Total Environ.* **573** 1089–102

[12] Srivastava G, Das C K, Das A, Singh S K, Roy M and Kim H *et al* 2014 Seed treatment with iron pyrite (FeS_2) nanoparticles increases the production of spinach *RSC Adv.* **4** 58495–504

[13] Aigbogun I E, Mohammed S S D and Orukotan A A 2017 The role of nanotechnology in food industries—a review *J. Adv. Microbiol.* **7** 1–9

[14] Sanguansri P A M 2006 Nanoscale materials development—a food industry perspective *Trends Food Sci. Technol.* **17** 547–56

[15] Ravichandran R 2010 Nanotechnology applications in food and food processing: innovative green approaches, opportunities and uncertainties for global market *Int. J. Green Nanotechnol. Phys. Chem.* **1** 72–96
[16] Otles S Y B 2010 Nano-biosensors as new tool for detection of food quality and safety *LogForum.* **6** 64–70
[17] Hwang S-H, Park Y-B, Han K and Suk D 2011 Smart materials and structures based on carbon nanotube composites *Carbon Nanotubes – Synthesis, Characterization, Applications* (Rijeka: InTech) Available from http://intechopen.com/books/carbon-nanotubes-synthesis-characterization-applications/smart-materials-and-structures-based-on-carbon-nanotube-composites
[18] Yu H, Park J-Y, Kwon C W, Hong S-C, Park K-M and Chang P-S 2018 An overview of nanotechnology in food science: preparative methods, practical applications, and safety *J. Chem.* **2018** 5427978
[19] Quintavalla S V L 2002 Antimicrobial food packaging in meat industry *Meat Sci.* **62** 373–80
[20] Shameli K, Bin A M, Jazayeri S D, Shabanzadeh P, Sangpour P and Jahangirian H *et al* 2012 Investigation of antibacterial properties silver nanoparticles prepared via green method *Chem. Cent. J.* **6** 73
[21] Manjumeena R, Duraibabu D, Sudha J and Kalaichelvan P T 2014 Biogenic nanosilver incorporated reverse osmosis membrane for antibacterial and antifungal activities against selected pathogenic strains: an enhanced eco-friendly water disinfection approach *J. Environ. Sci. Health Part* A **49** 1125–33
[22] Naraginti S and Sivakumar A 2014 Eco-friendly synthesis of silver and gold nanoparticles with enhanced bactericidal activity and study of silver catalyzed reduction of 4-nitrophenol *Spectrochim. Acta Part* A **128** 357–62
[23] Jo Y-K, Kim B H and Jung G 2009 Antifungal activity of silver ions and nanoparticles on phytopathogenic fungi *Plant Dis.* **93** 1037–43
[24] Owolade O and Ogunleti D 2008 Effects of titanium dioxide on the diseases, development and yield of edible cowpea *J. Plant Prot. Res.* **48** 329–36
[25] Niemeyer C M M C 2004 *Nanobiotechnology: Concepts, Applications and Perspectives* vol 1 (New York: Wiley)
[26] Oskam G 2006 Metal oxide nanoparticles: synthesis, characterization and application *J. Sol-Gel Sci. Technol.* **37** 161–4
[27] Marimuthu S, Rahuman A A, Rajakumar G, Santhoshkumar T, Kirthi A V and Jayaseelan C *et al* 2011 Evaluation of green synthesized silver nanoparticles against parasites *Parasitol. Res.* **108** 1541–9
[28] Dhakal R 2019 Biopesticides: a key to sustainable agriculture *Int. J. Pure Appl. Biosci.* **7** 391–6
[29] Fletcher S J, Reeves P T, Hoang B T and Mitter N 2020 A perspective on RNAi-based biopesticides *Front. Plant Sci.* **11** 51
[30] Hasaneen M and Abdel-aziz H M M O A M 2016 Effect of foliar application of engineered nanomaterials: carbon nanotubes NPK and chitosan nanoparticles NPK fertilizer on the growth of French bean plant *Biochem. Biotechnol. Res.* **4** 68–76
[31] Raliya R, Franke C, Chavalmane S, Nair R, Reed N and Biswas P 2016 Quantitative understanding of nanoparticle uptake in watermelon plants *Front. Plant. Sci.* **7** 1288
[32] Singh R, Manickam N, Mudiam M K R, Murthy R C and Misra V 2013 An integrated (nano-bio) technique for degradation of γ-HCH contaminated soil *J. Hazard. Mater.* **258–259** 35–41

[33] Shahid S A, Qidwai A A, Anwar F, Ullah I and Rashid U 2012 Improvement in the water retention characteristics of sandy loam soil using a newly synthesized poly(acrylamide-co-acrylic acid)/$AlZnFe_2O_4$ superabsorbent hydrogel nanocomposite material *Molecules* **17** 9397–412

[34] Omanović-Mikličanin E M M 2016 Nanosensors applications in agriculture and food industry *Bull. Chem. Technol. Bosnia Herzegovina.* **47** 59–70

[35] Munawar A, Ong Y, Schirhagl R, Tahir M A, Khan W S and Bajwa S Z 2019 Nanosensors for diagnosis with optical, electric and mechanical transducers *RSC Adv.* **9** 6793–803

[36] Willner M R and Vikesland P J 2018 Nanomaterial enabled sensors for environmental contaminants *J. Nanobiotechnol.* **16** 95

[37] Dubas S T P V 2008 Humic acid assisted synthesis of silver nanoparticles and its application to herbicide detection *Mater. Lett.* **65** 2661–3

[38] Pacioni N L and Veglia A V 2007 Determination of poorly fluorescent carbamate pesticides in water, bendiocarb and promecarb, using cyclodextrin nanocavities and related media *Anal. Chim. Acta* **583** 63–71

[39] Xia N, Wang Q and Liu L 2014 Nanomaterials-based optical techniques for the detection of acetylcholinesterase and pesticides *Sensors* **15** 499–514

[40] Zhao Y, Ma Y and Li H W L 2011 Composite QDs@ MIP nanospheres for specific recognition and direct fluorescent quantification of pesticides in aqueous media *Anal. Chem.* **84** 386–95

[41] Liu S, Yuan L, Yue X, Zheng Z and Tang Z 2008 Recent advances in nanosensors for organophosphate pesticide detection *Adv. Powder Technol.* **19** 419–41

[42] Shojaei T R, Salleh M A M, Sijam K, Rahim R A, Mohsenifar A and Safarnejad R *et al* 2016 Fluorometric immunoassay for detecting the plant virus Citrus tristeza using carbon nanoparticles acting as quenchers and antibodies labeled with CdTe quantum dots *Microchim. Acta* **183** 2277–87

[43] Chartuprayoon N, Rheem Y, Ng J C K, Nam J, Chen W and Myung N V 2013 Polypyrrole nanoribbon based chemiresistive immunosensors for viral plant pathogen detection *Anal. Methods* **5** 3497

[44] Ambika N, Meharoofa F, Menezes F E and Deepika M S K P K 2019 Agro-amigo: agro based web marketing and automated irrigation *Perspect. Commun. Embed Syst. Signal Process PiCES.* **11** 286–90

[45] Sadiq M T, Hossain M A M, Rahman K F and Sayem A S 2019 Automated irrigation system: controlling irrigation through wireless sensor network *Int. J. Electron Electr. Eng.* **7** 33–7

[46] Martinez O, Arguello C, León J and Carguacundo P D C D G E C 2019 Prototype of automated irrigation system improves the yield of potatoes (*Solanum tuberosum* L.) in Riobamba-Ecuador using wireless network sensors-WSN and 6LoWPAN *MASKA Y.* **9** 25–30

[47] Pandey S K, Jain A, Sharda R and Sharma P J A 2018 Economic analysis of automated drip irrigation system for production of tomato crop *Indian J. Econ. Dev.* **14** 513–20

[48] Millán S, Campillo C, Casadesús J, Moñino M and Vivas A P M 2019 Automated irrigation scheduling for drip-irrigated plum trees *Precision Agriculture'19* (Wageningen: Wageningen Academic Publishers) pp 59–66

[49] Ortiz D and Litvin A G F M 2018 A cost-effective and customizable automated irrigation system for precise high-throughput phenotyping in drought stress studies *PLoS One* **6** e0198546

[50] Nazir R, Ayub Y and Tahir L 2020 Green-nanotechnology for precision and sustainable agriculture *Biogenic Nano-Particles and their Use in Agro-ecosystems* (Singapore: Springer) pp 317–57

[51] Huang J, Huang G, An C, He Y, Yao Y and Zhang P S J 2018 Performance of ceramic disk filter coated with nano ZnO for removing *Escherichia coli* from water in small rural and remote communities of developing regions *Env. Pollut.* **238** 52–62

[52] He Y, Huang G, An C, Huang J, Zhang P and Chen X X X 2018 Reduction of *Escherichia coli* using ceramic disk filter decorated by nano-TiO_2: a low-cost solution for household water purification *Sci. Total Env.* 1628–37

[53] Ghahremani D, Mobasherpour I, Salahi E, Ebrahimi M and Manafi S K L 2017 Potential of nano crystalline calcium hydroxyapatite for Tin(II) removal from aqueous solutions: equilibria and kinetic processes *Arab J. Chem.* **10** S461–71

[54] Mahmoud M E and Abou Kana M T H H A A 2015 Synthesis and implementation of nanochitosan and its acetophenone derivative for enhanced removal of metals *Int. J. Biol. Macromol.* **81** 672–80

[55] Dahrazma B, Naghedinia A and Gorji H G S S 2019 Morphological and physiological responses of *Cucumis sativus* L. to water with micro-nanobubbles *J. Agric. Sci. Technol.* **21** 181–92

[56] Liu Y, Zhou Y, Wang T, Pan J and Zhou B M T *et al* 2019 Micro-nano bubble water oxygation: synergistically improving irrigation water use efficiency, crop yield and quality *J. Clean. Prod.* **222** 835–43

[57] Sang H, Jiao X, Wang S, Guo W and Salahou M K L K 2018 Effects of micro-nano bubble aerated irrigation and nitrogen fertilizer level on tillering, nitrogen uptake and utilization of early rice *Plant Soil Env.* **64** 297–302

[58] Singh K R and Singh R P 2021 Utility of nanobiosensors in environmental analysis and monitoring *Nanosensors for Environment, Food and Agriculture* ed V Kumar *et al* (Singapore: Springer) pp 229–46

[59] Gopinath A and Kadirvelu K 2018 Strategies to design modified activated carbon fibers for the decontamination of water and air *Environ. Chem. Lett.* **16** 1118–37

[60] Holt M S 2000 Sources of chemical contaminants and routes into the freshwater environment *Food Chem. Toxicol.* **38** S21–7

[61] Lu F and Astruc D 2018 Nanomaterials for removal of toxic elements from water *Coord Chem. Rev.* **356** 147–64

[62] Dar O A, Malik M A, Talukdar M I A and Hashmi A A 2020 Bionanocomposites in water treatment *Bionanocomposites* (Amsterdam: Elsevier) pp 505–18 Available from https://linkinghub.elsevier.com/retrieve/pii/B9780128167519000192

[63] Kumar N, Mittal H, Alhassan S M and Ray S S 2018 Bionanocomposite hydrogel for the adsorption of dye and reusability of generated waste for the photodegradation of ciprofloxacin: a demonstration of the circularity concept for water purification *ACS Sustain. Chem. Eng.* **6** 17011–25

[64] Qi X, Wei W, Su T, Zhang J and Dong W 2018 Fabrication of a new polysaccharide-based adsorbent for water purification *Carbohydr. Polym.* **195** 368–77

[65] Crane M, Watts C and Boucard T 2006 Chronic aquatic environmental risks from exposure to human pharmaceuticals *Sci. Total Environ.* **367** 23–41

[66] Falconer I R, Chapman H F, Moore M R and Ranmuthugala G 2006 Endocrine-disrupting compounds: a review of their challenge to sustainable and safe water supply and water reuse *Environ. Toxicol.* **21** 181–91

[67] Ukhurebor K E, Singh K R R, Nayak V and UK-Eghonghon G 2021 Influence of SARS-CoV-2 pandemic: a review from the climate change perspective *Environ. Sci. Process. Impacts* **23** 1060–78
[68] Devaraj M, Saravanan R, Deivasigamani R, Gupta V K, Gracia F and Jayadevan S 2016 Fabrication of novel shape Cu and Cu/Cu_2O nanoparticles modified electrode for the determination of dopamine and paracetamol *J. Mol. Liq.* **221** 930–41
[69] Dietrich D R, Webb S F and Petry T 2002 Hot spot pollutants: pharmaceuticals in the environment *Toxicol. Lett.* **131** 1–3
[70] Smith S C and Rodrigues D F 2015 Carbon-based nanomaterials for removal of chemical and biological contaminants from water: a review of mechanisms and applications *Carbon* N **91** 122–43
[71] Thines R K, Mubarak N M, Nizamuddin S, Sahu J N, Abdullah E C and Ganesan P 2017 Application potential of carbon nanomaterials in water and wastewater treatment: a review *J. Taiwan Inst. Chem. Eng.* **72** 116–33
[72] Jayaraman T, Murthy A P, Elakkiya V, Chandrasekaran S, Nithyadharseni P and Khan Z *et al* 2018 Recent development on carbon based heterostructures for their applications in energy and environment: a review *J. Ind. Eng. Chem.* **64** 16–59
[73] Chowdhury S and Balasubramanian R 2014 Recent advances in the use of graphene-family nanoadsorbents for removal of toxic pollutants from wastewater *Adv. Colloid Interface Sci.* **204** 35–56
[74] Khan A, Khuda F, Elseman A M, Aly Z, Rashad M M and Wang X 2018 Innovations in graphene-based nanomaterials in the preconcentration of pharmaceuticals waste *Environ. Technol. Rev.* **7** 73–94
[75] Yu J-G, Zhao X-H, Yang H, Chen X-H, Yang Q and Yu L-Y *et al* 2014 Aqueous adsorption and removal of organic contaminants by carbon nanotubes *Sci. Total Environ.* **482–3** 241–51
[76] Mashkoor F, Nasar A and Inamuddin 2020 Carbon nanotube-based adsorbents for the removal of dyes from waters: a review *Environ. Chem. Lett.* **18** 605–29
[77] Malik P, Katyal V, Malik V, Asatkar A, Inwati G and Mukherjee T K 2013 Nanobiosensors: concepts and variations *ISRN Nanomater.* **2013** 327435
[78] Talaiekhozani A and Rezania S 2017 Application of photosynthetic bacteria for removal of heavy metals, macro-pollutants and dye from wastewater: a review *J. Water Process Eng.* **19** 312–21
[79] Abdi G, Alizadeh A, Zinadini S and Moradi G 2018 Removal of dye and heavy metal ion using a novel synthetic polyethersulfone nanofiltration membrane modified by magnetic graphene oxide/metformin hybrid *J. Memb. Sci.* **552** 326–35
[80] Madima N, Mishra S B, Inamuddin I and Mishra A K 2020 Carbon-based nanomaterials for remediation of organic and inorganic pollutants from wastewater: a review *Environ. Chem. Lett.* **18** 1169–91
[81] Khezrianjoo S and Revanasiddappa H D 2013 Photocatalytic degradation of acid yellow 36 using zinc oxide photocatalyst in aqueous media *J. Catal.* **2013** 582058
[82] Shah V and Madamwar D 2013 Community genomics: isolation, characterization and expression of gene coding for azoreductase *Int. Biodeterior. Biodegrad.* **79** 1–8
[83] Mahmoodi N M and Arami M 2009 Numerical finite volume modeling of dye decolorization using immobilized titania nanophotocatalysis *Chem. Eng. J.* **146** 189–93
[84] Bandala E R, Stanisic D and Tasic L 2020 Biogenic nanomaterials for photocatalytic degradation and water disinfection: a review *Environ. Sci. Water Res. Technol.* **6** 3195–213

[85] Wypij M, Jędrzejewski T, Ostrowski M, Trzcińska J, Rai M and Golińska P 2020 Biogenic silver nanoparticles: assessment of their cytotoxicity, genotoxicity and study of capping proteins *Molecules* **25** 3022

[86] Komeili A 2006 Magnetosomes are cell membrane invaginations organized by the actin-like protein MamK *Science* **311** 242–5

[87] Bellan L M, Wu D and Langer R S 2011 Current trends in nanobiosensor technology *WIREs Nanomed. Nanobiotechnol.* **3** 229–46

[88] Long F, Zhu A and Shi H 2013 Recent advances in optical biosensors for environmental monitoring and early warning *Sensors* **13** 13928–48

[89] Shahbazi R, Salouti M, Amini B, Jalilvand A, Naderlou E and Amini A *et al* 2018 Highly selective and sensitive detection of *Staphylococcus aureus* with gold nanoparticle-based core–shell nano biosensor *Mol. Cell Probes* **41** 8–13

[90] Amini B, Kamali M, Salouti M and Yaghmaei P 2017 Fluorescence bio-barcode DNA assay based on gold and magnetic nanoparticles for detection of Exotoxin A gene sequence *Biosens. Bioelectron.* **92** 679–86

[91] Kaushal M and Wani S P 2017 Nanosensors: frontiers in precision agriculture *Nanotechnology* ed R Prasad, M Kumar and V Kumar (Singapore: Springer) pp 279–91

[92] Topete A, Alatorre-Meda M, Iglesias P, Villar-Alvarez E M, Barbosa S and Costoya J A *et al* 2014 Fluorescent drug-loaded, polymeric-based, branched gold nanoshells for localized multimodal therapy and imaging of tumoral cells *ACS Nano* **8** 2725–38

[93] Antonacci A, Arduini F, Moscone D and Palleschi G S V 2018 Nanostructured (bio) sensors for smart agriculture *TrAC, Trends Anal. Chem.* **98** 95–103

[94] DeRosa M C, Monreal C, Schnitzer M, Walsh R and Sultan Y 2010 Nanotechnology in fertilizers *Nat. Nanotechnol.* **5** 91

[95] Allen B L, Kichambare P D and Star A 2007 Carbon nanotube field-effect-transistor-based biosensors *Adv. Mater.* **19** 1439–51

[96] Britto P J, Santhanam K S V and Ajayan P M 1996 Carbon nanotube electrode for oxidation of dopamine *Bioelectrochem. Bioenerg.* **41** 121–5

[97] Modi A, Koratkar N, Lass E, Wei B and Ajayan P M 2003 Miniaturized gas ionization sensors using carbon nanotubes *Nature* **424** 171–4

[98] Mauter M S and Elimelech M 2008 Environmental applications of carbon-based nanomaterials *Environ. Sci. Technol.* **42** 5843–59

[99] Timmer B, Olthuis W and van den Berg A 2005 Ammonia sensors and their applications—a review *Sens. Actuators B Chem.* **107** 666–77

[100] Occupational Safety and Health Standards OSHA 1993 1910.1000 table Z-1]

[101] Lee K, Scardaci V, Kim H-Y, Hallam T, Nolan H, Bolf B E, Maltbie G S, Abbott J E and Duesberg G S 2013 Highly sensitive, transparent, and flexible gassensors based on gold nanoparticle decorated carbon nanotubes *Sens Actuators B.* **188** 571–5

[102] Cui S, Pu H, Lu G, Wen Z, Mattson E C, Hirschmugl C, GajdardziskaJosifovska M, Weinert M and Chen J 2012 Fast and selective room-temperature ammonia sensors using silver nanocrystal-functionalized carbon nanotubes *Interfaces ACS Appl Mater.* **4** 4898–904

[103] Yoon B, Liu S F and Swager T M 2016 Surface-anchored poly(4- inylpyridine)–single-walled carbon nanotube–metal composites for as detection *Chem. Mater.* **28** 5916–24

[104] Duc Hoa N, Van Quy N, Suk Cho Y and Kim D 2007 Nanocomposite of SWNTs and SnO_2 fabricated by soldering process for ammonia gas sensor application *Phys. Status Solidi* A **204** 1820–4

[105] Rasheed T, Nabeel F, Adeel M, Rizwan K, Bilal M and Iqbal H M N 2019 Carbon nanotubes-based cues: a pathway to future sensing and detection of hazardous pollutants *J. Mol. Liq.* **292** 111425

[106] Abu-Ali H, Nabok A and Smith T 2019 Development of novel and highly specific ssDNA-aptamer-based electrochemical biosensor for rapid detection of mercury (II) and lead (II) ions in water *Chemosensors* **7** 27

[107] Liu J and Lu Y 2004 Accelerated color change of gold nanoparticles assembled by DNAzymes for simple and fast colorimetric Pb 2+ detection *J. Am. Chem. Soc.* **126** 12298–305

[108] WHO 2017 *Guidelines for Drinking-Water Quality* https://who.int/water_sanitation_health/publications/drinking-water-quality-guidelines-4-including-1staddendum/en/

[109] Davydova A, Vorobjeva M, Pyshnyi D, Altman S, Vlassov V and Venyaminova A 2016 Aptamers against pathogenic microorganisms *Crit. Rev. Microbiol.* **42** 847–65

[110] Jin B, Wang S, Lin M, Jin Y, Zhang S and Cui X *et al* 2017 Upconversion nanoparticles based FRET aptasensor for rapid and ultrasenstive bacteria detection *Biosens. Bioelectron.* **90** 525–33

[111] EUdirective 2006 *DIRECTIVE 2006/7/EC of the European Parliament and of the COUNCIL of 15 February 2006 Concerning the Management of Bathing Water Quality and Repealing Directive 76/160/EEC* https://eur-lex.europa.eu/legal-content/EN/TXT/PDF/?uri=CELEX

[112] Soukarié D, Ecochard V and Salomé L 2020 DNA-based nanobiosensors for monitoring of water quality *Int. J. Hyg. Environ. Health* **226** 113485

[113] Singh K R, Nayak V, Sarkar T and Singh R P 2020 Cerium oxide nanoparticles: properties, biosynthesis and biomedical application *RSC Adv.* **10** 27194–214

[114] Nayak V, Singh K R, Singh A K and Singh R P 2021 Potentialities of selenium nanoparticles in biomedical science *New J. Chem.* **45** 2849–78

[115] Fernandes M, Singh K R, Sarkar T, Singh P and Singh R P 2020 Recent applications of magnesium oxide (MgO) nanoparticles in various domains *Adv. Mater. Lett.* **11** 20081543

IOP Publishing

Bionanomaterials for Environmental and Agricultural Applications

Ravindra Pratap Singh and Kshitij RB Singh

Chapter 2

Antimicrobial potentialities: special emphasis on metal and metal oxide-based bionanomaterials

Salmiah Jamal Mat Rosid, Sarina Mat Rosid, Nur Atiqah Nasir, Wan Nazwanie Wan Abdullah and Nurul Najidah Mohamed

The growing exploration in bionanomaterials has increased, especially as an antimicrobial agent. The antimicrobial properties of metal and metal oxide nanoparticles are correlated with physicochemical characteristics such as particle size, shape, charge and concentration of particle. Moreover, metal nanomaterials interact strongly with bacteria and biofilms and exert their bacterial effects over the generation of reactive oxygen species (ROS). Microbes, plants, animal cells and animal products have been used for the biogenesis of nanoparticles either intra- or extracellularly. Further explanation about the potential of synthesized metal and metal oxide on antimicrobial properties will be given in this chapter.

2.1 Introduction

Increasing population growth has resulted in high demand for the agricultural sector and an impact on the environment. This also leads to an increase in various types of microorganisms, either beneficial or harmful, which are causing serious problems by spreading disease, dominating or killing other important species [1]. To overcome the effects of these microorganisms, various organic and inorganic agents known as antimicrobial agents have been developed. Antimicrobial agents are chemical or physical drugs that can kill or stop the growth of microorganisms [2]. There are various types of antimicrobial agents classified according to their function and the microorganisms that mainly act upon. Table 2.1 shows some of the types of antimicrobial agents and their mechanisms of action [3].

doi:10.1088/978-0-7503-3863-9ch2

Table 2.1. Types of antimicrobial agents [3, 4]

Antimicrobial agents	Synthetization method	Action
B-lactam drugs	Chemically	Inhibitors for bacterial cell wall synthesis
Fosfomycin	Naturally	
Vancomycin	Naturally	
Tetracyclines	Naturally	Inhibitors for protein biosynthesis
Erythromycin	Naturally	
Macrolides	Naturally, Chemically	
Aminoglycoside antibiotics	Naturally	
Lipopeptide antibiotics	Naturally	
Streptogramins	Naturally	
Quinolones	Chemically	Inhibitors for DNA synthesis
Nitroimidazole (anti-protozoa)	Chemically	
Rifampicin	Naturally	Inhibitors for RNA synthesis
Ethambutol	Chemically	
Pyrazinamide (anti-tuberculosis)	Chemically	
Sulfonamides	Chemically	Inhibitors for the enzyme dihydropteroate synthetase (DHPS) to synthesis folate

The antimicrobial agents launched today are primarily either chemically or naturally synthesized [4, 5]. Most of today's antimicrobials are derived from organic synthesis because they are biocompatible, stable, highly selective and easy to synthesize [6]. Some synthetic agents include sulfonamides, cotrimoxazoles, quinolones, antivirals, antifungals, anticancer drugs, antimalarial drugs, antituberculosis drugs, antireprotic drugs, and antiprotozoal drugs. [6, 7]. Notwithstanding, microorganisms have mutated and become immune to the developing antimicrobial agents. Recent findings suggest that conventional antimicrobial agents are not effective against certain microbial strains due to mutation, change in absorption mechanism and metabolic pathways by microorganisms known as antibiotic-resistant microorganisms [3, 8]. This is because the long-lasting antimicrobial technology was based on either poisons or heavy metals, which may not have completely killed the microorganisms so that they survive, transform and become resistant to poisons and/or heavy metals. As a result, this poses a major threat, increasing the high risk of infectious diseases that are likely to be fatal to living organisms and, ultimately, humans. There is a need to find a stable, cost-effective, biocompatible and environmentally friendly alternative to avoid a major risk infectious diseases situation.

Antimicrobial technology is a recent alternative against disease-causing organisms that limit usage of heavy metals and toxins. Therefore, inorganic materials are highly regarded as effective antimicrobial agents. Nanomaterials, nanocomposites and metal–organic frameworks are the most favorable and far more reliable than the conventional agents and others. Due to a large number of metal oxides, the infinite number of metal and non-metal ratios, many metal–ligand combinations can be synthesized and used as antimicrobial agents to track, label, inhibit the activity of transcription factors and ultimately kill harmful microorganisms [9]. Notably, inorganic materials have more advantages than organic materials like reduced toxicity, improved durability, good selectivity and lower resistance.

Nanoscale materials are very effective due to their multidimensional properties such as small size (1–100 nm), different morphologies and shapes, and different particle charges for microbial destruction [10, 11]. Recent studies have shown that smaller sizes are generally more common because small sized bionanomaterials easily penetrate cell membranes to alter transport, interact with biomolecules in microbial cells, and interfere with communication even at the genomic level, forming decaying cells. It turned out to be an essential factor in microbial delay that causes the ultimate death of microorganisms. In addition, smaller bionanomaterials with larger specific surface areas provided greater permeability to cell membranes [4, 12, 13]. This indicates that antimicrobial activity increases as the size of the nanomaterial decreases. Morphology and shape are also crucial because as morphology changes, nanomaterial properties also change, especially the size of antimicrobial biomaterials. As reported by Cha *et al*, regarding the shape of the nanoparticles, various gradations of damage to bacterial cells occurred when the nanoparticles interacted with the periplasmic enzyme [14]. In addition, the shape of nanorods and nanowires is more efficient than other shapes such as hollow and spherical. For instance, cube-shaped Ag nanoparticles exhibit stronger anti-bacterial activity than spherical and wire-shaped Ag nanoparticles with similar diameters due to their specific surface area and facet reactivity [15], which have less of an effect on microbial susceptibility [16]. Figure 2.1 shows the type of morphologies of nanomaterials.

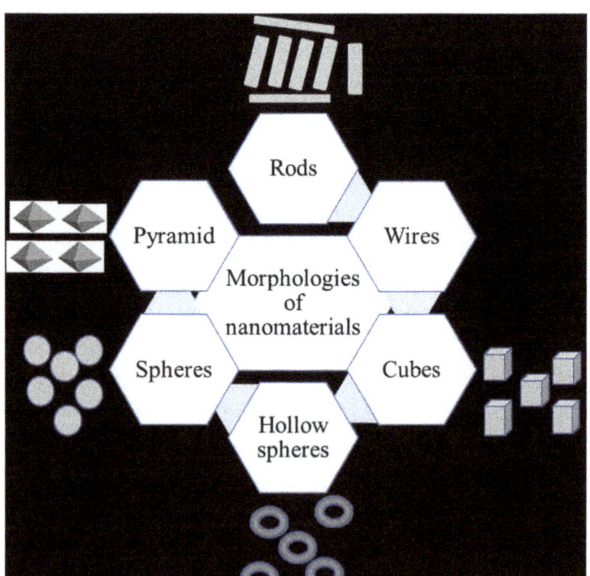

Figure 2.1. Type of morphologies of nanomaterials.

In addition to that, surface roughness also acts as a factor in antimicrobial activity. When the surface roughness of the bionanomaterial increases, the size and surface area-to-mass ratio also increases, which promotes more adsorption of microbial proteins and subsequently reduces microbial adhesion [17]. Moreover, morphology and the size of nanomaterials can be controlled with the help of different preparation techniques. The electrostatic attraction between positively charged nanomaterials and the negatively charged cell membranes of microorganisms has a positive surface charge and tends to be adsorbed on the surface of microorganisms. In contrast to negatively charged counterparts, they are meticulously connected to microorganisms [18] and increase vascular permeability [19] by limiting adhesion of microorganisms by ion exchange [20]. Compared to negatively charged nanomaterials and neutral nanomaterials, positively charged counterparts are thought to promote ROS formation, leading to interactions between nanomaterials and microbial surfaces [21].

Nanomaterials are synthesized by chemical, physical and biological approaches. The synthesis of nanomaterials by physical and chemical methods is very common, but the use of toxic chemicals puts a boundary to their use in agricultural, food, and environmental applications [22]. However, nanomaterials synthesized by biological approach have shown great polydispersity, size and stability. They are also cost effective, environmentally friendly and non-toxic. Due to that, bionanomaterials have become a major concern for researchers in mass production and are applied as antimicrobial agents in a variety of applications. There are various sources to produce bionanomaterials like plants and microorganisms (fungi, bacteria or yeast). The use of plants is preferred due to their non-toxicity, quiet scale-up and reproductivity in making them. In addition, the use of microbial cells to synthesize nanosized materials has emerged as a new approach for the production of metal and metal oxide nanomaterials. Several fungal strains have been used as promising resources for nanomaterial production, including *Fusarium*, *Aspergillus*, *Verticillium*, *Pencilium*, and yeast [22]. Table 2.2 shows some of the bionanomaterials produced from plant and microorganism sources. Hence, this chapter will discuss more about classification of metal and metal oxide bionanomaterials and their physicochemical properties in section 2.2. In section 2.3 we overview the synthesis method of metal and metal oxide based bionanomaterial for antimicrobial properties with different type of plants, algae, yeast and bacteria. Section 2.4 discuss in depth the mechanism of antimicrobial activity through metal and metal oxide-based bionanomaterials. Section 2.5 summarizes the future prospects and challenges, and finally, section 2.6 outlines the concluding remarks and recommendations.

2.2 Classification of metal and metal oxide-based bionanomaterials and their physicochemical properties

Over the years, various types of metal and metal oxide-based bionanomaterials have established promising antimicrobial activity [36]. The physico-chemical properties of metal and metal oxide-based bionanomaterials are mostly rated by the parameters

Table 2.2. List of synthesized nanomaterials from plant and microorganism.s

Sources	Types of bionanomaterials produced		Reference
Plant	Leaf extracts of *Annona muricata*	Au	[23]
	Pulp of *Aloe vera*	Ag	[24]
	Aloe vera	Au and Ag	[25]
	Leaf extracts of *Azadirachta indica*	Au and Ag	[26]
	Leaf extracts of *Mirabilis jalapa*	Au	[27]
	Plumeria rubra	Ag	[28]
Microorganism	*Aspergillus clavatus*	Ag	[29]
	Saccharomyces cerevisiae MTCC 2918	ZnS	[30]
	Penicillium decumbens	Ag	[31]
	Fusarium oxysporum	Ag, Fe_3O_4	[32]
	Verticillium luteoalbum	Au	[33]
	Trichoderma sp.	Au	[34]
	Bhargavaca indica	Au and Ag	[35]

like size, shape, and composition. In this section, some examples of promising antimicrobial activities on metal and metal oxide-based bionanomaterials and their physicochemical characteristics are reviewed.

2.2.1 Silver-based bionanomaterials

Currently, owing to the uniqness of chemical and physical properties of silver nanoparticles (AgNPs), this metal is progressively used in numerous fields, including medical, food, health, consumer, and industrial uses. Meanwhile, with the addition of high anti- microbial activity of the AgNPs, this makes it possible to use this metal in a wide range of applications including food, pharmaceutical, fabric, and packaging industries. Compared to other promising metal nanomaterials, AgNPs appear to be promising antibacterial agents because of their large surface-to-volume ratios and crystallographic surface structure [37]. AgNPs also function as an antibacterial agent by inducing the formation of reactive oxygen species. Due to the excellent anti-microbial property of AgNPs, they have also been combined with starch. This fabrication was achieved by utilizing β-d-glucose as a reducing agent and starch as a protective agent in a mild-heating system. By combining both materials, the attractive force between the AgNPs and starch is reversible and weak at high temperatures, contributing to the dissolution of the synthesized AgNPs [38].

Further, AgNPs also have been synthesized using Tollens method. Through this method, in the presence of ammonia, Ag^+ ions were reduced by saccharides, and synthesised AgNPs with various shapes and sizes in the range of 50–200 nm were obtained [39]. The main factors that influence the AgNP size are the ammonia concentration and the nature of the reducing agents. In this method, $Ag(NH_3)^{2+}$ is a stable complex ion, produced by the robust affinity of ammonia for Ag^+ ions, which helps to obtain a smaller size of AgNPs.

AgNPs can also been synthesized using polyoxometalates. Polyoxometalates are a family of metal oxide molecular clusters with unique structures. Due to its range structure, this polyoxometalate has the ability to be used in inactively synthesizing noble NPs through stepwise and multi-electron redox reactions [40]. The produced monodispersed AgNPs are spherical shaped with 38 nm diameter.

Meanwhile, research by Pal *et al* [41] has reported that the shape of the AgNPs has significant influence on the activity on gram (−) bacterium, *E. coli.* The finding shows that the triangular [111] facets of AgNPs show a better antibacterial activity in comparison to spherical AgNPs (cubo-octahedral, or quasi-spherical shapes) [100] facets. This finding concluded that the size and shape of the NPs have huge influence on *E. coli* activity, even for NPs possessing the same area-to-volume ratio properties [42].

In addition, the surface charge of NPs also seems to affect the antibacterial activity as demonstrated by Correa *et al* [43]. Their research showed that significant fluctuations in antibacterial activity occur at various surface charge of NPs. The AgNPs can have positive, neutral, or negative surface charge, depending on the method of synthesis. So, care should be taken when synthesizing AgNPs since it may affect their antibacterial activity. Besides, AgNPs are also unstable metal owing to the inherent instability of metal AgNPs, which tend to ionize the material when in solution. After being ionized, Ag ions penetrate the bacterial cell walls, condense DNA molecules and prevent replication or biofilm development. Due to the inhibition process, the bacterial cells no longer proliferate, thus the cell undergoes necrosis which make the AgNPs act as a good antimicrobial agent [44].

2.2.2 Copper- based bionanomaterials

CuNPs have also shown promising antimicrobial activity and have been clearly established too [45]. CuNPs offer good heat resistance, chemical stability and outstanding antimicrobial properties owing to their large surface area. Their outstanding anti-bacterial, antifungal, antiviral, and anti-inflammatory properties have given rise to their application in a wide field of industries, such as pharmaceuticals and food packaging [43].

For example, Palza *et al* in his reported paper stated that the efficacy in antimicrobial activity depends on the copper content, relative humidity and also temperature [46]. According to their findings, the major mechanism of the antimicrobial behaviour of this material is the release of copper ions from copper-based materials. Cu^{+} ions can generate vastly reactive hydroxyl groups through Fenton-like reactions, thereby increasing toxicity. However, by forming organic complexes with functional groups in the microorganisms, Cu^{2+} is also considered as a biocidal agent, which can destroy the conformational structure of proteins and nucleic acids, as well as promote oxidative phosphorylation reactions and changes in osmotic balance. Due to these properties, copper ions are hence the real active compounds, which can kill microorganisms (cidal effect) or inhibit the growth of microorganisms (static effect).

In other findings, the potential of antimicrobial properties of copper compounds has also been claimed to depend on the structure of the surface. For example, the surface of electroplated copper executed bacteria faster than polished copper or native rolled copper [47]. Furthermore, findings by Micheal *et al* have concluded that copper also has the advantages of being more active and reliable in indoor environments, such as in dry and cold conditions compared to silver, one of the other well-known antimicrobial metals [48].

Meanwhile, Vincent *et al* in their review paper concluded that the most fascinating commonality of the literature is that all bacteria exposed to copper survive for only a few minutes before dying [49].

There has been no complete resistance to copper in prolonged exposure. Several studies have also discussed the relationship between particle size and microbicidal activity, concluding that the smallest nanoparticles are the most efficient ones.

In addition, a stable CuNP is also formed when $CuSO_4{\cdot}5H_2O$ solution is treated with *Magnolia officinalis* leaf extract. The copper nanoparticles obtained are generally between 37 and 110 nm in diameter, according to electron microscopy studies. An antibacterial test was performed by counting live *E. coli* cells after they had grown for 24 h in a shake-flask filled with latex foam coated with CuNPs. The biosynthetic CuNP-coated foam demonstrated higher antibacterial activity than foams untreated with CuNPs or foams treated chemically with NPs synthesized from sodium borohydride [50].

In addition, pure CuNPs were produced by chemical methods in the presence of chitosan stabilizers. The findings show that the chitosan medium significantly affects the growth rate of the microorganisms. Besides, depending on the concentration of chitosan stabilizer, the resulting CuNPs range from 2–350 nm and, with 0.05% by weight of chitosan–CuNPs being relatively large, the particles obtained are in the range of 50–300 nm. Meanwhile, at a concentration of 0.1 wt%, it was observed that the obtained NPs are in the range of 50–270 nm. Similarly, the size range for chitosan–CuNPs at a concentration of 0.2 wt% is 5–50 nm. On the other hand, chitosan-CuNPs containing 0.5% chitosan appear to have a minimum particle size of 2 nm. This finding highlights the importance of chitosan in controlling CuNP size [51].

Rajeshkumar *et al*, have synthesized CuNPs from the rare medicinal plant species, *Cissus arnotiana* and tested its antibacterial activity opposed to Gram-negative and Gram-positive bacteria. The morphology and characterization of the synthesized CuNPs were investigated using UV-visible spectroscopy at 350–380 nm wavelength and XRD analysis was conducted to analyse the crystal properties. SEM, TEM and AFM were performed to evaluate the size of spherical shape with the range of 60–90 nm and the surface roughness. The biosynthetic CuNPs displayed improved antibacterial activity against the Gram-negative bacteria *Escherichia coli*, and the inhibition zone of 75 $\mu g\ mL^{-1}$ was 22.20 ± 0.16 mm [52].

2.2.3 Gold-based bionanomaterials

AuNP is a clustered or colloidal particle composed of a gold core and an inert, biocompatible compound. AuNPs offer synthetic versatility, with the advantage of being able to control particle size, shape and surface properties. In addition, AuNPs coating can be altered or adjusted to control particle stability, solubility and environment interaction [53]. Research conducted by Balakumaran *et al*, [54] has shown that AuNPs have an elegant antimicrobial activity against a diverse range of Gram-positive and Gram-negative bacteria, and multidrug-resistant human pathogens [55].

Mishra *et al* have reported on a method that is both environmentally friendly and quick for AuNPs biosynthesis derived from *Trichoderma sp* [34]. Their research was also successful in converting 4-nitrophenol to 4-aminophenol by AuNPs through this biocatalytic which provides new hope for green and sustainable bioremediation, while their ability to inhibit pathogenic bacteria demonstrates their suitability as a new generation antimicrobial.

2.2.4 Zinc-based bionanomaterials

Semiconductor metal oxide, ZnO has antimicrobial properties that can be strengthened due to its size, shape and surface-capping agents when used as a nanomaterial [56]. Aditya *et al*, in their article have stated that Zn^{2+} are ions simultaneously produced when ZnO NPs are adopted by bacterial cells, resulting in the yielding of large amounts of reactive oxygen species (ROS) [57]. The increased ROS levels will disrupt bacterial cell homeostasis, causing the cells to rupture. They also claimed that when exposed to light, ZnO NPs will produce electron–hole pairs. And, once these electron–hole pairs react with –OH groups on the ZnO surface, hydroxyl radicals, superoxide ions (O_2), and perhydroxyl radicals ($HO_2\cdot$) are formed. The formation of these species can also result in microbial cell degeneration.

In a recent study, Batool *et al*, used biosynthesised ZnO NPs to study adsorption, wound healing and antimicrobial activities. Based on their research, ZnO NPs exhibit potent antimicrobial activity against *B. licheniformis, B. subtilis* and *K. pneumonia*, *E. coli*, and two fungal strains, *C. albicans* and *A. niger* [58]. The study was conducted using ZnO-NP/silica gel (ZnO-NP/SG) as dressings for mouse wound. The healing efficiency resulting from microscopic investigations shows that the wound area is reduced significantly within 11 days of observation using ZnO-NP/SG-30 ppm dressings. These findings prove that ZnO NPs are effective and powerful in inhibiting bacterial and fungal strains and also a powerful wound healing agent.

Moreover, Awwad *et al*, have also successfully produced ZnO NPs through 'green synthesis' using *Ailanthus altissima* fruits aqueous extract [59]. The result showed that despite very low concentrations, the NPs were very active as antibacterial agent against pathogens such as *S. aureus* and *E. coli*. Moreover, Emamifar *et al*, have also used ZnO NPs combined with low-density polyethylene (LDPE) and developed orange juice packages [60]. Through their study, orange juice was sterilized and inoculated with *lactobacillus plantarum* at 8.5 log cfu mL^{-1}. The orange juice was then filled into packages made from nanocomposite films and stored at 4 °C. The microbial stability of the juice was assessed after storage for 7, 28, 56, 84, and 112 days. These results signified that the use of this nanocomposite packaging material significantly compressed the growth rate of microorganisms. ZnO NPs were also tested for antifungal activity against *Aspergillus niger* and *Aspergillus flavus* [61]. The results discovered that *Aspergillus flavus* inhibits at a higher rate than *Aspergillus niger*. ZnO NPs inhibited *Aspergillus flavus* with 75% inhibition at 500 ppm and 76% inhibition at 1000 ppm, while *Aspergillus niger* germinated with 57% and 63% inhibition, respectively.

At the same time, biological ZnO NPs were successfully synthesized using *Pandanus* leaf extract (POLE). The characterization has been conducted, which then proved that the obtained pure spherical NPs have a size of about 90 nm. Furthermore, biological ZnO NPs were investigated as an antibacterial agent against pathogens. The highest antibacterial activity observed was 50 g per well against Gram-positive *Bacillus subtilis* (26 nm) and Gram-negative *Escherichia coli* (24 mm). Complete bacterial growth (100%) was completely eliminated after treatment with 85 gml^{-1} of ZnO NPs. Overall, POLE-mediated biotype ZnO NPs can be used as important anti-cancer and antibacterial agents, and used to develop new drugs and skin care products [62].

2.2.5 Titanium-based bionanomaterials

Titanium dioxide (TiO_2) is an inorganic material with outstanding whitening, photocatalytic, and antimicrobial properties and is widely applied in a variety of products such as cosmetics, paints, orthodontic composites, and food [43]. Furthermore, TiO_2 has been used as a food additive, most commonly as an anticaking or whitening agent, or to improve the color and glazing of food [63]. Reduction of titanium dioxide to nanoscale (TiO_2 NPs) significantly improves its photocatalytic properties and increases reactive oxygen species (ROS). ROS will damage the DNA strands, bacteria cell and other cell structures because of oxidative stress. Owing to this capability, it has been also focused in water disinfection and food packaging [64]. de Dicastillo *et al* successfully synthesized hollow nanotubes and nanospheres of TiO_2 with outstanding antimicrobial activity via electrospinning coupled with atomic layer deposition techniques [65]. When the NPs were exposed to UVA radiation for 60 min, the results showed that the highest antimicrobial activity toward methicillin-resistant *Staphylococcus aureus* 97–7 and 622–4 was exhibited by nanospheres of TiO_2. The increase in bioactivity of nanometer-sized TiO_2 particles is because of an increase in contact area and/or volume caused by a decrease in particle size, especially the thickness (< 100 nm), resulting in better interaction between the molecules.

TiO_2 NPs, a biocomposite material, has been developed to prevent the growth of mold on the surface of cheese during antibacterial testing [66]. Through their research, the weight loss of coating Ras cheese with a prepared bio-nanocomposite material and its effect on the microbial, chemical and physical properties of Ras cheese compared to uncoated cheese during the maturation process was evaluated. Cheese coating reduces weight and water loss, but does not affect the normal aging process of Ras cheese. Its microbiology, chemistry and texture characteristics have changed. They also concluded that cheese coated with a film containing 2% TiO_2 NPs can eliminate growth of mold on the cheese surface.

Aditya *et al* have successfully biosynthesized TiO_2 NPs from a leaf extract of *Trigonella foenum-graecum*, producing spherical NPs with a size in the range of 20 to

90 nm, and evaluated their antimicrobial activity using standard disc diffusion method [57]. The NPs were tested for antimicrobial properties against *Yersinia enterocolitica* (10.6 mm), *Escherichia coli* (10.8 mm), *Enterococcus faecalis* (11.4 mm), *Staphylococcus aureus* (11.2 mm), and *Streptococcus faecalis* (11.4 mm) (11.6 mm). The findings show that TiO_2 NPs are a potent antimicrobial agent that can pave the way for the development of new antimicrobial agents. Aside from that, spherical shaped TiO_2 NPs were produced from plants, specifically using *Morinda citrifolia* leaf extract and advanced hydrothermal method. The developed TiO_2 NPs were an excellent quasi-spherical shape in the size range of 15 to 19 nm. Furthermore, antimicrobial activity against human pathogens such as *E. coli, B. subtilis, S. aureus, P. aeruginosa, A. niger* and *C. albicans* were tested. TiO_2 NPs exhibited interesting antimicrobial activity, especially against Gram-positive bacteria [65].

2.2.6 Nickel-based metal NPs

Due to their excellent properties and application in various fields, interest in the synthesis of NiO NPs has increased in recent years. NiO NPs are used not only in photocatalytic sensors, but also in medical applications such as antibiotics, drug release in target cells, and nanocarriers aimed at moving along the growth trends of cell imaging and therapy [67, 68].

Recently, Uddin *et al* have conducted a study on the green synthesis of NiO NPs from *Berberis balochistanica* stem for investigating bioactivities [69]. Through their study, the antibacterial activity of phytofabricated BBS-NiONPs (100, 500 and 1000 μg mL^{-1}) against *Proteus vulgaris* and *Staphylococcus aureus* bacterial strains were conducted. BBS-NiONPs showed a dose-dependent response against both selected bacterial strains. Moreover, 10 μg Ciprofloxacin (positive control) was found more reactive than all the applied doses of BBS-NiONPs during antibacterial activities. Antifungal analysis of BBS-NiONPs was also carried out using different concentrations against *Fusarium oxysporum, Aspergillus niger* and *Alternaria alternata*. Dose-dependent responses against all three analysed fungus strains were observed. *F. oxysporum* was found less vulnerable at high concentration (1000 μg mL^{-1}) while *A. alternata* was found highly susceptible with a percentage inhibition of 71.25% followed by *A. niger* with a percentage inhibition of 39.51% at 1000 μg mL^{-1}.

In addition, Sabouri *et al* have proposed a novel synthesis method for producing neat and single-phase NiO NPs via a simple and cost-effective method using okra plant extract and $Ni(NO_3)_2.6H_2O$ as an inorganic precursor [67]. The antibacterial properties of NiO NPs prepared by applying okra plant extract were evaluated using the disc diffusion method against Gram-positive *S. aureus* and Gram-negative *E. coli* and *P. aeruginosa.* The results obtained conclude that the antibacterial effectiveness of NPs depends on particle size, stability, and moderately growing concentration. Hence, NPe size plays a vital role in antibacterial effectiveness, considering how most of the findings show that antibacterial activity increases as particle size decreases.

2.3 Synthesis method of metal and metal oxide based bionanomaterial for antimicrobial properties

Biological synthesis, also known as green synthesis, has grown great interest among researchers to produce nanomaterials that excel in antimicrobial properties. This is because green synthesis is more cost effective, always available, non-toxic and environmentally friendly as well as showing great stability, biocompatibility, safety and size/morphology of produced nanomaterials. The nanomaterials produced through this method are known as biogenic nanomaterials. The advantages using green synthesis method are shown in figure 2.2.

Bionanomaterials can be produced through different sources like plants extract (leaves, pulp, stem and etc) and microorganisms (bacteria, yeast and fungi). Plants use offers a few benefits over other sources, as phytochemicals can act as protective and stabilizers and prevent particles agglomeration [37]. Moreover, plant-based processes are cost-effective, suitable for large-scale synthesis of nanomaterials and safe for human and the environment [70]. Recently, there is a new trend to synthesize nanomaterials with ecologically friendly, cost-efficient, sustainable and much easier method, which is from biological and agricultural waste. All of sources more preferable due to their non-toxicity and excellent antimicrobial properties towards various types of strains microorganism. The bionanomaterials production via green synthesis method can occur either by intracellularly or in extracellular cell depends on the location of biomolecules release.

2.3.1 Plants

Bio-reduction method is a cost-effective approach for green synthesis nanomaterials that involved one pot or single step method, which require low energy usage [71]. Biomolecules present in plant extracts, such as enzymes, amino acids, proteins, vitamins, polysaccharides and organic acid (such as citric acid), can reduce metal ions to form nanomaterials. Different parts of the plant, such as leaves, roots, flowers, seeds, stems and fruits have been used to synthesize bionanomaterials of different sizes and shapes.

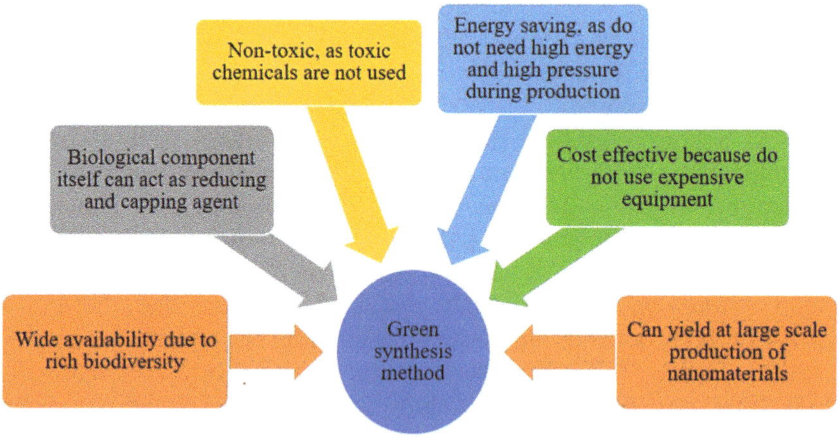

Figure 2.2. Highlight advantages using green synthesis method.

For instance, in study by Sentilkumar *et al*, cerium oxide (CeO_2) was synthesized using *S. acuta* leaf extract as reducing agent in cerium nitrate solution [72]. The resulting Chi-CeO_2 showed potential antimicrobial activities with inhibition zone of *E. coli* was 11 mm and 8 mm for *B. subtilis* at 50 µl. Meng and Sun also produced silver (Ag) NPs from $AgNO_3$ solution using *Rosa chinensis* flower extract and exhibited good antimicrobial activity, which showed zone of inhibition of *E. coli* (9.36 mm) and *S. aureus* (12.06 mm) as well as value of minimum inhibitory concentration (MIC), 0.016 µg mL^{-1} and 0.033 µg mL^{-1}, respectively [73].

Ag and Au NPs were also produced using *Euphrasia officinalis* leaf extracts and evaluated for antimicrobial properties. The result showed that zone of inhibition of *P. aeruginosa, E. coli, V. parahaemolyticus and S. aureus* at 1000 ppm were 15.3 mm, 11.7 mm, 14.7 mm and 13.7 mm respectively [74]. Study by Logeswari *et al* used 5 different types of plants namely *Syzygium cumini, Ocimum tenuiflorum, Centella asiatica, Solanum tricobatum* and *Citrus sinensis* to produce Ag NPs [75]. From the antimicrobial test, it showed that Ag NPs produced by *S. tricobatum* and *O. tenuiflorum* extracts exhibited the highest antimicrobial activity toward to *S. aureus* (30 mm) and *E. coli* (30 mm) respectively.

Geethalakshmi and Sarada obtained both Au and Ag nanomaterials from root extract of *Trianthema decandra* [76]. Both gold and silver nanoparticles showed outstanding activity resistant to *Y. enterocolitica, P. vulgaris, E. coli, S. aureus* and *S. faecalis*. For gold nanoparticles, an inhibition area of 13.5–15.5 mm was recorded, while the corresponding silver recorded an inhibition area of 15.5–20.5 mm. Karthiga used stem extract of *Garcinia mangostana* to yield silver nanomaterials and tested for its antimicrobial activity opposed to *E. coli, B. subtilis* and *K. planticola* [77]. The results showed that it is more effective against *E. coli* than *B. subtilis* and *K. planticola* because of the lower content of peptidoglycan and weaker cell wall.

On other hand, Prasanth *et al* used mangrove plant, *Rhizophora lamarckii* leaves to produce magnesium oxide (MgO) and tested with 4 microorganisms [78]. The obtained MgO found to have great antimicrobial activity with zone of inhibition of *S. aureus, S. pneumonia, S. typhi* and *E. coli*: 26.5 mm, 26.3 mm, 25.7 mm and 26.1 mm respectively. Meanwhile, Khan *et al* gained MgO through extraction of *Camellia sinensis* leaves with Mg $(NO_3)_2.6H_2O$ solution and have tested with three Gram positive and four Gram negative microorganisms [79]. With concentration of 25 µg mL^{-1}, S^{-1}. *mercescents* and *S. pyogenes* exhibited the highest zone of inhibition, 21 mm compared to *E. coli* (15 mm), *P. aeruginosa* (19 mm), *K. pneumoniae* (18 mm), *S. aureus* (18.8 mm) and *S. epidermidis* (19 mm).

In a study by Subhapriya and Gomathipriya, TiO_2 was produced from leaf extract of *Trigonella foenum-graecum* [80]. Various species of microoganisms have been tested, namely *S. faecalis, S. aureus, E. coli, E. faecalis, P. aeruginosa, P. vulgaris, Y. enterocolitica, K. pneumoniae, B. subtilis* and fungus *C. albicans*. Synthesized TiO_2 are effective toward to *Y. enterocolitica* (10.6 mm), *E. coli* (10.8 mm), *S. aureus* (11.2 mm), *E. faecalis* (11.4 mm) and *S. faecalis* (11.6 mm) shows significant antimicrobial activity. Table 2.3 summarized the metal and metal oxide nanomaterials from plants extract for antimicrobial test.

Table 2.3. The metal and metal oxide nanomaterials from plants extract from antimicrobial test

Materials produced	Part of plant used	Size and Morphology	Microorganisms tested	Antimicrobial activity Method	Reference
Chitosan-CeO_2	Leaf extract	3.61–24.40 nm (spherical shaped)	*E. coli* and *B. subtilis*	Disk diffusion	[72]
Ag	Flower extract	9–32 nm (spherical)	*E. coli and S. aureus*	Disk diffusion	[73]
Ag and Au	Leaf extract	Ag = 40.37 nmAu = 49.72 nm (spherical shaped)	*P. aeruginosa, E. coli, V. parahaemolyticus and S. aureus*	Disk diffusion	[74]
Ag	Plant extract	*O. tenuiflorum* (28 nm), *S. tricobatum* (22.3 nm), *S. cumini* (26.5 nm), *C. asiatica* (28.4 nm) and *C. sinensis* (65 nm) (Rod shaped)	*S. aureus, P. aerugina, E. coli and K. pneumoniae*	Disk diffusion	[75]
Au and Ag	Root extract	Au = 33.7–99.3 mm (spherical, triangular, hexagonal and cubic) Ag = 36–94 nm (spherical)	*S. aureus, S. faecalis, E. faecalis, E. coli, P. aeruginosa, P. vulgaris, B. subtilis, Y. enterocolitica*, and *C. albicans*	Disk diffusion	[76]
Ag	Stem extract	15–20 nm (spherical)	*E. coli, B. subtilis and K. planticola*	Disk diffusion	[77]
MgO	Leaf extract	A few nm to 80 nm (spherical shaped)	*E. coli, P. aeruginosa, S. mercescents, K. pneumoniae, S. aureus, S. epidermidis and S. pyogenes*	Disk diffusion	[79]
TiO_2	Leaf extract	20–90 nm (spherical shaped)	*S. aureus, S. faecalis, E. coli, P. vulgaris, E. faecalis, P. aeruginosa, Y. enterocolitica, K. pneumoniae, B. subtilis* and *fungus C. albicans*	Disk diffusion	[80]

2.3.2 Microorganism

In recent years, microorganisms have been utilized to synthesize a variety of nanomaterials in a manner that is not harmful to the environment. In this regard, there are many types of bacteria, fungi, algae and yeast species has been used as a source for nanomaterials production. The main similarity between them is the capability to reduce metal ions owing to the present of reductase enzyme.

2.3.2.1 Bacteria

Bacteria are a good choice because of their existence in many surroundings and the ability to adapt to extreme conditions. They are also rapid growing, cheap to develop and easy to control. Bacterial cell walls play a very crucial role. Essential metals need to penetrate the cytoplasm through the wall and be released back to the outside of the cell through the wall network. The cell wall component of peptidoglycan supports the interaction between metal and the chemically reactive group, followed by the polyanion for metal inorganic deposition. The walls with many possible metal bonding sites can be modified by chemical reaction (anhydride treatment) [81]. These chemical processes are carried out in specific groups, including amines and carboxyl groups that change positive charges to negative charges (during the critical procedural process of metal bonding). The crystalline and amorphous phases of particles are usually affected through the intracellular and extracellular environment and species with specific bacterial morphology.

Ramasamy *et al* attained Au–Ag NPs via green synthesis methods by mixing extracts of *Shewanella oneidensis* and $HAuCl_4.4H_2O$ and $AgNO_3$ [82]. The produced Au–Ag NPs have been tested with four microbials and showed great antimicrobial properties with zone of inhibition of *E. coli, P. aeruginosa, E. faecalis and S. aureus*: 21 mm, 23 mm, 20 mm and 22 mm, respectively. Meanwhile, in a study by Jayaseelan *et al* ZnO has been synthesized using *Aeromonas hydrophilia* dan tested with different types of bacteria and fungi [83]. Zone of inhibition result showed that 25 $\mu g\ mL^{-1}$ of ZnO had great antimicrobial activity toward *A. flavus* (19 ± 1.0 mm) and *P. aeruginosa* (22 ± 1.8 mm) compared to other species.

On the other hand, extract *Bacillus megaterium* has been introduced into $ZnNO_3.5H_2O$ solution to yield ZnO nanomaterials. ZnO shows significant antibacterial effects on multidrug resistant *H. pylori* strains. In addition, at a concentration of < 12.5 $\mu g\ mL^{-1}$, the biogenic ZnO is less toxic to mammalian cells [84]. ZnO nanomaterials with the size of 20–50 nm, synthesized with *Streptomyces* sp. showed as quite a promising candidate for antimicrobial properties with inhibition zone of *E. coli* (12 ± 5 mm) and *B. subtilis* (2 ± 5 mm) [85].

Bacteria of *Brevibacterium frigoritolerans* was also mixed with $AgNO_3$ solution to produce silver nanomaterials and tested with six types of microorganisms. Zone of inhibition result shows that synthesized Ag has excellent antimicrobial activity with *C. albicans* KACC 30 062 (22±0.7 mm), *V. parahaemolyticus* ATCC 33 844 (23±1.1 mm), *B. anthracis* NCTC 10 340 (14±0.2 mm), *B. cereus* ATCC 14 579 (14±0.3 mm), *S. enterica* ATCC 13 076 (12±0.4 mm) and *E. coli* ATCC 10 798 (10±0.7 mm) [86]. Meanwhile, *Enterococcus faecalis* was used to synthesize selenium

(Se) nanomaterials from sodium selenite (Na_2SeO_3) via extracellular method. Antimicrobial activity was tested against four types of microorganisms, *B. subtilis*, *P. aeruginosa*, *S. aureus*, and *E. coli*. However, the results showed Se nanomaterials only worked on *S. aureus* [87]. Table 2.4 summarises the bionanomaterial synthesized from bacteria species for antimicrobial properties.

2.3.2.2 Fungi

The synthesis of NPs by fungi is superior to that of bacteria. Compared to NPs synthesized by bacteria, fungi are capable of synthesizing NPs of nanometer size and have a more resistant monodisperse [88]. Fungi have potential strategies for synthesizing NPs extracellularly to improve their commercial viability. For example, *Fusarium acuminatum* releases the reducing agent into solution in the existence of an aqueous $AuCl_4$ ion solution with a reaction mediated by the NADH enzyme, forming Au NPs. Synthesized Au NPs exhibit long-term stability in solution because of protein binding via cysteine-lysine residue binding.

Manjunath *et al* obtained Au nanomaterials with the size of particles in the range of 30–60 nm of particles from fungus *Cladosporium cladosporioides* [89]. This study shows that NADPH-dependent reductase and phenolic compounds are involved in the bio-reduction of gold metal salts to nanomaterials. Synthesized Au exhibited promising antimicrobial activity with inhibition zone of *E. coli* (10.5 mm), *S. aureus* (12 mm), *B. subtilis* (9.5 mm), *S. epidermidis* (11.5 mm) and *A. niger* (10 mm). On the other hand, Ahmad *et al* produced Au and Ag nanomaterials using fungus *Candida albicans* [90]. The MIC of Au and Ag nanomaterials for Gram-negative bacteria was 128 g ml^{-1} and 8 g ml^{-1}, and for Gram-positive bacteria was 512 μg mL^{-1} and 32 μg mL^{-1}, respectively. The average zone of inhibition diameters of Au and Ag NPs for *Staphylococcus aureus* are 11 and 19 mm, respectively, while the average diameters of *E. coli* are 16 mm and 24 mm, respectively.

Silver nanomaterials synthesized using *Macrophomina phaseolina* fungus with particles size, 5–40 nm exhibited promising antimicrobial activity opposed to *A. tumefaciens* and *E. coli* with inhibition zone 29 and 27 mm, respectively. It was found that these NPs were naturally coated with protein. SDS-PAGE analysis also showed that there is an 85 kDa protein band, which is responsible for the capping and stabilization of Ag NPs [91]. *Aspergillus clavatus* was also used to synthesize Ag nanomaterials via extracellularly with particle sizes between 10–25 nm. The results show that Ag has a strong inhibitory effect on plates inoculated with *C. albicans*. A maximum inhibition zone of 16 mm compared to 9 mm was observed using *C. albicans* in the control group. The MIC range for *C. albicans* is 5.8 μg mL^{-1}, while the MFC is about 9.7 μg mL^{-1}. Ag was found to be effective toward *P. fluorescens* and *E. coli*, with their zone of inhibitions 14 and 10 mm, respectively [92]. Table 2.5 tabulates bionanomaterials synthesized by different kinds of fungi species for their antimicrobial properties.

2.3.2.3 Yeast

Yeast is recognized as a suitable substitute for nanomaterial production. Although there is no recognized bio-reduction mechanism, it is known that certain

Table 2.4. Bionanomaterials synthesized from bacteria species for antimicrobial properties

Bacteria species	Materials produced	Size and Morphology	Microorganisms tested	Synthesis Method	Reference
Shewanella oneidensis	Au-Ag	20 ± 3 nm (Spherical shaped)	*E. coli, P. s aeruginosa, E. faecalis* and *S. aureus*	Extracellular	[82]
Aeromonas hydrophila	ZnO	42–64 nm (Spherical shaped)	*A. Hydrophila, E. coli, P. aeruginosa, E. faecalis, S. pyogenes, A. flavus, A. niger*, and *C. albicans*	Extracellular	[83]
Bacillus megaterium	ZnO	45–95 nm (Rod and cubic shaped)	*H. pylari*	Extracellular	[84]
Streptomyces sp.	ZnO	20–50 nm (spherical shaped)	*E. coli and B. subtilis*	Extracellular	[85]
Brevibacterium frigoritolerans	Ag	50–150 nm (Spherical shaped)	*V. parahaemolyticus, S. enterica, B. anthracis, B. cereus, E. coli*, and *C. albicans.*	Extracellular	[86]
Enterococcus faecalis	Se	29–195 nm (Spherical)	*S. aureus, B. subtilis, E. coli* and *P. aeruginosa*	Extracellular	[87]

Table 2.5. Bionanomaterials synthesized by different kinds of fungi species for their antimicrobial properties

Fungi species	Materials produced	Size and Morphology	Microorganisms tested	Synthesis Method	Reference
Cladosporium cladosporioides	Au	30–60 nm (Spherical)	*E. coli, S. aureus, B. subtilis, S. epidermidis* and *A. niger*	Extracellular	[89]
Candida albicans	Au and Ag	Au = 4–10 nmAg = 30 nm (spherical)	*S. aureus* and *E. coli*	Extracellular	[90]
Macrophomina phaseolina	Ag	5–40 nm (Spherical)	*E. coli* and *A. tumefaciens*	Extracellular	[91]
Aspergillus clavatus	Ag	10–25 nm (Spherical and hexagonal)	*C. albicans, P. fluorescens* and *E. coli*	Extracellular	[92]

biochemical reducing agents in yeast cells are responsible for the formation of nanomaterial. *Saccharomyces cerevisiae* is a well-known yeast strain, which has been widely studied in recent years due to its ability to synthesis green nanomaterials through a biological reduction process. Certain special secreted proteins (such as cysteine) can ensure the biological reduction and capping of the nanomaterial produced [88]. For instance, Saravanan *et al* used Baker's yeast, *Saccharomyces cerevisiae* to synthesize Ag nanomaterials through extracellular method [93]. It was proven that synthesized Ag exhibit promising antimicrobial activity against *P. aeruginosa* (22 mm), *K. pneumoniae* (20 mm), *S. aureus* (15 mm), and *S. pyogenes* (14 mm), respectively. Asghari-paskiabi *et al* also used *Saccharomyces cerevisiae* to produce selenium sulfide (SeS) nanomaterials and the obtained SeS strongly inhibited the growth of pathogenic fungi; *A. fumigatus, A. flavus, A. alternata, M. canis, T. rubrum* and *C. Krusei*. In MTT assay, no cytotoxicity was observed [94].

On the other hand, Moghaddam *et al* studied the potential ability of a new yeast strain *Pichia kudriavzevii* to synthesize zinc oxide (ZnO) nanomaterials [95]. The synthesized ZnO has hexagonal wurtzite structure with particle size about 10–61 nm. The antimicrobial properties of ZnO have been tested against *B. subtilis, S. aureus, E. coli, S. epidermidis* and *S. marcescens* and maximum inhibition zone was achieved by *S. epidermidis* with 19 mm at 100 $\mu g\ mL^{-1}$ concentration.

2.3.2.4 Algae

Algae are eukaryotic aquatic autotrophs. Biological reduction of algae shows great possibilities for the development of green and clean synthesis of various metals and metal oxide NPs. It is known that a significant number of algae can synthesize different NPs, but the understanding is still incomplete in controlling the size and shape of the product. Hence, many researchers are concerned with exploring algae-mediated NP biosynthesis. Through evolution, algae have evolved the ability to produce complex inorganic salts in the intracellular or extracellular structure [96]. Metal ions are first seized on the surface of plant cells through electrostatic interactions between the ions and the negatively charged carboxylic acid groups existing on the cell surface. The ions are then reduced by cellular enzymes to form the nucleus, and then further reduced to form metal ions [81, 97].

Jena *et al* investigated the potential of algae, *Chlorococcum humicola* in production of Ag nanomaterials [98]. These findings indicate that the presence and binding of proteins and Ag nanomaterial may lead to their possible stabilization. The protein might be bonded to the nanomaterials via free amine or cysteine residues or via the electrostatic attraction of the negatively charged carboxylic acid groups of the enzyme present on the cell wall. The antimicrobial activity was tested with *E. coli* and showed promising potential as an antimicrobial agent with zone of inhibition 13 mm.

Urospora sp. is a green seaweed and has a variety of phytochemicals, including carbohydrates, phenols, steroids, alkaloids, saponins, and flavoids. Thus, reduction of Ag nanomaterials from seaweed is a green synthesis method with good antimicrobial activity. This is proven with the appearance of clear inhibitory zone of *S. aureus* (23 mm), followed by *B. subtilis* (20 mm) and *E. coli* (18 mm), as well as *P. aeruginosa* (14 mm) and lastly *K. pneumoniae* (12 mm) [99]. Meanwhile, study by Abboud *et al* used brown alga extract, *Bifurcaria bifurcate* to produce copper oxide nanomaterials and it is proven to have good antimicrobial properties with zone of inhibition of *E. aerogenes* and *S. aureus*, 14 and 16 mm, respectively [100].

On the other hand, in the literature, it is found that marine red algae are a rich source of phenolic compounds, especially bromophenol. According to reports, phenolic substances have a great range of biological effects such as antioxidants, anti-inflammatory drugs, vasodilators and antimicrobial agents. Therefore, Abdel-Raouf *et al* has used red algae, *Galaxaura elongate* to synthesize Au nanomaterials and showed quite promising antimicrobial properties against *S. aureus* (13 mm), *E. coli* (17 mm), *K. pneumoniae* (17 mm) and *P. aeroginose* (13 mm) [101]. A similar study has been also conducted on Ag nanomaterials derived from red algae, *Portieria hornemannii*. The synthesized Ag nanomaterials were found to be active against *V. parahaemolyticus, V. harveyi, V. anguillarum* and *V. alginolyticus*. It was found that the width of the inhibition zone of *V. harveyi* was 16 and 19 mm, while *V. anguillarum* was 23- and 28-mm. *V. parahaemolyticus* displayed that the diameter of inhibitory zone was about 20 mm, and the diameter of *V. vulnificus* was 18 and 22 mm, respectively [102]. Table 2.6 tabulates types of bionanomaterials synthesized from algae species for their antimicrobial properties.

2.4 Mechanism of antimicrobial activity through metal and metal oxide-based bionanomaterials

There are multiple mechanisms for metal NPs to attack the bacteria cell. According to Gold *et al* [44], there are three major pathways for metal NPs to interact physically with bacterial cells, as shown in figure 2.3.

Table 2.6. Type of bionanomaterials synthesized from algae species for their antimicrobial properties

Algae species	Materials produced	Size and Morphology	Microorganisms tested	Synthesis Method	Reference
Chlorococcum humicola (Green algae)	Ag	15–20 nm (Spherical)	*E. coli*	Extracellular	[98]
Urospora sp. (Green seaweed)	Ag	20–30 nm (Spherical)	*S. aureus, E. coli, P. aeruginosa, K. pneumoniae* and *B. subtilis*	Extracellular	[99]
Bifurcaria bifurcate (Brown algae)	CuO	5–45 nm (Spherical)	*E. aerogenes* and *S. aureus*	Extracellular	[100]
Galaxaura elongate (Red algae)	Au	3.85–77.13 nm (Spherical)	*S. aureus, E. coli, K. pneumoniae* and *P. aeruginosa*	Extracellular	[101]
Portieria hornemannii (Red algae)	Ag	60–70 nm (Spherical)	*V, harveyii, V. parahaemolyticus, V. alginolyticus and V. anguillarum*	Extracellular	[102]

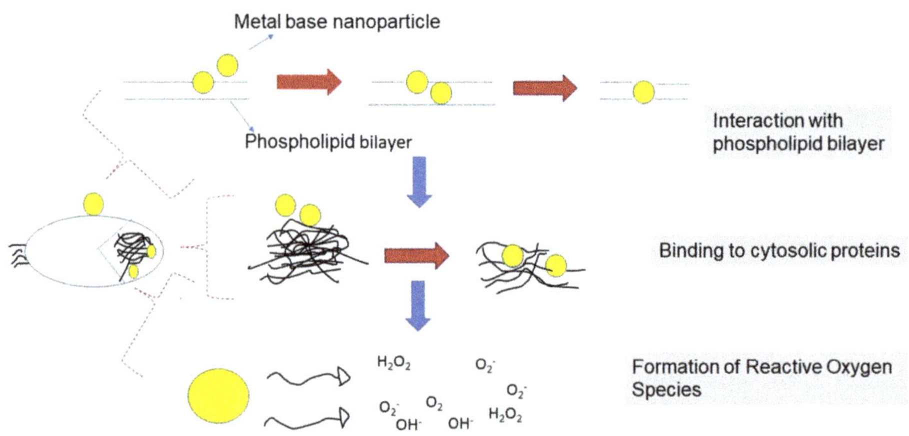

Figure 2.3. Metal nanomaterials interact with prokaryotic cell compromising cellular function.

2.4.1 General mechanistic pathway of the antimicrobial activity of nanoparticle-based metal and metal oxides

The majority of nanomaterials described in recent studies have antibacterial activity and undergo the following main mechanisms: inhibition of cell wall, inhibition of enzymes, reduced DNA production, and toxic ROS.

2.4.1.1 Interactions with phospholipid bilayer

As reported by Beyth *et al*, the cell membrane is destroyed by metal-based NPs through electrostatic binding to the bacterial cell wall and/or releasing metallic ions [103]. Via electrostatic contact, the negative charge of cellular parts and positive charge of the NPs interact on the surface. The membrane and bacterial proteins are destroyed due to this interaction and the disruption of cell barrier cause releasing water from cytosol. However, this imbalance of membrane stability and ions results in cell death and cells attempt to compensate for this loss. These effects have been demonstrated by gold, magnesium oxide, silver, zinc oxide, and titanium oxide NPs.

2.4.1.2 Cytosolic proteins binding

Enzymes and DNA are cytosolic proteins and the main mechanism that induces an antimicrobial response is through metallic binding to cytosolic protein. This interaction contributes to reduced activity, inhibiting the development of ATP, respiratory and metabolic pathways. Silver, for instance, binds to DNA and respiratory chain enzymes, retarding replication and division [104]. On the other hand, the interaction of Au with DNA inside the cell may cause alteration in genes. [105].

2.4.1.3 Production of reactive oxygen species

By generating oxygen free radicals, such as superoxide anions or hydrogen peroxide (H_2O_2) or ROS, NPs destroy bacteria. ROS formation is caused indirectly by NPs themselves. ROS leads to severe oxidative stress and destroys macromolecules in the cell that generally cause lipid peroxidation, protein modification, enzyme inhibition, and damage to RNA/DNA. Pits or holes within the bacterial membrane can also be formed by this serious oxidative stress, causing cell lysis [104]. With silver, hydroxyl radical (OH) formation has been observed [106]. ROS formation producing H_2O_2 from glucose oxidase is demonstrated by magnesium oxide, zinc oxide, and gold [105]. Titanium dioxide has triggered ROS production from both H_2O_2 and OH upon exposure to light. The antibacterial activity of NPs is summarized in figure 2.4.

2.4.2 Antimicrobial mechanism of various metals and metal oxides

TiO_2, Ag Pt, Ag, Au Cu, and ZnO are the main NPs that have been used in biomedical applications as antibacterial agents in recent years [107]. This is because of their long-term stability and outstanding biocompatibility characteristics.

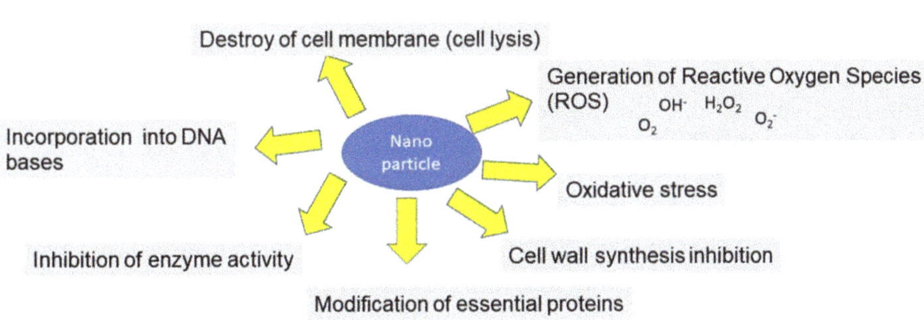

Figure 2.4. Summary of antibacterial activity of NPs.

2.4.2.1 Silver nanoparticles antimicrobial mechanism

AgNPs are capable of binding to the surface of the antimicrobial cell wall, penetrating and disrupting the cell membrane structure and eventually causing cell death. The bacterial cell surface shows formation of pits that indicate the damage of the cell membrane [104].

Another mechanism leading to cell death is believed to be the development of free radicals by AgNPs. Based on the study of electron spin resonance spectroscopy, it has been proposed that free radical formation occurs when AgNPs are in contact with bacterial cells. Therefore, the cell membrane can be weakened by these free radicals and made porous, ultimately leading to cell death [108]. ROS are highly reactive molecules that are capable of oxidizing cell structures and biomolecules, causing DNA denaturation, lipid peroxidation and protein modifications that contributes to the death of cells. Das *et al* have found that the expansion of ROS can lead to cytotoxicity in microbes induced by AgNPs [81].

As stated by Prabhu and Poulose, the silver ions are released from AgNPs via interaction of thiol group, thus inactivating several essential enzymes in the cell [109]. Another fact is that the essential components of DNA are phosphorous and sulfur. Since phosphorus and sulfur are soft bases, they have a natural propensity to react with each other, whereas silver is soft acid. The NP can kill DNA by interaction with these soft bases and cause targeted cell death. DNA replication problems in pathogens can also be caused by the interaction of AgNPs with sulfur and phosphorous DNA. Another biosynthesized AgNP antibacterial mechanism includes bacterial protein denaturation. Previous studies have revealed that the electrostatic forces among the microbial cell membrane and NPs are vital for antimicrobial effects [10, 110]. There are negatively charged surface membranes of microorganisms, including viruses, fungi and bacteria, while the NP has a positively charged surface. Kim *et al*, explained that the AgNPs can bind to protein functional group (-SH) and thus induce protein denaturation [111]. According to earlier studies, by harming the proton pump, AgNPs could inhibit fungal cell growth and cause fungal surface protein denaturation. This phenomena will destroy the cell membranes by changing permeability of protein lipid bilayer or membrane. The electron transport system can also be disrupted by AgNPs.

2.4.2.2 Antimicrobial mechanism of gold nanoparticles

Via two methods, AuNPs use their antimicrobial potentialities toward multidrug-resistant (Gram-negative) bacteria: degrading the membrane's ability to minimize ATP levels by constraining tRNA binding of the ribosome subunit and inhibiting ATPase activity. Cui *et al*, explained that AuNPs modified with 4,6-diaminopyrimidine thiol can inhibit the bacterial tRNA function by affecting their protein synthesis that eventually leads to microbial growth inhibition [112].

Gold (soft acid) has a higher probability to react to soft bases containing sulfur and phosphorus. Therefore, AuNPs can interact with bases containing phosphorus in DNA or proteins containing sulfur in the membrane. Such interactions eventually lead to cell death by disrupting the normal functioning of DNA, including the mechanism of repair, replication and synthesis of new strands.[113].

2.4.2.3 *Antimicrobial mechanism of zinc oxide nanoparticles*

The bactericidal behavior of ZnONPs can be affected by the size, shape and surface-capping agents of NPs. Xie *et al*, reported that higher antibacterial efficacy was correlated with the smaller size of ZnONPs [114]. ZnONPs can influence the function of cells by creating excessive amounts of nucleic acids. The increased nucleic acid level in fungal hyphae may be because of the stress response mediated by the NP. Similarly, the rise in the amount of carbohydrates can be viewed as a mechanism of self-protection from ZnONPs, and lead to deformed structures of hyphal cells and ultimately trigger cell death. Hence, these studies indicate that the mechanisms of action of ZnONPs vary from those of antibacterial activity. Thus, these studies indicate that the mechanisms of action of ZnONPs vary from those of antibacterial operation. Different fungi were also found to be known for their production of mycotoxins, such as patulin and fusaric acid. The inhibition rate of these mycotoxins has been determined to steadily decrease with an increased concentration of ZnONPs. Such inhibitions in the development of mycotoxins may be because of the inactivation by ZnONPs of enzymes involved in the biosynthesis process of these toxins.

2.4.2.4 *Antimicrobial mechanism of copper nanoparticles*

Based on preliminary study, the local conductivity and pH can be altered by Cu^{2+} ions release [110]. This recovery of Cu^{2+} ions into the solution will cause the disruption of bacteria membranes which disturb the function of cellular enzymes. In 2015, Shende *et al* found that a CuNP will change the fungal structure and function of the cell in a mechanistic study [115]. Furthermore, by affecting its DNA, it can induce cell death and disrupt its replication and transcription mechanisms. In addition, when CuNPs interacted with -SH (sulfhydryl) groups, it might inactivate proteins and cause growth inhibition. An alternative idea suggests that oxidative stress induced by CuNP within microbes may result in their cell death.

Recently, the synthesis of metal NPs using medicinal plants and microorganisms has been extensively studied and this is renowned as an environmentally friendly, nontoxic and efficient method for application in biomedicine and bioremediation. The mechanisms of antimicrobial activity of bionanomaterial are summarized in table 2.7.

2.5 Future prospects and challenges

Antibiotic drugs are one of the antimicrobial agents that give a huge benefit to the agriculture industry. It is widely used in the animals' food and welfare also as a control to plant diseases [122]. However, the consensus on the use of conventional antibiotic drug in agriculture leads to the rising of the antimicrobial resistance. Therefore, a shift towards the development of metal and metal oxide-based bionanomaterials as the antimicrobial agents is considered as a great move. Metal and metal oxide NPs can overcome drug resistance because they employ multiple mechanisms to kill the microbes concurrently. However, the stability and toxicity of the metal and metal oxide bionanomaterials should be studied in detail. Although it is reported to be less harmful towards mammalian cells compared to living things,

Table 2.7. Possible mechanism of antimicrobial activity of bionanomaterial.

Nanoparticle	Size (nm)	Plant	Properties	Antimicrobial Mechanism	References
AgNps	31	*Ocimum gratissimum*	Antibacterial	The plasma membrane is destroyed due to development of ROS, preventing the transport of electrons and distrupting enzymes in the respiratory chain, eventually causing cell death	[116]
AuNps	16	*Ananascomosus*	Antibacterial	As electrostatic interaction occurs between bacterial cells and AuNPs, Au ions are released and penetrate into cell wall of bacteria. These ions further disturb the metabolic processes that consequently lead to the death of cells.	[117]
ZnONPs	40–25	*Aloe*	Antimicrobial	Hydrogen peroxide, which can penetrate the cell wall of bacteria and cause bacterial death, is produced by ROS reacting with H ions. In addition, binding of NPs with presence bacterial protein groups on the cell wall to -SH, will reduce the cell permeability which leads to cell lysis. This disruption leads to leakage of mineral, protein and genetic material that causes cell death.	[118]
	Not given	*Azadirachtaindica (L.)*	Antimicrobial	Free radicals generated from H_2O_2 can penetrate the cell membrane, leading to the death of bacterial cells	[119]
CuONps	20	*Aloe vera*	Antibacterial	CuNPS release Cu ions and interact with the negatively charged cell wall and cause mutilation to the bacteria cell wall. Interaction of DNA molecule with Cu ions, forming nucleic acid cross-linking strands; hence, cell death is caused by disorganization of DNA helical strand (protein denaturation). Bacterial cell biochemical processes are also disrupted by these ions	[120]
CuNPs	20	*Citrus medica Linn. (Idilimbu)*	Antimicrobial	Ions are released by metal NPs and accumulate on the cell surface. The association of these ions with NPs creates pits in the membrane; induces cellular components leakage. In addition, CuNPs will penetrate into microbial cells and constrain their enzymes/proteins by increasing the production of cell death. The interaction of Cu ion with SH group generates ROS and distrupting the helical structure of DNA.	[121]

the slight impact of metal NPs especially the toxicity of silver metals should be considered before they can be widely used. Researchers should have explored the surface engineering techniques for the metal and metal oxide bionanomaterials to achieve biocompability.

Various kinds of different biological sources for synthesizing metal and metal oxide-based bionanomaterials were reported in the past decade and this is expected be improved in future. Plant extracts, algae and microorganisms such as fungi, bacteria and yeast are the probable sources for synthesizing metal and metal oxide nanomaterials. However, there are still ample potential biological sources that remain unexplored such as marine plants and plant gums. Therefore, the exploration of other biological sources for synthesizing of metal and metal oxide nanomaterials should be actively progressed. Bionanomaterial antimicrobial agents are considered as the important tools to reduce destructive associated with NPs that are commonly exploited in the laboratory and in industry. It has a great future as the antimicrobial agent in the agriculture industry. Therefore, these limitations open new and boundless opportunities in this emerging field of research.

2.6 Concluding remarks and recommendations

Metal and metal oxide-based bionanomaterials can reduce environmental impact, and are cost-effective and non-toxic to living things and the environment. The antibacterial properties of metal and metal oxide-based bionanomaterial are greatly influenced by their physicochemical properties, such as morphology and particle size. Triangular AgNPs showed an improved antibacterial activity in comparison to their spherical AgNPs. The morphology and size of the NPs can make a huge difference in their antibacterial activity. However, the reason behind this is not yet well understood. There is still a lack of understanding of the fundamental mechanism and modeling factors. It is recommended that future research and development of metal and metal oxide-based bionanomaterial as a antimicrobial agent should also be aligned towards extending laboratory-based to large-scale production by counting their present issues, especially the health and environmental effects. The properties and behavior of metal and metal oxide-based bionanomaterials should be upgraded to meet the aforementioned points before widely utilizing them in the agriculture industry.

Acknowledgment

The authors thank the Universiti Sultan Zainal Abidin, Universiti Teknologi Malaysia, Universiti Sains Islam Malaysia and Universiti Sains Malaysia for human capital assistance.

References

[1] Gupta A, Gupta R and Singh R L 2017 Microbes and environment *Principles and Applications of Environmental Biotechnology for a Sustainable Future. Applied Environmental Science and Engineering for a Sustainable Future* ed R Singh (Berlin: Springer)

[2] Moody V and Needles H L 2004 Antimicrobial agents *Tufted Carpet.* 145–53
[3] Reygaert W C 2018 An overview of the antimicrobial resistance mechanisms of bacteria *AIMS Microbiol.* **4** 482–501
[4] Varier K M, Gudeppu M, Chinnasamy A, Thangarajan S, Balasubramanian J, Li Y and Gajendran B 2019 Nanoparticles: antimicrobial applications and its prospects *Prog. Bot.* **80** 321–55
[5] Nussbaum V F, Brands M, Hinzen B, Weigand S and Häbich D 2006 Antibacterial natural products in medicinal chemistry—exodus or revival? *Angew. Chem. Int. Ed.* **45** 5072–129
[6] Fair R J and Tor Y 2014 Antibiotics and Bacterial Resistance in the 21st Century *Persp. Med. Chem.* **6** 25–64
[7] Ruddaraju L K, Pammi S V N, Guntuku G S, Padavala V S and Kolapalli V R M 2020 A review on anti-bacterials to combat resistance: from ancient era of plants and metals to present and future perspectives of green nano technological combinations *Asian J. Pharm. Sci.* **15** 42–59
[8] Breijyeh Z, Jubeh B and Karaman R 2020 Resistance of gram-negative bacteria to current antibacterial agents and approaches to resolve it *Molecules.* **25** 1–23
[9] Khairnar S D, Shinde S G and Shrivastava V S 2019 A short review on the improvement of antimicrobial activity by metal and nonmetal doping in nanoscale antimicrobial materials *J. Nanomed. Biotherap. Discov.* **9** 1–7
[10] Wang L, Hu C and Shao L 2017 The antimicrobial activity of nanoparticles: present situation and prospects for the future *Int. J. Nanomed.* **12** 1227–49
[11] Ali A S 2020 Application of nanomaterials in environmental improvement *Nanotechnology and the Environment* ed M Sen (Rijeka: Intechopen)
[12] Deplanche K, Caldelari I, Mikheenko I P, Sargent F and Macaskie L E 2010 Involvement of hydrogenases in the formation of highly catalytic Pd (0) nanoparticles by bioreduction of Pd (II) using *Escherichia coli* mutant strains *Microbiology.* **156** 2630–40
[13] Gurunathan S, Han J W, Kwon D N and Kim J H 2014 Enhanced antibacterial and anti-biofilm activities of silver nanoparticles against gram-negative and gram-positive bacteria *Nanoscale Res. Lett.* **9** 373
[14] Cha S H, Hong J, McGuffie M, Yeom B, VanEpps J S and Kotov N A 2015 Shape-dependent biomimetic inhibition of enzyme by nanoparticles and their antibacterial activity *ACS Nano.* **9** 9097–105
[15] Actis L, Srinivasan A, Lopez-Ribot J L, Ramasubramanian A K and Ong J L 2015 Effect of silver nanoparticle geometry on methicillin susceptible and resistant *Staphylococcus aureus*, and osteoblast viability *J. Mater. Sci.* **26** 215
[16] Talebian N and Zavvare H S H 2014 Enhanced bactericidal action of SnO_2 nanostructures having different morphologies under visible light: influence of surfactant *J. Photochem. Photobiol. B Biol.* **130** 132–9
[17] Sukhorukova I V, Sheveyko A N, Kiryukhantsev-Korneev P V, Zhitnyak I Y, Gloushankova N A, Denisenko E A, Filippovich S Y, Ignatov S G and Shtansky D V 2015 Toward bioactive yet antibacterial surfaces *Colloids Surf.* B **135** 158–65
[18] Pan X, Wang Y, Chen Z, Pan D, Cheng Y, Liu Z, Lin Z and Guan X 2013 Investigation of antibacterial activity and related mechanism of a series of nano-Mg $(OH)_2$ *ACS Appl. Mater. Interfaces* **5** 1137–42
[19] Maeda H 2010 Tumor-selective delivery of macromolecular drugs via the EPR effect: background and future prospects *Bioconjug. Chem.* **21** 797–802

[20] Fang B, Jiang Y, Nusslein K, Rotello V M and Santore M M 2015 Antimicrobial surfaces containing cationic nanoparticles: how immobilized, clustered, and protruding cationic charge presentation affects killing activity and kinetics *Colloids Surf.* B **125** 255–63
[21] Arakha M, Sweta P, Devyani S, Tapan K P, Bairagi C M, Krishna P, Bibekanand M and Suman J 2015 Antimicrobial activity of iron oxide nanoparticle upon modulation of nanoparticle-bacteria interface *Sci. Rep.* **5** 14813
[22] Chaudhry N, Dwivedi S, Chaudhry V, Singh A, Saquib Q, Azam A and Musarrat J 2018 Bio-inspired nanomaterials in agriculture and food: current status, foreseen applications and challenges *Microbial. Pathog.* **123** 196–200
[23] Folorunso A, Akintelu S, Oyebamiji A K, Ajayi S, Abiola B, Abdusalam I and Morakinyo A 2019 Biosynthesis, characterization and antimicrobial activity of gold nanoparticles from leaf extracts of *Annona muricata J. Nanostruct. Chem.* **9** 111–7
[24] Zhang Y, Yang D, Kong Y, Wang X, Pandoli O and Gao G 2010 Synergetic antibacterial effects of silver nanoparticles@aloe vera prepared via a green method *Nano Biomed. Eng.* **2** 252–57
[25] Chandran S P, Chaudhary M, Pasricha R, Ahmad A and Sastry M 2006 Synthesis of gold nanotriangles and silver nanoparticles using aloe wera plant extract *Biotechnol. Prog.* **22** 577–83
[26] Shankar S S, Rai A, Ahmad A and Sastry M 2004 Rapid synthesis of Au, Ag, and bimetallic Au core–Ag shell nanoparticles using Neem (*Azadirachta indica*) leaf broth *J. Colloid Interface Sci.* **275** 496–502
[27] Vankar P S and Bajpai D 2010 Preparation of gold nanoparticles from *Mirabilis jalapa* flowers *Indian J. Biochem. Biophys.* 47 157–60 Available from: ID: 20653286
[28] Patil C D, Patil S V, Borase H P, Salunke B K and Salunkhe R B 2012 Larvicidal activity of silver nanoparticles synthesized using *Plumeria rubra* plant latex against *Aedes aegypti* and *Anopheles stephensi Parasitol. Res.* **110** 1815–22
[29] Saravanan M and Nanda A 2010 Extracellular synthesis of silver bionanoparticles from *Aspergillus clavatus* and its antimicrobial activity against MRSA and MRSE *Colloids Surf.,* B **77** 214–18
[30] Mala J G S and Rose C 2014 Facile production of ZnS quantum dot nanoparticles by *Saccharomyces cerevisiae* MTCC 2918 *J. Biotechnol.* **170** 73–8
[31] Majeed S, Abdullah M S, Dash G K, Ansari M T and Nanda A 2016 Biochemical synthesis of silver nanoprticles using filamentous fungi *Penicillium decumbens* (MTCC-2494) and its efficacy against A-549 lung cancer cell line *Chin. J. Nat. Med.* **14** 615–20
[32] Mohammadian A, Shoja Alsadati S A and Habibi M 2007 *Fusarium oxysporum* mediates photogeneration of silver nanoparticles *Sci. Iran.* **14** 323–6 Available from http://scientiairanica.sharif.edu/article_2860.html
[33] Gericke M and Pinches A 2006 Biological synthesis of metal nanoparticles *Hydrometallurgy.* **83** 132–40
[34] Mishra A, Kumari M, Pandey S, Chaudhry V, Gupta K C and Nautiyal C S 2014 Biocatalytic and antimicrobial activities of gold nanoparticles synthesized by *Trichoderma* sp *Bioresour. Technol.* **166** 235–42
[35] Alghuthaymi M A, Almoammar H, Rai M, Said-Galiev E and Abd-Elsalam K A 2015 Myconanoparticles: synthesis and their role in phytopathogens management *Biotechnol. Biotechnol. Equip.* **29** 221–36

[36] Cheeseman S, Christofferson A J, Kariuki R, Cozzolino D, Daeneke T and Crawford R J 2020 Antimicrobial metal nanomaterials: from passive to stimuli-activated applications *Adv. Sci.* **7**

[37] Zhang X F, Liu Z G, Shen W and Gurunathan S 2016 Silver nanoparticles: synthesis, characterization, properties, applications, and therapeutic approaches *Int. J. Mol. Sci.* **17**

[38] Mochochoko T, Oluwafemi O S, Jumbam D N and Songca S P 2013 Green synthesis of silver nanoparticles using cellulose extracted from an aquatic weed; water hyacinth *Carbohydr. Polym.* **98** 290–94

[39] Korbekandi H and Iravani S 2012 Optimization of biological synthesis of silver nanoparticles using *Lactobacillus casei* subsp. *casei J. Chem. Technol. Biotechnol.* **87** 932–37

[40] Cauerhff A and Castro G R 2013 Bionanoparticles, a green nanochemistry approach *Electron. J. Biotechnol.* **16** 11

[41] Pal S, Tak Y K and Song J M 2007 Does the antibacterial activity of silver nanoparticles depend on the shape of the nanoparticle? A study of the gram-negative bacterium *Escherichia coli Appl. Environ. Microbiol.* **73** 1712–20

[42] Majumdar D, Jung K H, Zhang H, Nannapaneni S, Wang X and Ruhul Amin A R M 2014 Luteolin nanoparticle in chemoprevention: *in vitro* and *in vivo* anticancer activity *Cancer Prev. Res.* **7** 65–73

[43] Correa M G, Martínez F B, Vidal C P, Streitt C, Escrig J and de Dicastillo C L 2020 Antimicrobial metal-based nanoparticles: a review on their synthesis, types and antimicrobial action *Beilstein J. Nanotechnol.* **11** 1450–69

[44] Gold K, Slay B, Knackstedt M and Gaharwar A K 2018 Antimicrobial activity of metal and metal-oxide based nanoparticles *Adv. Ther.* **1** 1700033

[45] Porta E, Cogliati S, Francisco M, Roldán M V, Mamana N and Grau R 2019 Stable colloidal copper nanoparticles functionalized with siloxane groups and their microbicidal activity *J. Inorg. Organomet. Polym. Mater.* **3** 964–78

[46] Palza H, Nuñez M, Bastías R and Delgado K 2018 *In situ* antimicrobial behavior of materials with copper-based additives in a hospital environment *Int. J. Antimicrob. Agents* **6** 912–7

[47] Zeiger M, Solioz M, Edongué H, Arzt E and Schneider A S 2014 Surface structure influences contact killing of bacteria by copper *Microbiologyopen.* **3** 327–32

[48] Micheals H T, Noyce J O and Keevil C W 2009 Effects of temperature and humidity on the efficacy of methicillin- resistant *Staphylococcus aureus* challenged antimicrobial materials containing silver and copper *Lett. Appl. Microbiol.* **49** 191–5

[49] Vincent M, Duval R E, Hartemann P and Engels-Deutsch M 2018 Contact killing and antimicrobial properties of copper *J. Appl. Microbiol.* **124** 1032–46

[50] Lee H J, Song J Y and Kim B S 2013 Biological synthesis of copper nanoparticles using Magnolia kobus leaf extract and their antibacterial activity *J. Chem. Technol. Biotechnol.* **88** 1971–77

[51] Usman M S, El Zowalaty M E, Shameli K, Zainuddin N, Salama M and Ibrahim N A 2013 Synthesis, characterization, and antimicrobial properties of copper nanoparticles *Int. J. Nanomed.* **8** 4467–79

[52] Rajeshkumar S *et al* 2019 Antibacterial and antioxidant potential of biosynthesized copper nanoparticles mediated through *Cissus arnotiana* plant extract *J. Photochem. Photobiol.,* **197** 111531

[53] Sánchez-López E, Gomes D, Esteruelas G, Bonilla L, Lopez-Machado A L and Galindo R 2020 Metal-based nanoparticles as antimicrobial agents: an overview *Nanomaterials.* **10** 1–39
[54] Balakumaran M D, Ramachandran R, Balashanmugam P, Mukeshkumar D J and Kalaichelvan P T 2016 Mycosynthesis of silver and gold nanoparticles: optimization, characterization and antimicrobial activity against human pathogens *Microbiol. Res.* **182** 8–20
[55] Abdel-Kareem M M and Zohri A A 2018 Extracellular mycosynthesis of gold nanoparticles using Trichoderma hamatum: optimization, characterization and antimicrobial activity *Lett. Appl. Microbiol.* **67** 465–75
[56] Alavi M and Nokhodchi A 2020 An overview on antimicrobial and wound healing properties of ZnO nanobiofilms, hydrogels, and bionanocomposites based on cellulose, chitosan, and alginate polymers *Carbohydr. Polym.* **227** 115349
[57] Aditya A, Chattopadhyay S, Jha D, Gautam H K, Maiti S and Ganguli M 2018 Zinc oxide nanoparticles dispersed in ionic liquids show high antimicrobial efficacy to skin-specific bacteria *ACS Appl. Mater. Interfaces* **10** 15401–11
[58] Batool M, Khurshid S, Qureshi Z and Daoush W M 2021 Adsorption, antimicrobial and wound healing activities of biosynthesised zinc oxide nanoparticles *Chem. Pap.* **75** 898–907
[59] Awwad A M, Amer M W, Salem N M and Abdeen A O 2020 Green synthesis of zinc oxide nanoparticles (ZnO-NPs) using Ailanthus altissima fruit extracts and antibacterial activity *Chem. Int.* **6** 151–9
[60] Emamifar A, Kadivar M, Shahedi M and Soleimanian-Zad S 2011 Effect of nanocomposite packaging containing Ag and ZnO on inactivation of Lactobacillus plantarum in orange juice *Food Control.* **3–4** 408–13
[61] Al-Dhabi N A and Arasu M V 2018 Environmentally-friendly green approach for the production of zinc oxide nanoparticles and their anti-fungal, ovicidal, and larvicidal properties *Nanomaterials.* **8**
[62] Hussain A, Ovesb M, Alajmi M F, Hussain I, Amird S, Ahmede J, Rehmana M T, El-Seedif H R and Ali I 2019 Biogenesis of ZnO nanoparticles using *Pandanus odorifer* leaf extract: anticancer and antimicrobial activities *RSC Adv.* **9** 15357–69
[63] Dréno B, Alexis A, Chuberre B and Marinovich M 2019 Safety of titanium dioxide nanoparticles in cosmetics *J. Eur. Acad. Dermatol. Venereol.* **33** 34–46
[64] Radtke A 2019 Photocatalytic activity of nanostructured titania films obtained by electrochemical, chemical, and thermal oxidation of Ti_6Al_4V alloy�comparative analysis *Catalysts.* **9**
[65] de Dicastillo C L, Patiño C, Galotto M J, Vásquez-Martínez Y, Torrent C and Alburquenque D 2019 Novel hollow titanium dioxide nanospheres with antimicrobial activity against resistant bacteria *Beilstein J. Nanotechnol.* **10** 1716–25
[66] Youssef A M, Assem F M, Abdel-Aziz M E, Elaaser M, Ibrahim O A and Mahmoud M 2019 Development of bionanocomposite materials and its use in coating of Ras cheese *Food Chem.* **270** 467–75
[67] Sabouri Z, Akbari A, Hosseini H A, Hashemzadeh A and Darroudi M 2019 Eco-friendly biosynthesis of nickel oxide nanoparticles mediated by okra plant extract and investigation of their photocatalytic, magnetic, cytotoxicity, and antibacterial properties *J. Clust. Sci.* **6** 1425–34
[68] Saleem S, Ahmed B, Khan M S, Al-Shaeri M and Musarrat J 2017 Inhibition of growth and biofilm formation of clinical bacterial isolates by NiO nanoparticles synthesized from Eucalyptus globulus plants *Microb. Pathog.* **111** 375–87

[69] Uddin S, Safdar L B, Anwar S, Iqbal J, Laila S and Abbasi B A 2021 Green synthesis of nickel oxide nanoparticles from *Berberis balochistanica* stem for investigating bioactivities *Molecules.* **26** 1–14

[70] Keat C L, Aziz A, Eid A M and Elmarzugi N A 2015 Biosynthesis of nanoparticles and silver nanoparticles *Bioresour. Bioprocess.* **2** 47

[71] Singh J, Dutta T, Kim K H, Rawat M, Samddar P and Kumar P 2018 Green synthesis of metals and their oxide nanoparticles: applications for environmental remediation *J. Nanobiotechnol.* **16** 1–24

[72] Senthilkumar R P, Bhuvaneshwari V, Ranjithkumar R, Sathiyavimal S, Malayaman V and Chandarshekar B 2017 Synthesis, characterization and antibacterial activity of hybrid chitosan-cerium oxide nanoparticles: as a bionanomaterials *Int. J. Biol. Macromol.* **104** 1746–52

[73] Meng Y and Sun Y 2016 Development of biogenic silver nanoparticle using rosa chinensis flower extract and its antibacterial property *J. Nanosci. Nanotechnol.* **16** 3969–72

[74] Singh H, Du J, Singh P and Yi T H 2017 Ecofriendly synthesis of silver and gold nanoparticles by *Euphrasia officinalis* leaf extract and its biomedical applications *Artif. Cells, Nanomed. Biotechnol.* **46** 1163–70

[75] Logeswari P, Silambarasan S and Abraham J 2015 Synthesis of silver nanoparticles using plants extract and analysis of their antimicrobial property *J. Saudi Chem. Soc.* **19** 311–17

[76] Geethalakshmi R and Sarada D V L 2012 Gold and silver nanoparticles from *Trianthema decandra*: synthesis, characterization, and antimicrobial properties *Int. J. Nanomed.* 5375

[77] Karthiga P 2017 Preparation of silver nanoparticles by Garcinia mangostana stem extract and investigation of the antimicrobial properties *Biotechnol. Res. Innova.* **2** 30–6

[78] Prasanth R, Kumar S D, Jayalakshmi A, Singaravelu G, Govindaraju K and Kumar V G 2019 Green synthesis of magnesium oxide nanoparticles and their antibacterial activity *Indian J. Geo. Marine Sci.* **48** 1210–5 http://nopr.niscair.res.in/handle/123456789/49711

[79] Khan A, Shabir D, Ahmad P, Khandaker M U, Faruque M R I and Din I U 2021 Biosynthesis and antibacterial activity of MgO-NPs produced from *Camellia sinensis* leaves extract *Mater. Res. Express* **8** 1–10

[80] Subhapriya S and Gomathipriya P 2018 Green synthesis of titanium dioxide (TiO_2) nanoparticles by Trigonella foenum-graecum extract and its antimicrobial properties *Microbial. Pathog.* **116** 215–20

[81] Das R K, Pachapur V L, Lonappan L, Naghdi M, Pulicharla R, Maiti S, Cledon M, Dalila L M A, Sarma S J and Brar S K 2017 Biological synthesis of metallic nanoparticles: plants, animals and microbial aspects *Nanotechnol. Environ. Eng.* **2** 1–21

[82] Ramasamy M, Lee J H and Lee J 2016 Potent antimicrobial and antibiofilm activities of bacteriogenically synthesized gold–silver nanoparticles against pathogenic bacteria and their physiochemical characterizations *J. Biomater. Appl.* 1–13

[83] Jayaseelan C, Rahuman A A, Kirthi A V, Marimuthu S, Santhoshkumar T, Bagavan A, Gaurav K, Karthik L and Rao K V B 2012 Novel microbial route to synthesize ZnO nanoparticles using *Aeromonas hydrophila* and their activity against pathogenic bacteria and fungi *Spectrochim. Acta, Part* A **90** 78–84

[84] Saravanan M, Gopinath V, Chaurasia M K, Syed A, Ameen F and Purushothaman N 2018 Green synthesis of anisotropic zinc oxide nanoparticles with antibacterial and cytofriendly properties *Microbial. Pathog.* **115** 57–63

[85] Balraj B, Senthilkumar N, Siva C, Krithikadevi R, Julie A, Potheher I V and Arulmozhi M 2017 Synthesis and characterization of zinc oxide nanoparticles using marine *Streptomyces* sp. with its investigations on anticancer and antibacterial activity *Res. Chem. Intermed.* **43** 2367–76

[86] Kim Y J, Singh P, Yang D C, Singh H, Wang C, Farh M E A and Hwang K H 2015 Biosynthesis, characterization, and antimicrobial applications of silver nanoparticles *Int. J. Nanomed.* 2567

[87] Shoeibi S and Mashreghi M 2017 Biosynthesis of selenium nanoparticles using *Enterococcus faecalis* and evaluation of their antibacterial activities *J. Trace Elem. Med. Biol.* **39** 135–9

[88] Vaseghi Z, Nematollahzadeh A and Tavakoli O 2018 Green methods for the synthesis of metal nanoparticles using biogenic reducing agents: a review *Rev. Chem. Eng.* **34** 529–59

[89] Manjunath M H, Joshi C G, Danagoudar A, Poyya J, Kudva A K and BL D 2017 Biogenic synthesis of gold nanoparticles by marine endophytic fungus-Cladosporium cladosporioides isolated from seaweed and evaluation of their antioxidant and antimicrobial properties *Process. Biochem.* **63** 137–44

[90] Ahmad T, Wani I A, Manzoor N, Ahmed J and Asiri A M 2013 Biosynthesis, structural characterization and antimicrobial activity of gold and silver nanoparticles *Colloids Surf. B* **107** 227–34

[91] Chowdhury S, Basu A and Kundu S 2014 Green synthesis of protein capped silver nanoparticles from phytopathogenic fungus *Macrophomina phaseolina* (Tassi) Goid with antimicrobial properties against multidrug-resistant bacteria *Nanoscale Res. Lett.* **9** 365

[92] Verma V C, Kharwar R N and Gange A C 2010 Biosynthesis of antimicrobial silver nanoparticles by the endophytic fungus *Aspergillus clavatus Nanomedicine.* **5** 33–40

[93] Saravanan M, Amelash T, Negash L, Gebreyesus A, Selvaraj A, Rayar V and Dheekonda K 2013 Extracellular biosynthesis and biomedical application of silver nanoparticles synthesized from Baker's yeast *Int. J. Res. Pharm. Biomed. Sci.* **4** 822–28

[94] Asghari-Paskiabi F, Imani M, Rafii-Tabar H and Razzaghi-Abyaneh M 2019 Physicochemical properties, antifungal activity and cytotoxicity of selenium sulfide nanoparticles green synthesized by Saccharomyces cerevisiae *Biochem. Biophys. Res. Commun.* **516** 1078–84

[95] Moghaddam A B, Moniri M, Azizi S, Rahim R A, Ariff A B, Saad W Z, Namvar F, Navaderi M and Mohamad R 2017 Biosynthesis of ZnO Nanoparticles by a new *Pichia kudriavzevii* yeast strain and evaluation of their antimicrobial and antioxidant activities *Molecules.* **22** 872

[96] Sau T K and Murphy C J 2004 Room temperature, high-yield synthesis of multiple shapes of gold nanoparticles in aqueous solution *J. Am. Chem. Soc.* **126** 8648–9

[97] Parial D, Patra H K, Dasgupta A K R and Pal R 2012 Screening of different algae for green synthesis of gold nanoparticles *Eur. J. Phycol.* **47** 22–9

[98] Jena J, Pradhan N, Dash B P, Sukla L B and Panda P K 2013 Biosynthesis and characterization of silver nanoparticles using microalga chlorococcum humicola and its antibacterial activity *Int. J. Nanomater. Biostruct.* **3** 1–8

[99] Suriya J, Bharathi R S, Sekar V and Rajasekaran R 2012 Biosynthesis of silver nanoparticles and its antibacterial activity using seaweed Urospora sp *African J. Biotechnol.* **11** 12192–8

[100] Abboud Y, Saffaj T, Chagraoui A, El Bouari A, Brouzi K, Tanane O and Ihssane B 2013 Biosynthesis, characterization and antimicrobial activity of copper oxide nanoparticles (CONPs) produced using brown alga extract (*Bifurcaria bifurcata*) *Appl. Nanosci.* **4** 571–6
[101] Abdel-Raouf N, Al-Enazi N M and Ibraheem I B M 2017 Green biosynthesis of gold nanoparticles using Galaxaura elongata and characterization of their antibacterial activity *Arabian J. Chem.* **10** S3029–39
[102] Fatima R, Priya M, Indurthi L, Radhakrishnan V and Sudhakaran R 2019 Biosynthesis of silver nanoparticles using red algae Portieria hornemannii and its antibacterial activity against fish pathogens *Microbial. Pathog.* 103780
[103] Beyth N, Houri-Haddad Y, Domb A, Khan W and Hazan R E 2015 Alternative antimicrobial approach: nano-antimicrobial materials *Evid. Based Complement. Alternat. Med.* 246012
[104] Sondi I and Salopek-Sondi B 2004 Silver nanoparticles as antimicrobial agent: a case study on *E.coli* as a model for Gram-negative bacteria *J. Colloid Interface Sci.* **275** 177–82
[105] Zheng K, Setyawati M I, Leong D T and Xie J 2017 Antimicrobial gold nanoclusters *ACS Nano.* **11** 6904
[106] Kim J S *et al* 2007 Antimicrobial effects of silver nanoparticles *Nanomedicine* **3** 95
[107] Ling D and Hyeon T 2013 Iron oxide nanoparticles: chemical design of biocompatible iron oxide nanoparticles for medical applications *Small.* **9** 1450–66
[108] Danilczuk M, Lund A, Sadlo J, Yamada H and Michalik J 2006 Conduction electron spin resonance of small silver particles *Spectrochim. Acta Part A Mol. Biomol. Spectrosc.* **63** 189–91
[109] Prabhu S and Poulose E K 2012 Silver nanoparticles: mechanism of antimicrobial action, synthesis, medical applications, and toxicity effects *Int. Nano Lett.* **2** 32
[110] Nisar P, Ali N, Rahman L, Ali M and Shinwari Z K 2019 Antimicrobial activities of biologically synthesized metal nanoparticles: an insight into the mechanism of action *J. Biol. Inorg. Chem.* **24** 929–41
[111] Kim J, Lee J, Kwon S and Jeong S 2009 Preparation of biodegradable polymer/silver nanoparticles composite and its antibacterial efficacy *J. Nanosci. Nanotechnol.* **9** 1098–102
[112] Cui Y, Zhao Y, Tian Y, Zhang W, Lü X and Jiang X 2012 The molecular mechanism of action of bactericidal gold nanoparticles on *Escherichia coli Biomaterials.* **33** 2327–33
[113] Tan Y N, Lee K H and Su X 2011 Study of single-stranded DNA binding protein-nucleic acids interactions using unmodified gold nanoparticles and its application for detection of single nucleotide polymorphisms *Anal. Chem.* **83** 4251–57
[114] Xie Y, He Y, Irwin P L, Jin T and Shi X 2011 Antibacterial activity and mechanism of action of zinc oxide nanoparticles against *Campylobacter jejuni Appl. Environ. Microbiol.* **77** 2325–31
[115] Shende S, Ingle A P, Gade A and Rai M 2015 Green synthesis of copper nanoparticles by *Citrus medica* Linn. (Idilimbu) juice and its antimicrobial activity *World J. Microbiol. Biotechnol.* **31** 865–73
[116] Escárcega-González C E, Garza-Cervantes J, Vázquez-Rodríguez A, Montelongo-Peralta L Z, Treviño-González M, Castro E D B, Saucedo-Salazar E, Morales R C, Soto D R and González F T 2018 *In vivo* antimicrobial activity of silver nanoparticles produced via a green chemistry synthesis using *Acacia rigidula* as a reducing and capping agent *Int. J. Nanomed.* **13** 2349–63

[117] Kumar P V, Shameem U, Kollu P, Kalyani R and Pammi S 2015 Green synthesis of copper oxide nanoparticles using *Aloe vera* leaf extract and its antibacterial activity against fish bacterial pathogens *BioNanoScience.* **5** 135–9
[118] Jamdagni P, Khatri P and Rana J 2018 Green synthesis of zinc oxide nanoparticles using flower extract of Nyctanthes arbor-tristis and their antifungal activity *J. King Saud Univ. Sci.* **30** 168–75
[119] Gunalan S, Sivaraj R and Rajendran V 2012 Green synthesized ZnO nanoparticles against bacterial and fungal pathogens *Prog. Nat. Sci. Mater Int.* **22** 693–700
[120] Naika H R, Lingaraju K, Manjunath K, Kumar D, Nagaraju G, Suresh D and Nagabhushana H 2015 Green synthesis of CuO nanoparticles using *Gloriosa superba* L. extract and their antibacterial activity *J. Taibah Univ. Sci.* **9** 7–12
[121] Dobrucka R and Długaszewska J 2016 Biosynthesis and antibacterial activity of ZnO nanoparticles using Trifolium pratense flower extract *Saudi J. Biol. Sci.* **23** 517–23
[122] Durso L M and Cook K L 2014 Impacts of antibiotic use in agriculture: what are the benefits and risks? *Curr. Opin. Microbiol.* **19** 37–44

Chapter 3

Bionanocomposites for potential applications in agriculture

Elaine Gabutin Mission

As no single nanomaterial has exhibited superior properties and flawlessness, hybrid nanosystems have been developed. More importantly, the use of naturally derived bio-based polymers as a nanomaterial component has been proposed to address sustainability, safety and environmental concerns. This contribution is aimed at discussing the developments in bio-based nanocomposites and their potential application in the field of agriculture. This chapter will explore the design, fabrication, characterization and properties of bionanocomposites based on starch, chitosan, cellulose, lignin, protein and other natural polysaccharides. Finally, challenges and opportunities on bionanocomposite development will be discussed, including social and ethical aspects.

3.1 Introduction

Food shortage and environmental pollution are two of the fastest escalating global challenges we are facing nowadays. These two are intertwined, such that over-exploitation of one can also negatively affect the other. For instance, the Food and Agriculture Organization of the United Nations has identified the occurrence of a food chain crisis that would affect a huge portion of the globe due to various man-made (farming approaches, deforestation, overgrazing etc) and natural causes (pestilence, droughts, weather systems etc) [1]. This is further exacerbated by a steady increase in population, which means more mouths to feed, with finite available resources. On the other hand, the International Energy Agency (2010) has estimated global carbon dioxide (CO_2) emission likely growing by 21% during the period 2008–2035 [2]. So, if we do not improve the way we valorize the environment as we ramp up our food supply production, we will be caught in the cycle of struggling and living through worsening environmental conditions. Amidst

doi:10.1088/978-0-7503-3863-9ch3

these global challenges, aggravated by the incidence of the coronavirus pandemic since 2019, scientists and researchers have joined hands to intensify further their approach in addressing these issues, striving for a balance on how we can sustainably increase food quantity without damaging the environment.

Having innovation at the forefront of every scientific discipline, breakthrough technologies have allowed scientists to work at nanoscale, leading to the manipulation, fabrication, and development of molecules. It is said that the more we look into the fundamental behavior of molecules, the more we understand them and the more power we have to harness their properties. In this context, the field of nanotechnology has become a popular mechanism to develop a class of materials that possess interesting properties that can be tailored to every possible application. However, not all nanomaterials developed have desirable properties. In fact, two major issues for the acceptability of nanomaterials into mainstream usage is bioaccumulation and nanotoxicity. Engineered nanoparticles have been found in environmental compartments (soil, water, air) and their detrimental impact to humans and other living organisms is fundamentally unknown, primarily due to their size and ease of transmission. Therefore, their acceptance is tainted with doubts. This brought to light the huge demand for sustainable, functional materials and massive interest in their properties and applications.

In order to address the weaknesses in individual nanoparticles, they need to be assembled, combined and blended together. When inorganic and inorganic materials are assembled, this leads to the formation of hybrid materials, more popularly known as nanocomposites. Nanocomposites are an emerging class of nanosized materials containing at least two constituent materials of heterogeneous properties, at least one of which has a dimension in the range of 1–100 nm. Typically, these nanomaterials include an organic (dendrimers, liposomes, polymers), inorganic based (metal and metal oxides, quantum dots) or carbon portion (carbon nanotubes, fullerenes, graphenes). Syntheses of these hybrid nanoparticles were largely driven by the necessity to capitalize on individual component materials' strengths, enhance existing properties, and minimize *individual* defects. Published materials regarding nanocomposites first appeared in PubMed and ScienceDirect in 1987 and 1998, respectively. Since then, there has been a surge in interest that yielded a combined 175 000 articles in both scientific databases since the first report appeared until March 2021, as shown in figure 3.1. The versatility of the developed materials is widely accepted but is still plagued with issues of end-user safety and environmental compatibility. Therefore, a huge effort has been exerted by scientists in order to address this issue which gave rise to an emerging field that lies at the forefront of biology, nanotechnology and materials science.

The name bionanocomposites has been coined to refer to nanocomposites containing a component of biological origin and biodegradable nature [3]. The first articles regarding this topic appeared in PubMed and ScienceDirect in the years 2004 and 2003, respectively. The prefix 'bio' is understood as that the materials originated from living organisms or microorganisms; hence, proteins, starch, cellulose, lignin, chitosan and alginate would first come to mind, although derivatives that were obtained via synthetic routes such as oils, lipids, amino acids

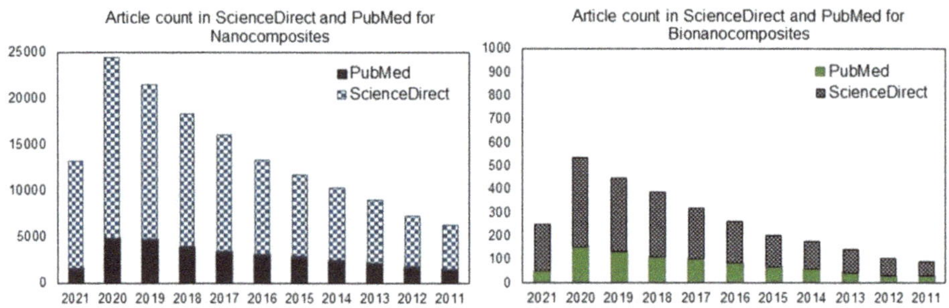

Figure 3.1. Comparison of the number articles published in PubMed and ScienceDirect in the last 10 years using keywords 'nanocomposites' [left] and 'bionanocomposites' [right].

and fats are also included in this category. Often, naturally occurring biopolymers or polysaccharides and their corresponding monomers, unfortunately, lack mechanical properties such as strength, processability, and stiffness found in their petroleum-based counterparts. Some are relatively sensitive to moisture that limits their application in humid environments, thus necessitating their integration within or into materials that could enhance their poor properties. Therefore, the biobased materials will have to be admixed and reinforced with organic or inorganic fillers, such as clays, polymer nanofibers or carbon particles. The inclusion of these nature-derived materials is perceived to impose renewability, nontoxicity, biodegradability and biocompatibility into the nanomaterial matrix. In contrast to artificial polymers, materials of biological origin have more complex structures and random configuration that makes their exploration interesting. Their shapes, structure and composition would affect how they would specifically bind or coordinate within the bionanacomposite structure. Despite the potential of these materials for development, the number of published articles has not picked up since their first introduction. As shown in figure 3.1, the total combined articles published in both PubMed and ScienceDirect is only about 3000, representing about 2% of nanocomposites article volume.

In this chapter, the recent developments in bionanocomposites for potential applications in the field of agriculture will be presented. First, the major constituents of bionanocomposites will be stated. The design, synthesis, properties, characterization, and application of the bionanocomposites will be discussed, including the trends, opportunities and challenges on bionanocomposites design and synthesis.

3.2 Major components of a bionanocomposite-natural polymers

Natural polymers are abundant and can be sourced from lignocellulosics, algae and marine products. The most commonly used are starch, cellulose, lignin, chitin, alginate and proteins. Due to their complex and interesting structure, natural polymers have been widely studied for a wide variety of applications. What is interesting in them is their flexible and tunable structural capacity for chemical adaptation to elicit unique properties that cannot be achieved with synthetic molecules. More importantly, they possess biodegradability, biocompatibility, cost-effectiveness, and nontoxicity.

3.2.1 Starch

Starch is one of the most common natural polysaccharides derived from food-based crops, such as rice, potato, maize, cassava, wheat and tapioca. It consists of carbon, hydrogen and oxygen in the formula $C_6H_{10}O_5$. It is a polysaccharide of amylose and amylopectin with glucose as its monomeric units connected by an $\alpha -1,4$ and $\alpha -1,6$ glycosidic bond. Starch is often found as an ingredient in bread, biscuits, pastries and cakes, where it improves food stability and stiffness. It has also found application in non-food items such as glue, textiles, stationery, laundry stiffener and flocculants. It has been widely accepted for the production of thermoplastic component of packaging films, however, it cannot be utilized on its own due to poor mechanical and physical properties [4]. Starch also includes features such as the ability to improve stiffness and stability of nanocomposites.

3.2.2 Cellulose

Lignocellulosics have been established as a sustainable bioresource originating vastly from terrestrial biomass and from aquatic biomass in smaller quantities. Cellulose, $(C_6H_{10}O_5)_n$, co-exists with hemicellulose and lignin in plants, but on its own is accorded as the most abundant biopolymer. It is a linear polysaccharide of glucose connected with $\beta-1,4$ glycosidic bonds, that provides a rich content of surface hydroxyl groups. The molecular weight of cellulose can be estimated through various methods and formulae correlating the degree of polymerization with the viscosity, the most common of which is $DP = 190\ [\eta]$ [5]. In fact, the hydroxyl group attributed as the primary alcohol is key to the modifications, some of which involve interfacial interactions and functionalization [6]. Cellulose can be extracted from plant biomass through a variety of techniques that include, but are not limited to, chemical (sulfuric acid, ionic liquid), subcritical//supercritical fluid, enzymatic, and mechanical fractionation that can lead to nanosized cellulose families, namely cellulose nanocrystals and nanofibrils [7–11]. Cellulose can also be synthesized by some acetic acid bacteria (*Gluconaacetobacter xylinus*, *Agrobacterium*, *Rhizobium* and *Sarcina*), yielding bacterial cellulose [12]. Bacterial cellulose has been used as a matrix and loaded with graphene oxide for a potential drug delivery system.

3.2.3 Lignin

Lignin is a complex, amorphous biopolymer found in cell walls mostly of woody biomass and serves as a binder for cellulose fibrils. It consists of three aromatic monolignols or aromatic alcohols [ρ-coumaryl, coniferyl and sinapyl alcohols] in random connections and varying proportions with molecular weight reaching as high as 25 000 g mol^{-1} [13, 14]. It is a major by-product of the paper industry and can be extracted as charged lignosulfonates or kraft lignin via sulfite pulping or alkaline kraft pulping. Either of these extraction processes imparts sulfonate and carboxylate functionalities onto lignin. Lignin is characterized by a high degree of crosslinking between units, hydrophobic structure and abundant functional groups (carbonyl, carboxylic and phenolic groups) that allows it to establish a strong interfacial interaction with other polymers. When lignin was grafted onto polyacrylamide, it was found to decrease aggregation due to the hydrophobicity of lignin [13]. In this case, the incorporation of lignin was found to impact colloidal and emulsion properties, which can therefore be controlled by adjusting lignin to polyacrylamide ratio and grafting tendencies. Lignin was added to starch to improve its physical and chemical properties, particularly decreasing water solubility of the starch/lignin film [15]. Lignin was also utilized as a barrier for the slow-release of urea fertilizers in conjunction with clay [16]. It was positioned as a potential replacement to sulfur coatings that would enable release rates matching that of plant uptake.

3.2.4 Chitin/chitosan

Chitin, $(C_8H_{13}O_5N)n$, is a high molecular weight macromolecule abundant in exoskeletons of sea-animal shells and crustaceans and internal structures of invertebrates. It is synthesized from N-acetyl-D-glucosamine and resembles cellulose structure except that, instead of hydroxyl groups, acetyl amine groups are attached in each monomer. On the other hand, chitosan is a heterogeneous group of polymers bearing 10 000 to 1 000 000 Da derived from the acetylation of chitin and is the second most abundant natural biopolymer. As it consists of binary linear chains of β −1,4 –D- glucosamine and N-acetyl-D-glucosamine linked with a β (1–4), it contains a huge amount of amino ($-NH_2$) and hydroxyl (-OH) groups [17]. Treating crab and shrimp shells with alkali sodium hydroxide is a common way of isolating chitosan. Chitosan has been widely employed in coatings due to its antimicrobial properties that are dependent on its type, polymerization degree and other environmental factors, hence its application for salami wrapper, sliced turkey meat deli, and seafood and fish wrappers [18]. Chitosan and alginate are widely used for encapsulation and controlled release of active ingredients [19].

3.2.5 Alginate

Alginate is an anionic, water-soluble linear polysaccharide extracted from brown algae. It is composed of 1–4 linear copolymers of β-D-manuronic and α-L-guluronic acids. Depending on the source, the manuronic and guluronic units may have huge variations, thus affecting the alginate composition and characteristics. The increase in guluronic units chain length is attributed to improvement of alginate mechanical properties. Meanwhile, when manuronic units chain length are increased, the alginate demonstrates immunogenicity, allowing for applications in biological systems. Alginate easily bonds with cations, such as Ca^{2+}, Ba^{2+}, Zn^{2+} and Al^{3+}, owing to the anionic carboxylate groups present in its structure. In its sodium form, it has been useful in extending shelf life of manufactured products [20].

3.2.6 Protein

Fibrous protein exists in nature in the form of silk, collagen, keratin and gelatin. Collagen is the main structural protein in human connective tissue and comprises 90%–95% of organic bone matter. Gelatin is mainly composed of alpha chains of amino acids derived from collagen. It has superb swelling and water uptake properties. Myoglobin is a 16.7 kDa water soluble heme-protein with a single polypeptide chain containing an iron heme as its prosthetic group and is found in mammals and vertebrates. It has the ability to store oxygen and enhance its diffusion rate [21]. The electrochemistry of proteins is significant for biosensors and bio-reactors. Enzymes which are biologically catalytic agents that can be secreted from a particular source, were also used in bionanocomposite applications. In principle, enzymes consist of amino acids and may sometimes have some carbohydrate or lipid fragment that they use when binding or connecting to a substrate. Some reported examples include lacasse, a multicopper enzyme, and horseradish peroxidase, found in horseradish roots [22–25].

3.2.7 Other biopolymers and synthetically-derived polymers

Aside from the major biopolymes discussed in the preceding sections, there are many other naturally occurring biopolymers in nature that are seldom heard about. Some of them, such as biopolyesters found in plant cuticles, namely suberin and cutin [26, 27], are not produced in huge quantities, so their exploitation for further application is also limited. There is also pullulan, a water soluble, exopolysaccharide that is extracellularly produced by the fungus *Aureobasidium pullulans*, in starch and sugar cultures, that is not yet heavily exploited. Pullulan is a water-soluble biopolymer with good film-forming abilities and adhesiveness properties [28–31].

Some polymers have natural sources, such as polylactic acid that can be sourced from starch or sugarcane. However, despite having a natural origin and renewable constituents, the biodegradability of polylactic acid is limited to industrial composting facilities. Simpler composting facilities will not suffice and are deemed to emit some harmful fragments during decay. The use of nanoclays into a polylactic acid biopolymer increases barrier properties to oxygen and water vapour and extends the shelf life of food products.[32]

3.3 Bionanocomposites synthesis

The development of bionanostructures is benchmarked from naturally occurring hierarchical structures, such as the human bone, tooth enamel, seashells and nacre in pearls. Nanobiocomposite synthesis can be achieved via two approaches namely top down approach or bottom up approach, as illustrated in figure 3.2. Top down approach is considered as the traditional processing, wherein, existing materials are precisely fabricated based on available and existing materials, whereas the bottom up approach is considered as the process of nature, where particles are able to recognize and bind onto specific materials eventually assembling on its own. In recent years, this has been known as 'self-assembly'. The resulting

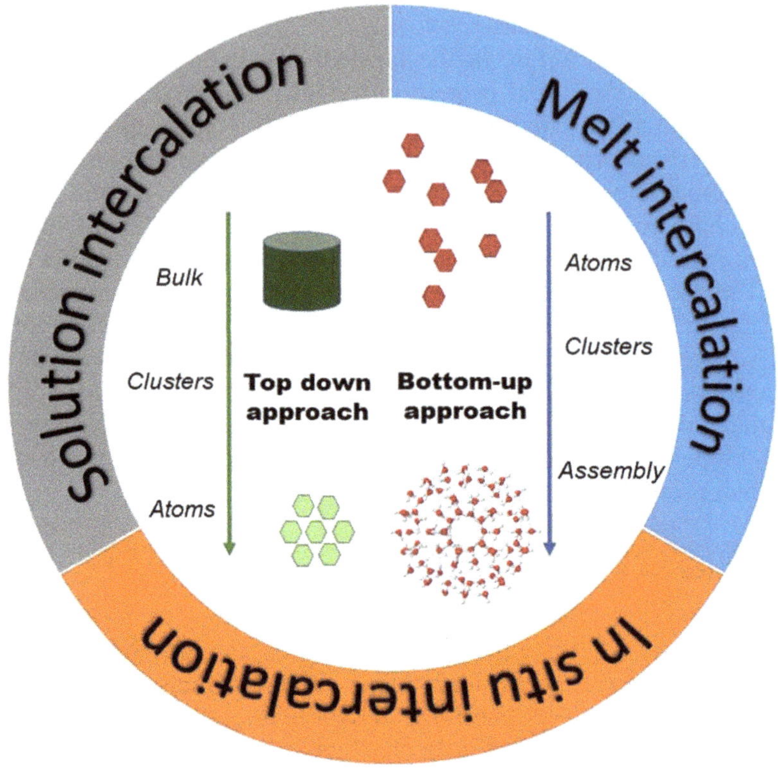

Figure 3.2. General techniques and approaches for the synthesis of bionanocomposites.

bionanocomposites can take various forms or shapes such as spherical, cubic, tubular, fibrous, rodlike, whiskers, platelets and sheets but could also be shapeless or irregular. In terms of preparative techniques, there are three broad classifications being utilized today [3].

3.3.1 Solution intercalation

This technique relies on the ability of the biopolymer to dissolve in a solvent system and the property of the inorganic matrix to swell. In such case, when the matrix has swollen in the solvent system, it allows the biopolymer chains to intercalate the interlayers of the matrix. Then, the solvent is removed and the intercalated structure remains, yielding a bionanocomposite formation. The formation of bacterial cellulose/graphene BNC is usually carried out through this process [12]. Chitosan-based composite film incorporated with grapefruit seed extract was successfully fabricated using this technique [33]. Gluten-based nanobiocomposites reinforced with cellulose nanofibrils and nanocrystals are prepared according to this technique [7]. Nanoencapsulation by emulsion cross linking are the most elaborate approaches of this technique [19, 20, 34].

3.3.2 *In situ* intercalation polymerization

This technique requires the monomer form of the biopolymer to exist in liquid form. The inorganic matrix can be swollen in the liquid monomer, then heat or radiation can be applied to initiate the polymerization. An initiator or catalyst may also be employed alternatively. In this case, the biopolymer forms within the interlayers of the inorganic matrix. In some other works, this method is also known as 'grafting'. Silmore *et al* demonstrated that lignin can be grafted onto polyacrylamide, producing a bionanocomposite with tunable amphiphilicity and hydrophobicity and that can form stable emulsions [13].

3.3.3 Melt intercalation

This technique is suitable for materials not suitable for solution or *in situ* intercalation. It involves the annealing of the biopolymer/inorganic matrix above the softening point of the polymer. During the annealing process, the biopolymer chains diffuse from the bulk polymer melt into the interlayer. The degree of penetration can be controlled resulting in intercalation or exfoliation. It has been reported as more economical and more environmentally friendly, but may result in uneven dispersion in contrast to the other two methods [35]. *In situ* template process or sol–gel technology also falls under this category. Cooper *et al* utilized this method for the hybrid nanocomposite resulting from poly-butylene succinate, biocarbon and graphene [35].

These preparation techniques are carried out using extruders, mixers, or ultrasonicators. Biomaterials bond with their inorganic matrix through a multitude of ways, some which are via non-covalent interactions (Van der Waals, π-π interaction); hydrogen and ionic bonds, inorganic metal–ligand formation, electrostatic interactions, hydrophobic interactions, charge transfer and atomic pi-stacking [36]. What favors the reaction between the biopolymer and matrix would be the presence of highly reactive functional groups such as carboxyls, amides, amines, hydroxyls, phenolics and aromatic rings that serve as bonding sites between the polymer, fillers or active substances for further chemical modifications [37]. Macroscale operating conditions such as processing temperature and pressure, reaction time, formulation ratios, nature of solvents and additives and pH will have a significant impact on the formation and stability of the bionanocomposite.

3.4 Bionanocomposite characterization

To the naked eye, the nanoparticles appear as ordinary powder or suspension or solution in a solvent system. Therefore, the confirmation of the actual structure requires the utilization of advanced microscopy and imaging techniques. Since not all laboratories are equipped with this advanced equipment, a variety of classical and instrumental analyses can also be used to characterize and confirm the success of bionanocomposite formation [19]. The most widely employed characterization techniques are enumerated below. Several references have been analyzed and it is worth noting that the combination of various approaches would be suitable to provide other scientists with a common ground in describing the structure and characteristics of resulting bionanocomposites [38–43]:

Fourier transform infrared spectroscopy (FTIR) can be used to confirm the available functional groups of the constituent or bionanocomposite as whole. The stretching or bending vibration of functional groups are reflected in FTIR spectra and can be used to confirm successful intercalation or functionalization. As such, FTIR provides qualitative insight into the bionanocomposite structure.

Scanning electron microscope/transmission electron microscope (SEM/TEM) are used to visualize specimens and have the ability to zoom in on the surface morphology by up to two million times. This is useful to observe structural changes and determine constituents in the formed bionanocomposite. Careful consideration should be taken since a non-conducting or poorly conducting specimen would require coating with electrically conducting metal such as gold, platinum or silver, which could lead to obscuration of the biological component of the bionanocomposite. This would particularly affect bionanocomposites with carbon or organic-based matrix. **Energy dispersive X-ray microscopy (EDX)** aside from combustion elemental analysis, this technique allows for the determination and estimation of constituent elements or characterization of chemicals on the area of a specimen. It is typically attached as an accessory to an SEM or TEM. Caution should be exercised as this technique may not be suitable for trace levels of elements or inhomogeneous samples.

Dynamic Light scattering (DLS) is used to ascertain the average particle sizes of synthesized bionanocomposites (hydrodynamic diameter) as well as polydispersion and aggregation. The technique can be configured for powdered form or stable suspensions, and can be coupled with microscopy in order to assess very asymmetric as-synthesized bionanoparticles.

Differential scanning calorimetry (DSC) and thermogravimetric analysis (TGA) are used for thermal stability and durability analyses. **Zeta potential measurements** are used to determine colloidal stability and surface charges. An increase or decrease in the value of the surface charge indicates interactions between material interfaces.

X-ray diffraction (XRD) is used for phase identification or confirm crystallinity or amorphous structure of the constituent or bionanocomposite as a whole. Meanwhile, **x-ray photoelectron spectroscopy (XPS)** is useful in evaluating chemical states and dominant elements, signalling homogeneity and confirming binding energies, energy bands specie incorporation and conformational changes. Whilst it is suitable for a broad range of materials, it also requires training for the interpretation of the spectra and is one of the most complex amongst the techniques stated herein.

Ultraviolet–visible spectrophotometry (UV–vis) is used to investigate absorption of light by the substance and can aid in determining the chemical purity and characteristics of the sample. Another spectroscopy technique called **nuclear magnetic resonance (NMR)** spectroscopy also determines purity and content of samples, particularly geared towards molecular structure and conformational changes. This is also an advance technique requiring knowledge in interpretation of the spectra.

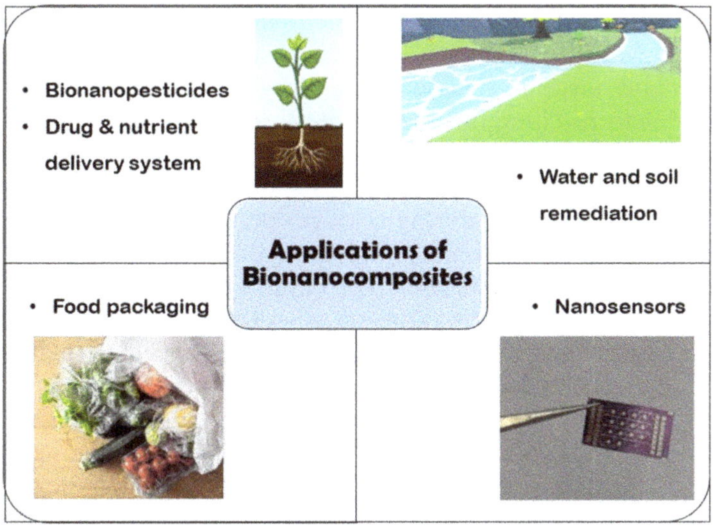

Figure 3.3. Application of bionanocomposites.

3.5 Agricultural applications of bionanocomposites

Most of the bionanocomposites developed are based on the need and target application for which they were originally intended. Therefore, bionanocomposites have found application in agriculture on major avenues such as nanopesticides/drug delivery, food packaging, remediation, and nanobiosensor development, which will be discussed in the following sections and are illustrated in figure 3.3. It is quite interesting to note that the bulk of the literature reviewed for this chapter indicated a huge interest in food packaging, followed by nanopesticides and drug delivery. There was quite an absence of bionanocomposites developed for direct application, presumably due to the nonbiodegradability of the inorganic matrix.

3.5.1 Bionanopesticides

Pests remain a major burden for farmers as their presence in agricultural fields would mean damaged crops, loss of yields and consequently, low productivity. Therefore, the use of pesticides became rampant. At the beginning, these pesticides worked wonders which drove excessive application. The excess pesticides eventually reached environmental and ecological matrices including soil, surface and groundwater and hydrological systems, through erosion, run-off, leaching, volatilization. Most of these substances are characterized as carcinogenic, toxic, mutagenic and teratogenic. Eventually, there were reports of adverse health effects arising from the presence of pesticides in the environment. The use of traditional agrochemicals has long been recognized as an approach to boost farming yields. However, it was also found out that the agrochemicals can have a negative effect on the soil quality and on the ecological health. In agriculture, bionanocomposites are popular for controlled release applications.

Nanotechnology offers a promising potential to act as carriers of pesticides, drugs, growth enzymes and catalysts [44]. Encapsulation has become an interesting approach to enable targeted release of the drugs or pesticides and prevent their degradation prior to reaching the desired site for release. Another approach is the development of coatings [16, 45]. This not only minimizes the amount of pesticides being used, but it also ensures that the pesticides reach the appropriate site for release at a specified timing or during a longer period of time. In terms of design, active ingredients are encapsulated within hydrophobic organic or inorganic particles that serve as diffusion barrier. The natural polymers are usually structured as the carrier in nanohollow formations or spherical configurations.

Figueroa and her team synthesized a composite hydrogel of poly(acrylamide) and starch as for the encapsulation of amoxicillin for controlled drug release and antibacterial application [48]. Poly(acrylamide)/starch formulation at volumetric ratios of 60/40 and 50/50 was found to be efficiently loaded with amoxicillin as 78% and 98%, respectively, of the 0.2 g ml^{-1} therapeutic dose. The presence of starch within the poly(acrylamide) network contributed to an increase in water uptake and swelling degree. For the 50/50 ratio, the amoxicillin release was found to be at 48% within the first hour and has shown inhibition effect against *Escherichia coli* ATCC 25922 and *Staphylococcus aureus* ATCC 25923 and *Pseudomonas aeruginosa* strain. The antibiotic release was also found to be influenced by optimum pH and temperature (pH = 7.4 and temp = 37 °C). Kumar *et al* utilized the polysaccharide sodium alginate as nanocarrier for the imidacloripid active ingredient to address leafhopper sucking pests in Okra crops. Their developed nano-encapsulates have a 150 nm range [20]. The encapsulation efficiency was found to be between 2.5%–98.7%, corresponding to about 148 mg bound active ingredient per 6000 mg of the bionancomposite formulation. The field evaluation demonstrated that the nanoencapsulates were effective in reducing the leafhopper population infesting the plant. The effect has been notable for 15 days, which outperformed the control experiment.

Santos-Silva *et al* [19] loaded paraquat in an alginate/chitosan system. Paraquat is an herbicide intended for weed control and is classified as very dangerous and highly toxic. In fact, it has a 15 mg l^{-1} toxicity limit towards aquatic organisms. Therefore, it merits to be encapsulated in order to reduce excessive usage. At an encapsulation efficiency of about 74%, the resulting bionanocomposite was found to be 635 ± 12 nm small. The release profile of the herbicide was found to reach near 100% in 8 h, which is about 2 h longer than when free herbicide was used. Meanwhile, Falsini and co-workers developed polymeric nanocapsules based on lignin as carriers of bioactive compounds, specifically gibberellic acid (GA) [34]. GA is a plant growth regulator and has been useful for enhancing seed germination. The synthesized lignin-based nanocapsules have an average of 300 nm size, which reduces to 200–250 nm when loaded with GA. The lignin nanoparticles were observed to be absorbed by the seed's tissues and penetrate within the vascular system of the germinating seed via fluorophore signalling. This finding highlights the need for biocompatible nanoparticles, otherwise, germinating seeds and developing plants might be intoxicated and die.

Aside from pesticide application, increasing plant productivity can also be achieved by means of biological control agents. These substances would induce resistance of plants to diseases and improvement of soil properties. Chitosan-immobilized in silica having a crystallite size of 5.8 nm was developed by Gatahi *et al* to bring about resistance of tomato plants to pathogens (i.e. the *Ralstonia solanacearum* bacterial wilt). They reported that the bionanocomposite supported formation of chitinase and glucanase, are effective in addressing bacterial wilt. An increase in treatment concentration from 0% to 100% led to higher inhibition rates, with a maximum of 92%. Germination rates of tomato seedlings, chitinase levels and shelf life generally improved as treatment concentration increased. Moreover, they observed that chitosan has a better impact on soil pH than chitin, since the former more easily breaks down than the latter [45]. Chitosan was also embedded in polyacrylic acid hydrogel via ionic gelation and applied onto onion plants. The formed particles were less than 90 nm, depending on the amount of copper loaded to address micronutrient deficiency. The treatment of the onion plants with the chitosan-polyacrylic acid hydrogel-copper bionanocomposite resulted in higher plant length, enhanced bulb diameter and chlorophyll content and improved number of roots and leaves per plant. Since copper is essential on photosynthesis and plant development and metabolism, the improvement in the plant health and productivity is attributed to the proper release of copper from the nanobiocomposite. It was determined that 75 ppm is the best copper loading concentration to achieve the most enhancement [46].

Sometimes, more complex molecules and combinations are explored. For instance, urea has been coated with biochar and polyvinyl alcohol (PVA) and polyvinylpyrrolidine (PVP) for slow-release application [47]. Originally, the PVA is responsible for film coating but due to its limited ability as diffusion barrier, its properties need to be enhanced by cross linking with PVP. Meanwhile, biochar was added to serve as support material for the delivery of urea and impart mechanical strength. The biochar was even sourced from rice, maize and forest litter to further highlight the use of natural components. Slow release performance has been successful and was found to reach around 65% in 22 days.

3.5.2 Food packaging

One of the major avenues by which bionanomaterials found application is in food packaging, by virtue of improved food safety, superior shelf life, oxygen and moisture barrier, protection from harmful toxins that may develop due to storage conditions and to minimize the use of chemical preservatives. It has been reported by FAO in 2011 that food intended for human consumption could be wasted without proper handling, storage and cooling facilities [37]. Moreover, with over 300 million tons of plastics produced globally each year that has huge possibility to be improperly disposed of, it would be fitting to have these materials replaced with biodegradable counterparts [49].

Starch has been a popular choice since the 1920s, but since it disintegrates in water due to the dissolution of amylose and swelling of amylopectin, it has to be reinforced with other nanoparticles for utilization in packaging films. An example was the graphene oxide reinforced lignin/starch bionanocomposite developed by Aqlil *et al* [15]. Starch/lignin bionanocomposite was prepared with 5:1 ratio, followed by the incorporation of graphene oxide of up to 0.7 wt%. The resulting nanocomposite showed better properties in terms of water swelling, moisture uptake, hydrolytic degradation, water vapour permeability, thermal stability and mechanical properties, indicating the success of this synthesis procedure.

Tan *et al* fabricated a chitosan-based composite film which was further incorporated with grapefruit seed extract to elicit antimicrobial and antifungal properties for bread [33]. The grapefruit seed extract enriched-chitosan composite film showed a 2 day delay in fungal development for bread stored at 24 °C, as opposed to low density polyethylene, highlighting the benefits of bionanocomposites over petroleum based plastic. Moreover, increasing the content of the grapefruit seed extract up to 1.5% v/v of chitosan-based film further afforded an additional 7 days of delay in fungal development, further prolonging the shelf life of the packed food. Essential oils, extracted from plants such as oregano, lemongrass, thyme and cinnamon, that contain various volatile compounds including terpenes, terpenoids, and alkaloids have also been explored for antimicrobial effects [18, 50].

Proteins isolated from spent hen (that were processed including skin, bones, breast, leg and gizzard) were cross-linked with chitosan using glycerol as plasticizer and reinforced with nanoclay for the preparation of bionanocomposite films [49]. The resulting bionanocomposite, comprised of 20% glycerol, 3% chitosan and 5% nanoclay with respect to the protein content. The as-prepared bionancomposite showed mechanical strength doubling that of plasticized films without nanoclay reinforcement (protein/chitosan/glycerol only).

Meat spoilage takes place due to the high value of unsaturated lipids, susceptibility to oxidation processes and presence of aerobic spoilage bacteria. To address these concerns, Souza *et al* developed bionanocomposite film based on chitosan and up to 2.5 wt% montmorillonite (37) for packaging fresh poultry meat and further reinforced with rosemary oil (0.5%–2% v/v) to impart antibacterial and antioxidant properties. The resulting film was found to be yellowish in color, flexible and transparent with an average thickness of 51 μm. Whilst the montmorillonite improved the water vapor permeability and oxygen permeability of the bionanocomposite film, it also confined the rosemary oil which avoided its diffusion onto the food. Regardless, it was capable of extending the shelf life of poultry meat until the 7th day, compared with control samples that perished on the 3rd day.

3.5.3 Remediation

Heavy metals, when released onto soil and aquatic resources, have the tendency to accumulate leading to irritation, abnormalities or even death of microorganisms and higher ordered living beings. This would mean economic devastation for the agricultural industry and human exposure with detrimental impacts resulting to disability, diseases or even death. Whilst massive heavy metal contamination of agricultural resources could mean several years or decades of clean up, minor contamination which may be due to a natural phenomenon such as volcanic eruption can be addressed by employing adsorption. Therefore, it is a must to develop highly adsorptive materials. Base materials for these applications include, graphene, activated carbon, carbon nanotubes, zeolites and polymeric materials. However, these materials at individual levels suffer from low adsorption capacities, slow adsorption rate and difficulty of regeneration. Qin *et al* developed a lignin-based nanocomposite intended to capture copper (II), zinc (II) and nickel (II) from wastewater. In order to improve the adsorption ability of polyethylene amine, lignin was used as s supporting material. Through the Mannich reaction, the fabricated bionanocomposite adsorbent was found to have many pores and functional groups. Removal efficiency for copper II, zinc II and nickel II reached 94%, 76% and 65%, respectively, at a dosage of 20 mg per 50 ml wastewater. Adsorption was found to take place in 20 min while the percentage removal slightly decreased from 94% to 81% after 5 cycles.

In the works of Mallakpour and Madani, they reported on the hydrothermal synthesis of an α-MnO2 and valine amino acid as matrix for chitosan that was utilized for adsorption of lead ions [51]. As chitosan bears amino and hydroxyl groups along its backbone, it is structurally conducive as a sorbent for dyes and heavy metals, which is useful for remediation of contaminated soils and waterways. They have observed that α-MnO_2/valine content has improved the thermal stability of the synthesized bionanocomposite. Bacterial cellulose/graphene nanocomposite has also been employed for metal ions removal [52].

3.5.4 Bionanosensors

Due to the rising costs of land, it is not impossible to find industrial complexes near agricultural zones. Then, there is the possibility that wastewater streams from industrial facilities get released, either intentionally or unintentionally, onto agricultural fields. As soil health and quality is critical in producing quality crops with high yields, develop smart agriculture techniques involve the utilization of sensors to assess soil health or water irrigation quality.

Various carbon based materials, the most prominent of which include carbon nanotubes and graphene and its derivatives have been the based materials for biosensors due to their ability to promote and facilitate reduction–oxidation reactions or electron transfer. The most common advances would be to develop bionanocomposite films that can be attached or incorporated onto the existing glassy carbon electrode surface. The most common requirements for biosensors are high sensitivity, high selectivity, working stability, anti-fouling ability and quick response.

Korkut *et al* reported a biosensor for the determination of phenol that can originate in wastewater streams of paint, oil, and pharmaceutical or polymer industries. The biosensor is a nanobiocomposite film electrode of carbon nanotube (CNT)/polypyrrole film with horseradish peroxidase on a gold surface [53]. The horseradish peroxidase is expected to induce good sensitivity, whereas the CNT is responsible for enhancing electron transfer in supporting the bionanocomposite architecture so that the bioactivity of the enzyme is maintained and in maintaining the stability of the working electrode. Pyrrole has relatively poor oxidation potential that enabled horseradish peroxidase immobilization, amounting to 2.7 μg. It should be noted that this value was further increased to 6.1 μg with the addition of CNT, as the CNT provided high surface area to which the enzymes have anchored themselves. The biosensor was found to be highly sensitive to the phenol derivatives tested in the study, reaching as high as 99% maximum response in roughly 2 seconds. The nanobiosensor was also found to be quite stable since it still exhibited up to 70% of its initial activity after storage of one month at 4 °C in the dark.

Collagen extracted from animal skin trimming wastes of a local tannery was blended with guar gum and incorporated onto carbon nanotubes in various formulations such as [100/100/0], [100/10/0.25], [100/10/1],[100/10/2] wt% collagen/guar gum/carbon nanotubes to develop electrically conducting nanobiocomposite films [54]. The as-synthesized nanobiocomposite films showed the capability to facilitate electron transfer by lighting a light emitting diode when inserted in between three coin cell batteries. Moreover, it was observed that the brightness of the emitted light was dependent on the carbon nanotube loading, with the 2 wt% loading having the highest potential to actually be utilized in biosensors or actuators.

Bolder research works involving the incorporation of haemoglobin and myoglobin in carbon nanotubes has been reported. For instance, Vilian *et al* immobilized myoglobin [21] in carbon nanotube/polytyramine composite functionalized with gold nanoparticles for detection of nitrite (NO^{2-}). Nitrite is widely utilized in the food and beverage industries and a precursor in the formation of N-nitrosamines, which were demonstrated to have potent carcinogenic effects in humans. Due to the poor electrocatalytic activities of protein molecules, the use of elaborate materials is necessary to improve their properties. The ionic liquid 1-ethyl-3-methylimidazolium tetrafluoroborate ($EMIMBF_4$) was utilized to modify a carbon paste electrode, after which it was incorporated with multiwalled carbon nanotubes. Thereafter, it was drop coated with myoglobin that had been entrapped in a Nafion film. The modified electrode has shown excellent ability towards the electrocatalytic reduction of trichloroacetic, peroxide and sodium nitrite [55].

3.6 Challenges and opportunities in the usage and design of bionanocomposites

Being a fairly developing field, it is not impossible for the challenges and opportunities to surface and somehow, provide a negative connotation. On the brighter side, these challenges and opportunities provide doors and pathways to which aspects would be necessary to develop this field further and push the disruptive change that bionanocmposites could bring in the scientific, technological and economic world.

First, since most of the studies are in their infancy, the best properties of each constituent material may not easily reflect on the resulting bionanocomposite. Substantial investment must be put into studying and developing bionanocomposites, much more on the analytical techniques that require massive capital to be laid out. It seems this will be worth the investments since eventually, very small quantities of materials and chemicals would be necessary to bring production to commercial scale. Having a lot of considerations and potential combinations by which the small particles react and interact with their matrix or reaction conditions, there would be countless trial-and-error approaches along the way. Therefore, the use of artificial intelligence and computational approaches in the design and synthesis of bionanocomposites would be advantageous. But before this can happen, there should be an appropriate extent of data and literature available for predictive models to be useful. For instance, cellulose nanocrystals (CNC) embedded in gluten resulted to the reduction in the permeability of gases, suggesting their potential for food packaging applications. However, further study needs to be conducted since CNC could agglomerate during fabrication, which has been shown to negatively impact the barrier properties [7]. These types of occurrence can be immediately captured in predictive models.

While the resources for biopolymer are abundant, they necessitate isolation and purification procedures. Normally, these involve the extraction of biopolymers and polysaccharides from biomass waste or industrial by-products. It is worth noting that the source of these biopolymers significantly affect their structure and intrinsic properties, thus, would also affect the resulting bionanocomposite [47, 54]. The occurrence of contamination has been rarely accounted for in the published literature.

Most of the reports on bionanocomposites centered on the synthesis, characterization and application. But some questions float and matters too, such as confirmation of the structural configuration, robustness and integrity and recyclability and whether biodegradation holds true for the biological part only and not the full material. The majority of the papers reviewed for the purpose of this chapter claimed that the synthesized bionanocomposites were biodegradable and environment friendly. However, very few papers have actually gone to the length of confirming the biodegradability of their synthesized materials. It should also be highlighted whether the biological material is the main active component, or was a support or filler. Sometimes, even if the biological materials were used as a filler in minute amounts, the bionanocomposite category is being assigned. There should be a common convention when the bionanocomposite should be used. For example, if the biological component is not the main active part, or if it acts as a filler or is found to be less than 10% of the whole bionanocomposite, it cannot be labelled as such. In addition, the extent of application should be tested beyond the walls of the laboratory. Field testing should be carried out using actual environmental conditions [20].

One major concern regarding the synthesis of bionanocomposites is the synthesis procedure: many successive steps, heavy labour and tediousness, requiring several hours of waiting and often entailing many chemicals. Can it still be considered green and sustainable if the synthesis process takes many chemicals and a significant amount of energy, often for a few grams of bionanocomposite that would be sufficient for characterization and not scalable for widespread utilization? There is also the difficulty in obtaining standardized particle sizes. The presence of aggregation has been reported when pesticide was loaded in sodium alginate [20].

3.7 Conclusion

Nanotechnology is truly a valuable approach by which the world can face emerging issues and concerns. The fields of life sciences, engineering, biology and nanomaterials have further integrated themselves to bring up a masterpiece, bionanocomposites, that holds the promise of improving our food production while sustainably accessing the world's resources. At present, they are not yet picking up pace when compared to nanocomposites, but it is possible to achieve bionanocomposites with tunable properties. However, there is much to know about the individual strengths of component materials and how these strengths can be harnessed when they are paired up or embedded in a mixture of other substances. Much work needs to be done in identifying their synergistic effects as well as how to further mask or diminish the inadequacies of individual components. Structures and configurations of these small structures can be verified by a variety of analytical and instrumental analyses, which can be costly and limiting in terms of developing the synthesized materials. In recent years, studies have become bolder and more disruptive by attempts to synthesize new materials, develop facile synthesis techniques and maximize the use of biomass waste.

Acknowledgment

The author wishes to acknowledge her postdoctoral contract under the Marie Skłodowska-Curie grant agreement No 894088.

References

[1] Food and Agriculture Organization of the United Nations Global report on food crises http://fao.org/emergencies/resources/documents/resources-detail/en/c/1272014/

[2] International Energy Agency 2020 World Energy Outlook https://iea.org/reports/world-energy-outlook-2020

[3] Zhao R, Torley P and Halley P J 2008 Emerging biodegradable materials: starch- and protein-based bio-nanocomposites *J. Mater. Sci.* **43** 3058–71

[4] Nafchi A M, Moradpour M, Saeidi M and Alias A K 2013 Thermoplastic starches: properties, challenges, and prospects *Starch/Staerke.* **65** 61–72

[5] Zhou Y, Zhang X, Zhang J, Cheng Y, Wu J and Yu J *et al* 2021 Molecular weight characterization of cellulose using ionic liquids *Polym. Test* **93** 106985

[6] Xiong R, Grant A M, Ma R, Zhang S and Tsukruk V V 2018 Naturally-derived biopolymer nanocomposites: interfacial design, properties and emerging applications *Mater Sci. Eng.* R **125** 1–41

[7] Fortunati E, Luzi F, Jiménez A, Gopakumar D A, Puglia D and Thomas S *et al* 2016 Revalorization of sunflower stalks as novel sources of cellulose nanofibrils and nanocrystals and their effect on wheat gluten bionanocomposite properties *Carbohydr. Polym.* **149** 357–68

[8] Martínez C M, Adamovic T, Cantero D A and Cocero M J 2019 Ultrafast hydrolysis of inulin in supercritical water: fructooligosaccharides reaction pathway and Jerusalem artichoke valorization *Ind. Crops Prod.* **133** 72–8

[9] Martínez C M, Cantero D A and Cocero M J 2018 Production of saccharides from sugar beet pulp by ultrafast hydrolysis in supercritical water *J. Clean Prod.* **204** 888–95

[10] Gao A, Chen H, Tang J, Xie K and Hou A 2020 Efficient extraction of cellulose nanocrystals from waste *Calotropis gigantea* fiber by SO42-/TiO2 nano-solid superacid catalyst combined with ball milling exfoliation *Ind. Crops Prod.* **152** 112524

[11] Badgujar K C, Dange R and Bhanage B M 2021 Recent advances of use of the supercritical carbon dioxide for the biomass pre-treatment and extraction: a mini-review *J. Indian Chem. Soc.* **98** 100018

[12] Troncoso O P and Torres F G 2020 Bacterial cellulose—graphene based nanocomposites *Int. J. Mol. Sci.* **21** 1–17

[13] Silmore K S, Gupta C and Washburn N R 2015 Tunable Pickering emulsions with polymer-grafted lignin nanoparticles (PGLNs) *J. Colloid Interface Sci.* **466** 91–100

[14] Abad-Fernández N, Pérez E and Cocero M J 2019 Aromatics from lignin through ultrafast reactions in water *Green Chem.* **21** 1351–60

[15] Aqlil M, Nzenguet A M, Essamlali Y, Snik A, Larzek M and Zahouily M 2017 Graphene oxide filled lignin/starch polymer bionanocomposite: structural, physical, and mechanical studies *J. Agric. Food Chem.* **65** 10571–81

[16] Zhang S, Fu X, Tong Z, Liu G, Meng S and Yang Y *et al* 2020 Lignin-clay nanohybrid biocomposite-based double-layer coating materials for controllable-release fertilizer *ACS Sustain. Chem. Eng.* **8** 18957–65

[17] Hernández-Téllez C N, Plascencia-Jatomea M and Cortez-Rocha M O 2016 Chitosan-based bionanocomposites: development and perspectives in food and agricultural applications *Chitosan in the Preservation of Agricultural Commodities* (New York: Academic) pp 315–38

[18] Sánchez-Ortega I, García-Almendárez B E, Santos-López E M, Amaro-Reyes A, Barboza-Corona J E and Regalado C 2014 Antimicrobial edible films and coatings for meat and meat products preservation *Sci. World J* **2014** 248935

[19] Silva M, dos S, Cocenza D S, Grillo R, Melo N F S, de, Tonello P S and de Oliveira L C *et al* 2011 Paraquat-loaded alginate/chitosan nanoparticles: preparation, characterization and soil sorption studies *J. Hazard. Mater.* **190** 366–74

[20] Kumar S, Bhanjana G, Sharma A, Sidhu M C and Dilbaghi N 2014 Synthesis, characterization and on field evaluation of pesticide loaded sodium alginate nanoparticles *Carbohydr. Polym.* **101** 1061–7

[21] Vilian A T E, Veeramani V, Chen S M, Madhu R, Kwak C H and Huh Y S *et al* 2015 Immobilization of myoglobin on Au nanoparticle-decorated carbon nanotube/polytyramine composite as a mediator-free H_2O_2 and nitrite biosensor *Sci. Rep.* **5** 1–10

[22] Jankowska K, Zdarta J, Grzywaczyk A, Degórska O, Kijeńska-Gawrońska E and Pinelo M *et al* 2021 Horseradish peroxidase immobilised onto electrospun fibres and its application in decolourisation of dyes from model sea water *Process. Biochem.* **102** 10–21

[23] Melo M N, Pereira F M, Rocha M A, Ribeiro J G, Diz F M and Monteiro W F *et al* 2020 Immobilization and characterization of horseradish peroxidase into chitosan and chitosan/PEG nanoparticles: a comparative study *Process. Biochem.* **98** 160–71

[24] Mahmoodi N M, Saffar-Dastgerdi M H and Hayati B 2020 Environmentally friendly novel covalently immobilized enzyme bionanocomposite: from synthesis to the destruction of pollutant *Compos. Part B Eng.* **184** 107666

[25] Li J, Chen X, Xu D and Pan K 2019 Immobilization of horseradish peroxidase on electrospun magnetic nanofibers for phenol removal *Ecotoxicol. Environ. Saf.* **170** 716–21

[26] Heredia A, Matas A and Dominguez E 2000 Investigating plant lipid biopolymers *Biochem. Educ.* **28** 50–1

[27] Graça J and Pereira H 1997 Cork suberin: a glyceryl based polyester *Holzforschung.* **51** 225–34
[28] Silva N H C S, Vilela C, Almeida A, Marrucho I M and Freire C S R 2018 Pullulan-based nanocomposite films for functional food packaging: exploiting lysozyme nanofibers as antibacterial and antioxidant reinforcing additives *Food Hydrocoll.* **77** 921–30
[29] Hassannia-Kolaee M, Khodaiyan F, Pourahmad R and Shahabi-Ghahfarrokhi I 2016 Development of ecofriendly bionanocomposite: whey protein isolate/pullulan films with nano-SiO2 *Int. J. Biol. Macromol.* **86** 139–44
[30] de Lima I A, Pomin S P and Cavalcanti O A 2017 Development and characterization of pullulan-polymethacrylate free films as potential material for enteric drug release *Brazil. J. Pharm. Sci.* **53** 1–9
[31] Kraśniewska K, Pobiega K and Gniewosz M 2019 Pullulan-biopolymer with potential for use as food packaging *Int. J. Food Eng.* **15** 1–14
[32] Lagaron J M, Cabedo L, Cava D, Feijoo J L, Gavara R and Gimenez E 2005 Improving packaged food quality and safety Part 2: Nanocomposites *Food Addit. Contam.* **22** 994–8
[33] Tan Y M, Lim S H, Tay B Y, Lee M W and Thian E S 2015 Functional chitosan-based grapefruit seed extract composite fi lms for applications in food packaging technology *Mater. Res. Bull.* **69** 142–6
[34] Falsini S, Clemente I, Papini A, Tani C, Schiff S and Salvatici M C *et al* 2019 When sustainable nanochemistry meets agriculture: lignin nanocapsules for bioactive compound delivery to plantlets *ACS Sustain. Chem. Eng.* **7** 19935–42
[35] Cooper C J, Abdelwahab M A, Mohanty A K and Misra M 2019 Hybrid green bionanocomposites of bio-based poly(butylene succinate) reinforced with pyrolyzed perennial grass microparticles and graphene nanoplatelets *ACS Omega.* **4** 20476–85
[36] Xiong F, Han Y, Wang S, Li G, Qin T and Chen Y *et al* 2017 Preparation and formation mechanism of size-controlled lignin nanospheres by self-assembly *Ind. Crop Prod.* **100** 146–52
[37] Souza V G L, Pires J R A, Vieira É T, Coelhoso I M, Duarte M P and Fernando A L 2019 Activity of chitosan-montmorillonite bionanocomposites incorporated with rosemary essential oil: from *in vitro* assays to application in fresh poultry meat *Food Hydrocoll.* **89** 241–52
[38] Gopi S, Amalraj A, Jude S, Thomas S and Guo Q 2019 Bionanocomposite films based on potato, tapioca starch and chitosan reinforced with cellulose nanofiber isolated from turmeric spent *J. Taiwan Inst. Chem. Eng.* **96** 664–71
[39] Mission E G, Quitain A T, Sasaki M and Kida T 2017 Synergizing graphene oxide with microwave irradiation for efficient cellulose depolymerization into glucose *Green Chem.* **19** 3831–43
[40] Mission E G, Quitain A T, Hirano Y, Sasaki M, Cocero M J and Kida T 2018 Integrating reduced graphene oxide with microwave-subcritical water for cellulose depolymerization *Catal. Sci. Technol.* **8** 5434–44
[41] Youssef F S, El-Banna H A, Elzorba H Y and Galal A M 2019 Application of some nanoparticles in the field of veterinary medicine *Int. J. Vet. Sci. Med.* **7** 78–93
[42] Singh V and Ahmad S 2014 Carboxymethyl cellulose-gelatin-silica nanohybrid: an efficient carrier matrix for alpha amylase *Int. J. Biol. Macromol.* **67** 439–45
[43] Tarannum N and Gautam Y K 2019 Facile green synthesis and applications of silver nanoparticles: a state-of-the-art review *RSC Adv.* **9** 34926–48

[44] Lang C, Mission E G, Ahmad Fuaad A A and Shaalan M 2021 Nanoparticle tools to improve and advance precision practices in the Agrifoods Sector towards sustainability – a review *J. Clean. Prod.* **293** 126063
[45] Gatahi D, Wanyika H, Kavoo A, Kihurani A and Ateka E 2016 Enhancement of bacterial wilt resistance and rhizosphere health in tomato using bionanocomposites *Int. J. Hortic. Sci. Technol.* **3** 129–44
[46] Abd El-Aziz M, Morsi S, Salama D, Abdel-Aziz M, Abd Elwahed M and Shaaban E *et al* 2019 Preparation and characterization of chitosan/polyacrylic acid/copper nanocomposites and their impact on onion production *Int. J. Biol. Macromol.* **123** 856–65
[47] Chen S, Yang M, Ba C, Yu S, Jiang Y and Zou H *et al* 2018 Preparation and characterization of slow-release fertilizer encapsulated by biochar-based waterborne copolymers *Sci. Total Environ.* **615** 431–7
[48] Torres-figueroa A V, Cinthia J P, Castillo-castro T, Bolado-mart E, Corella-madueño A G and Garc A M *et al* 2020 Composite hydrogel of poly (acrylamide) and starch as potential system for controlled release of amoxicillin and inhibition of bacterial growth *J. Chem.* **2020** 5860487
[49] Zubair M, Wu J and Ullah A 2019 Hybrid bionanocomposites from spent hen proteins *ACS Omega.* **4** 3772–81
[50] Quitain A T, Ozturk B, Mission E G and Sasaki M 2017 Microwave-assisted pressurized hot water extraction of alkaloids *Water Extraction of Bioactive Compounds: From Plants to Drug Development* (Amsterdam: Elsevier Inc) pp 269–89
[51] Mallakpour S and Madani M 2016 Use of valine amino acid functionalized α-mno2/chitosan bionanocomposites as potential sorbents for the removal of Lead(II) Ions from aqueous solution *Ind. Eng. Chem. Res.* **55** 8349–56
[52] Mensah A, Lv P, Narh C, Huang J, Wang D and Wei Q 2019 Sequestration of Pb(II) ions from aqueous systems with novel green bacterial cellulose graphene oxide composite *Materials (Basel).* 12
[53] Korkut S, Keskinler B and Erhan E 2008 An amperometric biosensor based on multiwalled carbon nanotube-poly(pyrrole)-horseradish peroxidase nanobiocomposite film for determination of phenol derivatives *Talanta* **76** 1147–52
[54] Meiyazhagan A, Thangavel S, Daniel P H, Pulickel M A and Palanisamy T 2015 Electrically conducting nanobiocomposites using carbon nanotubes and collagen waste fibers *Mater. Chem. Phys.* **157** 8–15
[55] Sun W, Li X, Wang Y, Li X, Zhao C and Jiao K 2009 Electrochemistry of myoglobin in Nafion and multi-walled carbon nanotubes modified carbon ionic liquid electrode *Bioelectrochemistry* **75** 170–5

Chapter 4

Utility of nanobiosensors in agriculture

Merve Çalışır, Ilgım Göktürk, Deniz Türkmen, Fatma Yılmaz and Adil Denizli

The increasing world population increases the demand for food, but the land that can be farmed is decreasing. So, food shortage is the most important emerging issue in today's conditions. In this respect, it is essential to focus on new developments in agricultural systems. The primary purpose of agriculture is to obtain a large amount of product with the least possible effort. Nanobiosensors are widely used for cheap, rapid and on-site detection methods. Nanobiosensor applications with high sensitivity, as highlighted in this chapter, are needed to ensure agricultural and consequently food security while conserving the natural habitat.

4.1 Introduction

Throughout the history of agriculture, biotechnological advantages have been used in the creation of agricultural products. Improved agricultural practices play a major role in maintaining the ecological balance [1, 2]. More than half of the world's population relies on agricultural income and agriculture is one of the most critical issues to be focused on in the future as it is now. Innovations in the agricultural systems are always based on obtaining the crops in the most efficient way possible, in other words, harvesting the maximum amount of produce with minimum energy and expense. For this purpose, each method of obtaining the product should be examined one by one, and attempts should be made to improve each step. During all these investigations, environmental impacts should also be considered in terms of sustainability. It is predicted that the world population will approach ten billion within 30 years. So it is now necessary to develop effective agricultural techniques for the increasing food demand. The agricultural sector is currently facing many global challenges such as climate change, low productivity, decreasing soil amount, low nutritional efficiency, and reducing water resources. On top of this, increased soil pollution caused by an excess amount of pesticide usage poses other dangers

doi:10.1088/978-0-7503-3863-9ch4

besides world hunger, so it is now necessary to develop effective techniques in agricultural systems with the increasing food demand [3, 4].

Considering all these reasons, it is concluded that other methods should be followed than traditional applications. Herein, nanotechnology offers promising ways for the development of agricultural systems and solutions to these existing problems. Nanotechnology provides control nanometer-scaled material production and can thus lead to many effects in agricultural applications. These developments include controlled release of micronutrients and fertilizers, biofertilizers, and nanobiosensors for sensitive pathogen and pesticide detection [5, 6]. Soil quality used in agriculture is directly related for determining the crop yield and thanks to nanotechnological applications, monitoring these crucial soil conditions has become much easier. In the last 50 years, the yield of grain crops has grown significantly and has a vital part in meeting the amount of food humankind needs. The reason for this yield was the increase in the use of chemical fertilizers. Although the preference for artificial fertilizers usage increases the amount of product, it has caused environmental pollution due to the leakage into the natural sources long term. Many of these disadvantages of chemical fertilizers have been eliminated with nanofertilizers [7]. Nanofertilizers are nanomaterials that consist of essential nutrients for the plant. Nanofertilizers increase product quality and quantity with high nutrient use and negligible waste, thus providing a low-cost and sustainable application [8]. Nanomaterials used during fertilization have become more economical and environmentally friendly with the application of controlled-release fertilizers. The yield obtained in conventional nitrogen fertilizers is between 30% and 60%, and in traditional phosphate fertilizers, between 8% and 90%. These percentages are highly affected by chemical bonding occurring in the soil, and the plant can not use much of these fertilizers. Applications applied directly to the plant root with controlled-release nanofertilizers increased the production efficiency by lowering the cost considerably [9, 10].

Nanopesticides are another promising application method for sustainable agricultural systems. Pesticides are chemical and biological materials that are frequently and specifically applied for the target in agricultural areas. Like chemical fertilizers, low amounts of pesticides reach the target pests and the rest cause pollution that is highly toxic to the food chain and human health. In addition, the increasing number of pesticides used in each new application causes the development of resistance to these compounds [11–13]. Nanopesticides have valid potential to overcome these limitations. In long-term applications, controlling of pesticides can be achieved with slow degradation and controlled release of nanopesticides. For this reason, the use of nanopesticides is essential for long-lasting, sustainable pest control as it prevents the accumulation of toxic chemicals [14]. The working principle of nanopesticides differs from conventional pesticides as nano-sized particles are based on transport in dissolved and colloidal states. The fact that harmful chemicals are soluble in water can undermine the natural habitat as they can infiltrate groundwater and microorganisms [15]. Nanoparticle-based pesticides, which increase the solubility of active compounds, are less harmful to the environment than traditional pesticides [16]. Although they can be used in many such areas, there are few commercial examples of nanotechnology applications in the current market. However, radical innovations are expected as a result of the cooperation of the agricultural sector with nanotechnology in the coming years. In particular, the sensor part of nanotechnology applications aims to end many current problems in agricultural systems. Hence, this chapter will present a comprehensive overview to the electrochemical, optical and piezoelectric based nanobiosensors focusing on the applications of these nanosensors in agriculture and the key points associated with them. For this purpose, detection studies of pesticides, microbial, food and toxin contaminants by nanobiosensors will be discussed by surveying important aspects in detail.

4.2 Nanobiosensor applications in agriculture

Biosensors are systems that can detect biological material using recognition elements that consist of biological substances [17]. Nanobiosensors are more sensitive and compact versions of biosensors designed to detect a biological analyte at much lower concentrations. Water, soil, fertilizer and pesticide precise determinations performed with nanobiosensors increase agricultural crop yield and offer more economical and environmentally friendly applications. The speed of electron transition, appropriate and sufficient surface area and sensitivity have put nanobiosensors ahead of traditional sensor systems [18]. One of nanosensors' great features is that they can design biological components to be mostly identical to the natural ones. Nanobiosensors use a signal-providing recognition element and a nanoscale material-based transducer. Figure 4.1 shows the parts of nanobiosensors. Among these biorecognition elements, DNA, aptamers, amino acids, nucleic acids, and sometimes even pathogens and bacteria can be used. Transducers have many types according to their signal reading methods, for instance optical sensing, electrochemical, and piezoelectric. In this section,

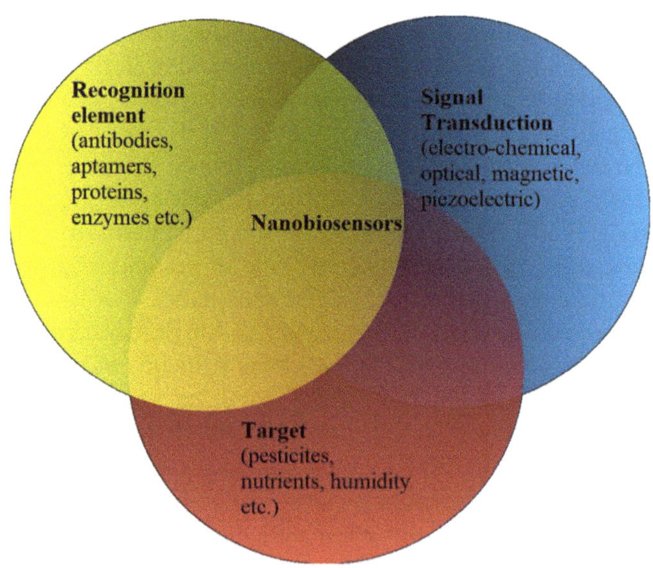

Figure 4.1. The main scheme for nanobiosensors.

changes in the analyte's sensory properties after binding to various bio-receptors will be examined. Naming is made according to different reading techniques used in sensor systems [19]. There are different types of the electrochemical biosensors such as voltammetric, amperometric, conductometric, potentiometric and impedimetric.

The term 'nano' originated from examining systems based on nanoscaled merits and molecules at the size of 1–100 nm [20]. Systems with controlled adjustment in such small sizes become open to many applications such as disease diagnosis and biomolecule detection. Nanoparticles are also used to increase microbial monitoring detection sensitivity, degradation and chemical recovery efficiency [21]. Sensors consist of different parts such as software, processing electronics and can detect and measure a physical property such as the mass of compounds [22]. The specificities of biological reactions are very advantageous to be used in biosensor systems. Biological recognition is the conversion of the signal (optical, piezoelectric and electrochemical) generated by the target analyte into a measurable effect. Biosensors consisting of biological recognition elements (nucleic acids, DNA, amino acids, enzymes) and signal transduction elements have a scheme in figure 4.2. In this chapter, the applications and developments of optical, electrochemical, and piezoelectric nanobiosensors formed by the marriage of nanotechnology and biosensor devices will be examined in the agriculture field. Aptamers and molecularly imprinted polymers developed to be used in the determination of target molecules were also mentioned. The application of pesticide and herbicide determinations required in agricultural applications with optical, electrochemical and piezoelectric nanobiosensors is presented with examples.

Optical detection methods including surface plasmon resonance (SPR), colourimetry, chemiluminescence and surface-enhanced Raman scattering (SERS) used for pesticide or herbicide determination represent the highest percentage of flow

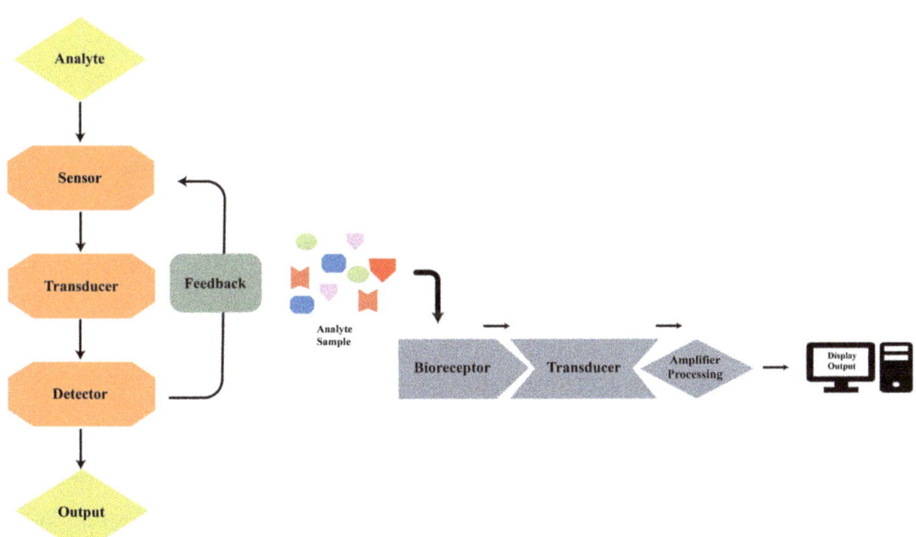

Figure 4.2. A typical sensor scheme consists of a transducer–amplifier and display unit.

methods designed [23–26]. Electrochemical sensor types are classified as potentiometric and amperometric sensors. These sensors are based on potential alteration and are practical devices used to detect biological or inorganic target molecules [27]. Electrochemical techniques have many remarkable properties and are preferred over other analytical methods. These portable, low-cost, rapid biosensors are classified according to signals measured in impedance, current, and potential [28]. The superior properties of the electrochemical impedance spectroscopy (EIS) technique such as high sensitivity, specificity, fast response time and easy preparation make it a commonly used technique. Electrochemical impedance spectroscopy can be categorized into two subgroups which are faradic and non-faradic EIS. Piezoelectric quartz crystal-based detection is quite simple. The detection of a target sample commonly relies on selective binding of adsorbate, leading to a mass change identified by a frequency change in the quartz crystal [29]. The advantages such as low cost, high sensitivity, real-time output and ease of use make piezoelectric sensors an alternative method compared to commonly used immunoassay tools to detect pesticides. Piezoelectric sensors can be designed using quartz crystals that resonate when an external alternating electric field is applied, without the need for expensive or hazardous labels [30, 31]. The frequency of the oscillation occurs as a function of the mass of the crystal. Therefore, the total mass will increase as a result of an analyte interaction with the corresponding molecule in a sample previously immobilized to the quartz crystal, which will be measured as a change in oscillation frequency [32]. Nanosensors and nanobiosensors are used in medicine, agriculture, food, packaging, and their development is crucial for these demanding research fields. They can be used as a valuable tool for agricultural applications in terms of soil fertility, moisture analysis, microbiological environment suitability, control mechanisms in fertilizer pesticide and herbicide applications. In this chapter, the studies conducted with electrochemical, optical and piezoelectric nanobiosensors used in different agricultural applications are given in detail.

4.2.1 Electrochemical nanobiosensors

Nanosensors are capable of 'nano-tuning', show superior performance levels with the desired detection limit levels and sensitivity. Nanobiosensors are preferred to monitor pesticides, herbicides, food allergens, additives, toxins, mycotoxin, microbial contamination, genetically modified organisms (GMOs), plant diseases and environmental parameters. Electrochemical sensors functioning according to the electrochemistry principle are the most widely used sensors. Electrochemical signals generate electrons that are consumed or produced during biointeraction. Voltage, current or impedance value measured due to chemical reactions generating ions or electrons are used in electrochemical nanosensors [33]. High-sensitive electrochemical transducers compatible with nanofabrication technologies are used in sensing applications. The produced electrochemical signal is proportional to the sample mass. Electrochemical nanosensor devices can be grouped as potentiometry, voltammetry and amperometry according to the working principles. The redox reaction taking place in an electroactive species creates a current and the working mechanism of the amperometric sensors is based on measuring this current continuously. When the potential is kept constant, the faradaic current determines the concentration of electroactive species. Amperometric biosensors have become cheaper, and highly sensitive via nanofabrication techniques. They find application areas in clinical, environmental, food and agriculture industries. Detection of sulfonamides, organophosphates, salbutamol and ractopamine, fructose content and hydrogen peroxide is an example of the use of amperometric-based nanosensors [34–38].

Voltammetry is the current determination by changing the potential. The oxidation and reduction rate potential of an electrochemical reaction are measured by cyclic voltammetry (CV). It depends on the measurement of voltage change between the working and standard electrode, and current occurs between the counter and working electrodes. CV has been applied in the food and agriculture fields to detect carbosulfan, *Salmonella typhi* and heavy metals [39–41]. Potentiometric sensors are based on potential created by the working electrode and the reference electrode. The complex biological matrices have ions such as Ca^{2+}, Na^+, H^+, NH_4^+ and K^+, and ISE identifies these ions using the change in potential [42]. Nanobiosensor technology can help increase crop yields by ensuring proper water, soil, fertilizers and pesticides. Nanobiosensors have high sensitivity and surface-to-volume ratio, increased lifetime stability and fast electron transfer dynamics [43]. These performed studies with potentiometric nanobiosensors, which are a type of electrochemical sensors, are promising in terms of being open to development and can be used in a wide range. In addition to this, it is one step ahead of other nanobiosensor systems since these systems can be designed much more easily and are ideal for reducing the cost of analysis.

4.2.1.1 Pesticide detection

The most significant number of inventions were made since the day humanity started using technology. They aimed to improve agriculture more than any other sector where technology is used. However, the growing hunger of the increasing world population today is being met. As the supply of traditional technological ammunition in agriculture is almost exhausted, the only way forward is to adopt innovative farming practices sustainably. Modern electronics and materials science to produce crops without further increasing fertilizers or pesticides can be termed innovative and sustainable agriculture. Scientists have made giant leaps in 'biology at the nanoscale' in the current century. Nanobiosensors contribute to smart agriculture through growth monitoring, real-time pest detection and monitoring of the environment continuously. Pesticides or plant protection products preventing disease and infestation are used to keep crops healthy [44]. Herbicides, insecticides, fungicides, acaricides, plant growth regulators and repellents belong to this class. Pesticides can be categorized according to chemical types such as organochlorines, organophosphates, carbamates, pyrethroids and pyrethrins [45]. Pesticides contain non-plant biocides such as wood preservatives, disinfectants used to keep disease-carrying pests away [46]. Pesticides leaching from soil into groundwater are potential drinking water contaminants [47]. Also, they can mix into surface water either by flow or by being dragged by emissions that pose a risk to human health and the ecosystem [48]. Gas chromatography (GC) and liquid chromatography (LC) coupled with mass spectroscopy (MS) are the conventional methods used in pesticide analysis. In addition to chromatographic techniques for pesticide detection, electrochemical biosensors have been used successfully [49].

The reason for the widespread use of organophosphorus pesticides (OP) is to increase agricultural productivity. However, excessive use of OP poses a potential threat to human health by causing pesticides to remain in the food and environment. For this reason, it is of great importance to quickly determine the trace level of the OPs in terms of environmental protection and food safety. Zheng *et al* offered a simple, environmentally friendly, sensitive and stable AChE biosensor for detecting OPs [50]. Bhardwaj *et al* prepared silica oxide modified copper-based metal organic framework (MOF) thin films for detecting a dangerous pesticide, atrazine [51]. Electrochemical immunosensors have become valuable tools for detecting pesticides, as limit of detection (LOD) values are achieved. Electrically conductive nanomaterials used to modify the working electrode surface enhance the analytical signals by allowing more antibodies to be immobilized, thus increasing their performance [52]. Sensitive and easy methodologies developed for fabricating nanobiosensors to detect and monitor pesticides without the need for extensive sample pretreatment are summarized in table 4.1. However, there are drawbacks in using immunosensors due to the expensive antibody obtaining and their instability using on-site environmental toxic substances detection [104]. Aptamer and MIP-based biosensors are promising as dependable affinity biosensors for detecting pesticides and can be applied in addition to immunosensors [105]. MIPs are biomimetic synthetic receptors formed by the reaction of functional monomers, template molecules and crosslinker to create specific cavities for target molecules [106]. AChE based biosensor was prepared to detect organophosphate pesticide dichlorvos using multiwalled carbon nanotubes (MWCNTs) as electron intermediator [34]. The results of the developed biosensor are shown in figure 4.3.

Nanocrystals having unique properties such as high photostability, complete wavelength absorption, controlled fluorescence emission are called quantum dots. ZnSe quantum dots functionalized with mercaptophenyl boronic acid was bonded on graphene–chitosan nanocomposites electrostatically and they were used for the determination of methyl parathion, and the limit of detection was determined as 0.2 nM [107]. Biosensors in which enzyme inhibition is monitored refer to indirect mode biosensors, while sensors, where the analyte or product formation concentration during reactions is measured, are called direct mode enzymatic biosensors. AChE inhibition has been used in indirect mode biosensor to detect organophosphate group pesticides [108, 109]. AChE biosensors used in electrochemical detection of organophosphates were reviewed by Dhull *et al.* Enzyme-linked biosensors have many advantages. However, there are problems in detecting multianalyte, and efforts are continuing to address these shortcomings [110]. Borah *et al* offered a system based on glutathione-S-transferase (GST) enzyme inhibition. The critical point in developing an enzyme-based biosensor is to preserve the enzyme properties such as active site and geometry while immobilizing it to the surface of the electrode

Table 4.1. Electrochemical biosensors used in environmental sensing.

Biosensor type	Recognition element	Material type	LOD	References
Electrochemical	PtNPs/CS	GCE	7 μg.L^{-1} and 20 μg.L^{-1}	[53]
Electrochemical	MWCNT/zirconia	Carbon paste electrode	9.0×10^{-9} mol. L^{-1}	[54]
Electrochemical	CuNPs	GCE modified with MWCNT	0.33 nmol.L^{-1}	[55]
Electrochemical	WO_3 center dot $0.33H_2O$ nanorod	Carbon paste electrode	233 nM	[56]
Electrochemical	Enzyme	Indium tin oxide (ITO) electrode with ZIF-8/MB composites	1.7 ng.mL^{-1}	[57]
Electrochemical	p-phenylenediamine	Pencil graphite electrodes	2.75 μg.mL^{-1}	[58]
Electrochemical	Hb	Reduced graphene oxide-chitosan (ERGO-CS/Hb/FTO)	79.77 nM	[59]
Electrochemical	$NiCo_2S_4$	Graphitized carbon nanofiber	20 nM	[60]
Electrochemical	CNT-H-IMZ	GCE	120 nM^{-1}	[61]
Electrochemical	NiO nanoparticles	GCE	0.5 μM	[62]
Electrochemical	SiO_2NP-CNT-μPs	GCE	0.092 μM	[63]
Electrochemical	Nanosilver/dodecane	GCE	0.6 nM	[64]
Electrochemical	MXene/CNHs/β-CD-MOFs)	Graphene	1.0 nM	[65]
Electrochemical	Stearic Acid/nanosilver composite	GCE	0.1 nM	[66]
Electrochemical	Enzyme	ITO electrode	8.8 ng.L^{-1}	[67]
Electrochemical	Aptamer	Au-TAN	0.077 pM	[68]
Electrochemical	NiAl-LDH	Carbon paste electrode	1×10^{-9} M	[69]
Electrochemical	Poly-L-cystein-gold nanoparticles (PLC-AuNPs/PGE)	Pencil graphite electrode	1.83×10^{-10} M	[70]
Electrochemical	SnS_2/NS–RGO)	GCE	0.17 nM	[71]
Electrochemical	Enzyme	MnO_2-Thi based electrode	3×10^{-10} M	[72]
Electrochemical	The Ag-ZnO/s-SWCNT	Au probe	0.27×10^{-16}M	[73]
Electrochemical	PtNPs-MoS_2-rGO	GCE	0.07 μM	[74]
Electrochemical	Anti-fiproni	GCE	112 ag.mL^{-1}	[75]
Electrochemical	Silver(Ag)-doped ZnO nanorods	GCE	0.34 nM L^{-1}	[76]

Electrochemical	ZnO	Ni-foam electrode	$1.2.10^{-15}$ M	[77]
Electrochemical	Nylon 6,6	Graphite HB pencil electrode	$0.94.10^{-8}$ M	[78]
Electrochemical	Nitride (g-C_3N_4)/ Graphene oxide (GO)-(Fc-TED)	GCE	8.3 nM	[79]
Electrochemical	(L-Cit)- (beta-CD)	GCE	10 nM	[80]
Electrochemical	BSA/AChE-Glu-s-SWCNTs/GCE	GCE	$3.75.10^{-11}$ M	[81]
Electrochemical	SWCNTsAu-ZrO2-GNs	GCE	1 $ng.mL^{-1}$	[82]
Electrochemical	ACC-HNFs	SPCE	6 $fg.mL^{-1}$	[83]
Electrochemical	GO	TRGOPE	0.015 $\mu mol.L^{-1}$	[84]
Electrochemical	N-methyl-2-pyrrolidone	GCE	0.78 nM	[85]
Electrochemical	Aptamer	GCE	0.35 fM	[86]
Electrochemical	Au NPs (3DG-Au)	GCE	0.0012 μM	[87]
Electrochemical	Enzyme	GCE and gold nanorods	0.7 nM	[88]
Electrochemical	Enzyme	SPE with carbon black nanoparticles	5 $\mu g.L^{-1}$	[89]
Electrochemical	Enzyme	SPE with Fe_3O_4 and gold nanoparticles	0.1 $ng.mL^{-1}$	[90]
Electrochemical	Enzyme	Carbon paste electrode and reticulated spheres structures of $NiCo_2S_4$	0.42 $pg.mL^{-1}$	[91]
Electrochemical	Enzyme	Carbon paste electrode with chitosan, gold nanoparticles, and Nafion	5 $fg.mL^{-1}$	[92]
Electrochemical	Aptamers	Carbon black and GO/Fe_3O_4	94 pM	[93]
Electrochemical	Enzyme	Platinum electrode with ZnO	12:00 PM	[94]
Electrochemical	Enzyme	Ionic liquids-gold nanoparticles porous carbon composite	0.3 pM	[95]
Electrochemical	Aptamers	Gold nanoparticles, MWCNT and rGO nanoribbons	17 fM	[96]
Electrochemical	Aptamers	Silver nanoparticles and nitrogen-doped GO	33 fM	[97]
Electrochemical	Aptamers	Platinum nanoparticles	1:00 PM	[98]
Electrochemical	Antibodies	SWCNT	0.01 $ng.mL^{-1}$	[99]
Electrochemical	Phage and antibody complex	Protein G functionalized Magnetic beads	0.2 $pg.mL^{-1}$	[100]
Electrochemical	Enzyme	GCE with GO and MWCNT	136 pM	[101]
Electrochemical	Enzyme	MWCNT and GO nanoribbons structure	1.7 nM	[102]
Electrochemical	Enzyme	Porous GCE with GO network	0.74 nM	[103]

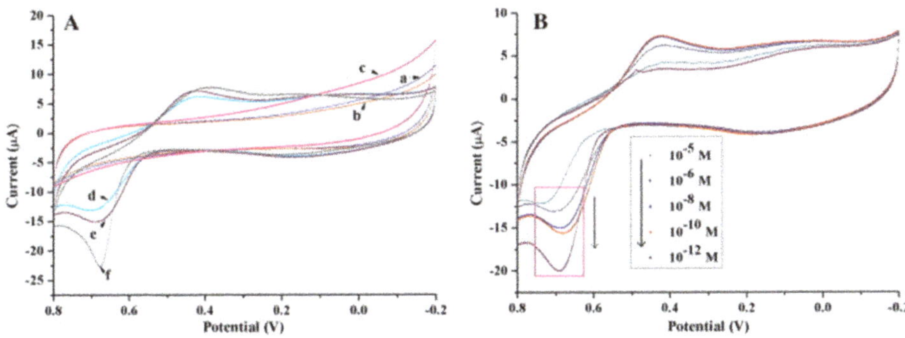

Figure 4.3. (A): CV of $(CS/ALB)_6$/GCE (a), bare glassy carbon electrode (b), $(MWCNTs/ALB)_6$/GCE (c) in pH 7.0 PBS, bare glassy carbon electrode (d), $(CS/ALB)_6$/GCE (e) and $(MWCNTs/ALB)_6$/GCE (f) in pH 7.0 PBS with 1 mM ATCl at a 50 mV s^{-1} scan rate; (B): Different concentrations of dichlorvos on $(MWCNTs/ALB)_6$/GCE [34].

[111]. The whole-cell containing electrochemical biosensors for pesticide detection has been developed as an alternative to enzymes. Since a cell is suitable for genetic modification, the detecting area can be quite wide and specific [112]. Artificial mediators such as potassium ferricyanide, menadione and benzoquinone overcome the slow transfer of electrons between the cell wall and the surface of the electrode by replacing the electron acceptor oxygen [113]. Triazines, organochlorines, or organophosphate types are pretty suitable for microbial biosensors detection [114, 115]. Another biosensor was designed to detect herbicide using cyanobacterial species [116]. Algal cells applied on an indium tin oxide (ITO) electrode in an electrochemical biosensor designed by Pabbi *et al* The biosensor measured alkaline phosphatase inhibition (ALP) in the cell wall exposed to the organophosphate pesticide acephate [115].

4.2.1.2 Microbial contamination

E. coli (Escherichia coli), *Salmonella, Campylobacter, Staphylococcus* etc, are common foodborne microbes causing outbreaks [117–119]. Most food pathogens transmission occurs through unhealthy water sources, raw meat, milk, fruit and vegetables. Gram-positive bacilli, usually found in foods such as dairy products, rice, vegetables, and various environmental conditions such as soil, were detected electrochemically in a milk sample using gold nanoparticles. A pathogenic biosensor using amperometric, voltammetric, and impedimetric measurements was examined [120, 121]. An electrochemical immunosensor for detecting *Brucella abortus* has been produced using copper-doped magnesium oxide-based matrix [122]. A sensor model was created for detecting *E. coli* with the help of glassy carbon electrode (GCE) modified with bismuth nanofilm [123]. *Salmonella* is one of the widespread food poisoning pathogens. A wide linear range was obtained using differential pulse voltammetry (DPV). Vaisocherová-Lísalová *et al* detected *Salmonella* with remarkably low concentrations successfully [124]. Gold nanoparticles are one of the most used particles in diagnosis because they are suitable for different methods. By using

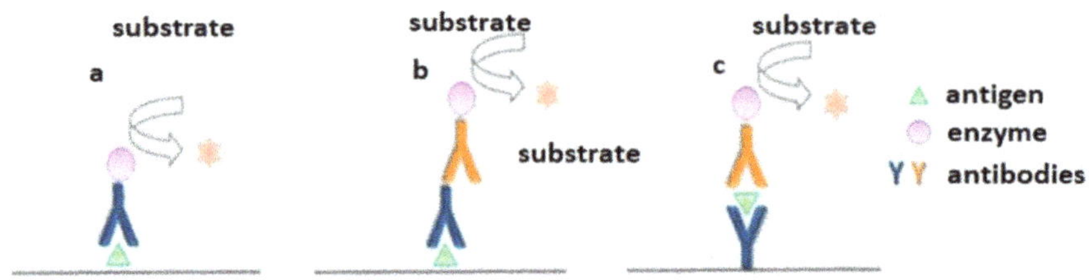

Figure 4.4. Schematic representation showing ELISA.

an electrochemical sensor that uses DNA as a recognition biological agent, Kuan *et al* were able to detect fungi at a lower detection level than PCR [125]. Wang *et al* detected *Sclerotinia sclerotiorum* fungus in oilseed crop by modifying the copper–gold electrode with nanoparticles [126].

The logic of immunosensors relies on the popular enzyme-linked immunosorbent assays (ELISAs) and which have two types of sandwich method namely, direct and indirect. Firstly, the direct ELISA method immobilizes antigens on an assay plate (figure 4.4(a)). Secondly, in an indirect ELISA, an enzyme-secondary antibody added after an unconjugated primary antibody, enhancing the signal and increasing sensitivity (figure 4.4(b)). Thirdly, in the sandwich ELISA method, two epitopes including antigens interact with again two antibodies that are enzyme-labelled (figure 4.4(c)). Electrochemical biosensors are used as enzymatic sensors, and they are low-cost and easily applicable for detecting plant pathogens [127, 128]. However, biosensing devices are susceptible to limited enzyme accessibility and for that they are significant disadvantaged sensors [129]. Xiang and Fei developed label-based electrochemical immunosensors with electrodes modified with graphene nanoplatelets (GNPs) to detect *Salmonella.* Xiang *et al* detected *Salmonella* using an ultrasensitive electrochemical immunosensor [130, 131]. GNPs dispersed in chitosan hydrogel (GNPs/Chi) were used to capture antibodies and amplify signals. Such a sandwich electrochemical immunosensor prepared by Fei *et al* was used for detection of *S. Pullorum* and *Gallinarum* in the concentration range of 10^4–10^9 CFU mL^{-1} with a LOD of 3×10^3 CFU mL^{-1} [131]. *Salmonella* was detected with label-based electrochemical immunosensors using antibody-modified magnetic particles. Afonso *et al* detected another pathogen bacterium in less than 2 h with a sandwich-type GNP immunosensor [132]. Wang *et al* used a two-step strategy involving immunomagnetic preconcentration and increased electrochemical signal [133].

Punbusayakul and colleagues fabricated a label-free immunosensor using double-walled carbon nanotube bundles for *S. Typhimurium* detection [134]. Chunyan Chai *et al* detected the *S. Pullorum* in chicken samples with a similar approach by using an MWCNT-based portable amperometric biosensor [135]. *Enterobacter sakazakii* and *E. coli* are detected by an electrochemical immunosensor including immobilized horseradish peroxidases-labelled antibodies. For enhancing the electrode sensitization, MWCNT–chitosan composite films are applied on the working electrodes [136]. DNA-based biosensors hybridize the target to the probe and DNA movement towards electrons occurs between electrodes. The DNA biosensor is much more sensitive than traditional PCR by detecting certain pathogens in a plant [137].

Similarly, by utilizing protein–ligand interactions, ultra-sensitive unique protein detection can be realized. DNA and protein detection biosensors can be used to detect plant pathogens. Amperometry, impedimetric cyclic voltammetry were used to identify picomolar DNA levels in the plum blossom virus by hybridization of label-free DNA [138, 139]. *Xylella fastidiosa* (Xf) pathogen affects economically essential crops and a Xf-DNA-electrochemical biosensor, prepared by considering the bacterial DNA of Xf, was used *in situ* analysis based on Cu (II)-Xf-dsDNA interaction [140]. These studies prove the effectiveness of nanobiosensors in terms of determination of pathogens and are also an indicator of their the excellent width of working range.

4.2.1.3 Food contaminants

Global food safety is essential as food-borne allergies affect patients badly and sometimes cause death, and there is no known treatment other than not eating allergic food [141]. Fang *et al* detected *Phytophthora cactorum* causing infection in strawberry fruits, electrochemically using TiO_2 and SnO_2 nanoparticles [142]. An electrochemical nanosensor to recognize *P. syringae* infection by Lau *et al* was also developed [143]. The trace level of a carcinogenic food additive quinoxaline-2-carboxylic acid was determined by functionalized MWCNTs [144]. The iron oxide nanoparticles-embedded electrochemical biosensor was used to detect H_2O_2 in milk samples presenting remarkably low detection limits [145]. In another study, detection of *staphylococcal enterotoxin B* in milk was performed using an electrochemical immunosensor with a detection limit of 0.017 ng mL^{-1} [146]. Duran and Marcato applied electrochemical nanosensors to detect common contaminants in the food and agriculture sectors [147]. An electrochemical sensor using cubic gold nanoparticles involved in a functionalized graphene oxide modified glassy carbon electrode was designed to sensitively determine tyrosine in milk by imprinting method [148]. Dong prepared an Ag-substituted ZnO nano-flower modified glassy carbon electrode for tyramine detection [149]. Nanobiosensors for vitamin C, isoflavones, vanilla flavours, and polyphenols were presented using an MWCNT screen-printed carbon electrode through amperometric detection [150, 151]. Detection of biotin (Vit B7) using gold nanoparticles was performed in an immunosensor electrochemical assay [152].

4.2.1.4 Toxin contamination

Chemicals released into water systems threaten human health and the ecosystem by causing pollution. Developing fast, reliable, and precise analytical detection systems is vital to comply with these limits [153]. The EU Directive 2013/39/EU boosts such methods' improvements to ensure reliable detection of situations where surface water environmental quality standard (EQS) limits are exceeded [154]. Brennan *et al* designed portable fluid biochips using rainbow trout gill epithelial cells (RTgill-W1) [155]. Many researchers have used electrochemical immunosensors to achieve greater selectivity by combining the antibody–antigen interaction's unique specificity with the sensitivity of electrochemical detection [156]. Sharma *et al* detected aflatoxin B1 with poly(3,4-ethylene dioxythiophene)

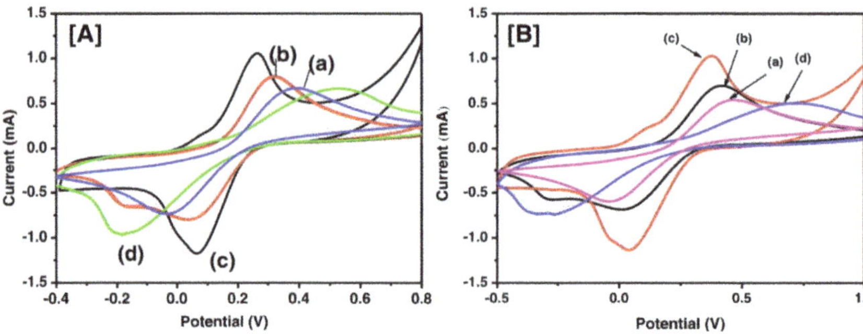

Figure 4.5. Cyclic voltammogram of aflatoxin B1 electrochemical immunosensor by Sharma *et al* [157].

modified with gold nanoparticles [157]. As shown in figure 4.5, after adjusting PEDOT/ITO with AuNPs, the values I_{pa} and I_{pc} increase to 1.10 mA and 1.12 mA (I_{pa}/I_{pc}~1) respectively (curve c). Kundu *et al* prepared a screen-printed electrode-based biosensor exhibiting excellent sensitivity for the detection of carcinogenic formaldehyde with a low detection limit, 0.03 mg.L^{-1}. The study results could help develop a range of biosensors for quantifying adulteration and contaminants along the entire food supply chain [158]. Zhang *et al* degraded perfluorinated compounds through electrochemical oxidation in wastewater treatment using nano-ZnO coated electrodes and achieved a cleaning efficiency of 39% to 66% within 40 min [159]. Further, using the MWCNT modified electrode, carcinogenic Sudan I was detected with a detection limit of 5.0 μg.L^{-1} [160].

Heavy metals, whose accumulation in soil and water becomes very dangerous, impair the ecosystem by completely affecting the food chain. [161–165]. Therefore, analytical techniques used in the heavy metal analysis should be able to fall below these limits. Currently, many analytical techniques are used to analyze heavy metals at billion concentration levels (ppb) [166–168]. Enzymatic biosensors are capable of detecting Pb^{2+}, Cu^{2+}, Cd^{2+}, Cr^{6+} and Hg^{2+}, well known heavy metals, and inhibition of the enzyme activity is the underlying logic of these biosensors [169–173]. Many researchers produce enzyme-based biosensors using nanomaterials. [174, 175]. The strategy of these biosensors is to use living cells as a cheaper biological sensing element than purified enzymes or antibodies [176–179]. Chemical energy contained in organic matter is metabolically converted into electrical energy by Microbial fuel cells (MFCs) [180]. An example of this kind of biosensor is designed for simultaneous monitoring of toxicity caused by mining drainage. Furthermore, maintaining the selectivity and sensitivity of cell activity remains a challenge, especially under unnatural environmental conditions [181]. Genetically modified microorganisms are used to improve selectivity to certain toxic substances such as Pb^{2+} and Hg^{2+} [182, 183]. With the heavy metal accumulation of past misapplications and the introduction of genetically modified crops, monitoring toxic contamination has become vital and as can be seen from these studies, nano-biosensors have proven themselves to be a useful option.

4.2.2 Optical nanobiosensors

The working principle of biosensors in biochemical reactions is based on internal reflection, absorption, fluorescence, luminescence, light scattering spectroscopy, or SPR [184, 185]. Spectroscopy systems contain sub-parts that detect changes in parameters like phase, wavelength, and polarization. Various optical methods based on SPR, fluorescence spectroscopy, and interferometry have been used in biosensor systems [186]. Luminescence consisting of fluorescence and phosphorescence is the emission of light from any substance. When incident light excites the electron in the ground state, after returning from the excited state to the ground state, excited single electrons emit photons. Fluorescence-based nanosensing is a vigorous technique due to its extreme sensitivity. Gold and silver nanoparticles with natural fluorescence mainly owe this specialty to the metallic nanoparticle's size and morphology. Stability of antioxidants is essential for packaged foods and cosmetic items. Hu *et al* prepared gold nanoclusters (Au-NCs) as a quantitative and straightforward spectroscopic method to detect antioxidants. The mechanism was explained as the quenching of gold nanoclusters' fluorescence of immensely reactive oxygen species (ROS) (peroxides, hydroxyl radical etc) that could be prevented by antioxidants [187]. Nitrite, a highly carcinogenic substance, is an essential food preservative. The reaction of nitrite with hydrogen peroxide produces peroxynitrous acid inducing aggregation of silver nanoparticles under acidic conditions, and fluorescent quenching is associated with nitrite concentration (LOD = 100 nM). Thus, nitrile detection was carried out using hyperbranched polyethyleneimine, which preserved the silver nanoparticles [188].

The change in the behavior of nanometer-sized semiconductor quantum dots with light causes them to exhibit extraordinary properties. Their optical absorption and emission energies which are significant for adjusting their luminescence stimulated by photons (photoluminescence), and the electric field (electroluminescence), can be shifted by quantum states and the confinement of their excitations [189]. The surface of CdSe and ZnS quantum dots can be shown as an example of this behavior in a study that includes a molecularly imprinted polymer with methyl acrylate functional monomer to detect dicyandiamide in dairies [190]. In a sensor study, arising from the combination of quantum dots with graphene, the moisture content of the soil could be measured precisely and accurately [191]. Uzek *et al* synthesized nanocomposite coating for the semiconductor quantum dots by molecularly imprinted shell via mini-emulsion polymerization to form the specific recognition sites for melamine detection with an LOD value of 0.14 ± 0.08 ppm [192]. Localized surface plasmon resonance (LSPR) and SPR based approaches, with high sensitivity among the several sensing techniques, have been used to detect biological and chemical compounds. Polarized light plays a vital role in the SPR technique and satisfies a specific resonance condition. A sharp dip monitors the resonance in the output optical signal at either resonance wavelength or resonance angle. The Kretschmann configuration is frequently utilized in which a thin layer of metals such as gold or silver is directly placed on the base of a coupling prism [193]. Figure 4.6(a) shows a schematic representation of SPR in which free conduction electrons in the metal

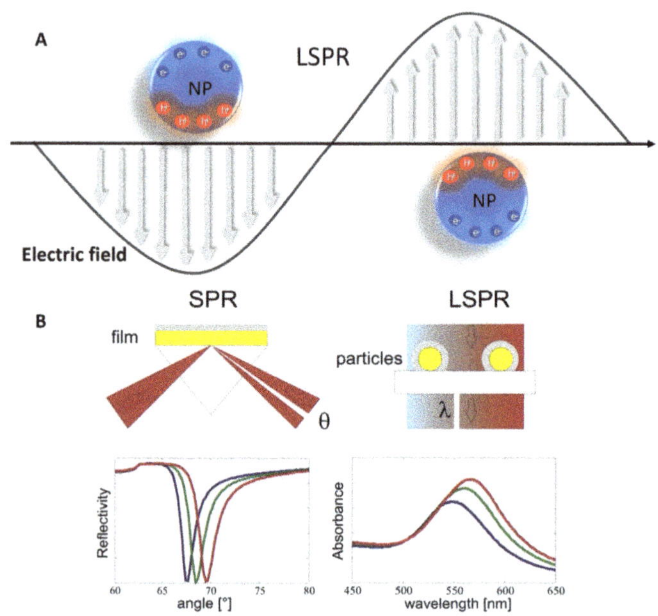

Figure 4.6. (a) Schematic representation of SPR with the free conduction electrons in the metal nanoparticles. (b) The methodology for the pSPR (left) and LSPR (right) systems [194].

nanoparticle are oscillated due to strong coupling with incident light. Schematic representation of the sensing principle for the propagating surface plasmon resonance (pSPR) and LSPR mechanism are presented in figure 4.6(b) [194].

In agriculture, fluctuations in the refractive index on which SPR sensors are based are insufficient in determining trace amounts of pesticides or herbicides, and sensor systems can be upgraded with nanoparticles to close this gap. [194–199]. Saylan *et al* prepared molecularly imprinted nanofilms-based SPR sensors for rapid, selective, and sensitive detection of cyanazine (SNZ), simazine (SMZ) and atrazine (ATZ) pesticides [200]. In order to increase the sensitivity of detection, Yılmaz *et al* offered an SPR-based affinity sensor system using molecularly imprinted plastic antibodies [201]. Figure 4.7(a) represents the polymerization illustration of SNZ imprinted nanofilm preparation. Besides, effect of imprinting is given with cross sensitivity values with three different molecules in figure 4.7(b). Surface-enhanced Raman scattering (SERS) is a promising analytical approach for overcoming problems with the sensitivity of detection. It has been used in an extensive range of physics and chemistry and life and also material science with unique characteristics such as rapid response, ultrasensitive detection, narrow peak bandwidth, and non-destructive detection. Therefore, SERS is used to utilize the detection of pesticides in food [202]. Pesticide residues, metal ions, mycotoxins, allergens, drug residues, contaminants from food processing and packaging, and preservatives are food contaminants [203, 204]. For the development of rapid and sensitive colourimetric sensors to efficiently detect food contaminants in foods and water sources, nanotechnology advances such as nanomaterials' modifications are enormously utilized. For example,

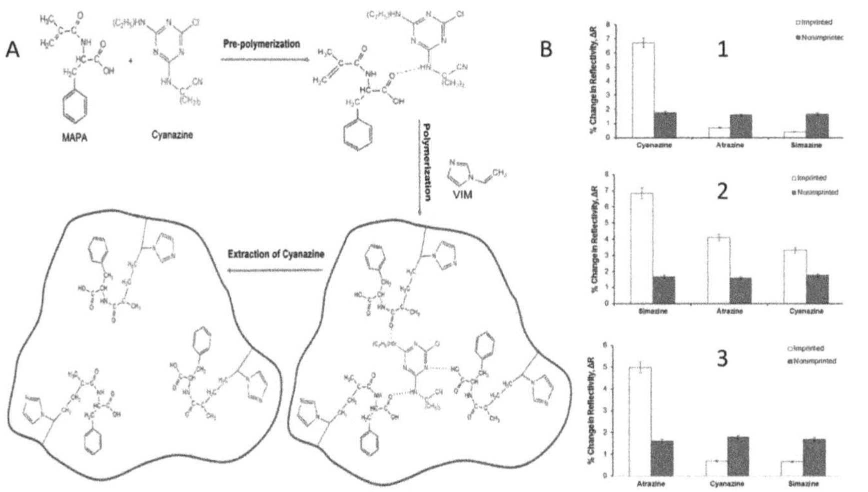

Figure 4.7. (a) Polymerization steps for SNZ imprinted nanofilm preparation. (b) Effect of imprinting on SPR sensors for different sensor molecules [200].

aptamers that can be folded into three-dimensional conformations are single-stranded oligonucleotides. They are unique because of the binding specifically to a target molecule. Wu *et al* reported an aptamer-based flow assay described for the mycotoxin ochratoxin A (OTA) visual and instrumental determination with LOD 1.86 ng mL^{-1}. This aptamer-based test strip was applied successfully for OTA's determination in spiked wheat and beer samples [205]. In another study, aptamer conjugated silver nanoparticles prepared by Divsar *et al* specifically interact with arsenic. This colourimetric Apt-SNP-based optical sensor can detect arsenic with a detection limit of 6 μg L^{-1} [206].

4.2.3 Piezoelectric nanobiosensors

The piezoelectric effect is the electrical voltage that some materials generate under mechanical stress. These piezoelectric materials generate voltage and vibration in an external electric field, and this effect is called reverse piezoelectric. In piezoelectric sensors, the compound (ligand) mass covering the measurement surface is inversely proportional to the resonance frequency that formed. This ratio is the basis for calculating binding kinetics and concentrations. The receptors in piezoelectric biosensors consist of some biological agents. These receptors, which include antibodies, DNA, nucleic acids, proteins, and amino acids, are fixed on the quartz crystal in the piezoelectric sensor. The mass increase that occurs during bonding decreases the oscillation frequency. Thus, information about the concentration of the bonded material can be obtained. The accuracy and precision of these measurements have placed piezoelectric biosensors in the category of reliable analytical tools. The piezoelectric sensor system provides real-time information about many biomaterials desired to be detected. The range of analysis with these sensors can be expanded by replacing ligands (e.g. nanoparticles) and modified surfaces. Quartz crystal microbalance (QCM) is a system that can perform repeatable and real-time analysis at a low cost. In QCM sensors, oscillation depends on the mass change in its content, sensitive quartz crystal. Differences in frequency vary with the ligands used and are thus highly discriminatory and liable in biomarkers' analysis [207].

Pesticides and herbicides are widely applied in practical agriculture, and their excessive usage places environmental integrity and human health in danger. It is vital to develop sensitive, fast and low-cost systems to analyse these toxic chemicals that accumulate in soil and water. With the integration of nano-sized materials into sensor systems, susceptible and reliable paths can be found. The study of Wu *et al* is an excellent example of combining the piezoelectric effect with sensor systems. A sensor system was designed and developed using polyvinylidene fluoride (PVDF) piezoelectric material in the study, which was used to increase the efficiency of the corn harvest [208]. As a specific agricultural study, the determination of α-terpinyl acetate in cardamom was made by Debabhuti *et al* with a QCM gas sensor [209]. The sensitivity of the sensor was compared with other volatile compounds and recorded as 0.3876 Hz ppm^{-1}. Good reversibility and long-term stability have been proved in calculations using QCM frequency shift. In a QCM sensor designed by Bakhshpour *et al*, pesticide detection was successfully performed using nanoparticles. Sensitive determinations were made with the sensor specially designed for chlorpyrifos (Cps), even at concentrations as low as 0.015–2.9 nM [210]. Another approach is developed for pesticide determination in honey produced under beekeeping practices. This sensor design, which also uses monoclonal antibody-based recognition agents, aims to determine the carbaryl pesticide with high accuracy and recovery percentage [211].

4.3 Conclusion and prospects

Nanotechnology has touched every field by its scientific novelties and shows great potential in various fields related to all public and industrial sectors, including materials, energy, agriculture, healthcare, communication, and information technologies. Nanomaterials have many potential applications in agriculture to enhance crop productivity and to improve the soil health, which has been highlighted in this section. The food industry and agriculture have an essential role also in the economic growth of a country. Therefore, process control and quality monitoring are important. Nanosensors and nanobiosensors determine the toxic chemicals in wastewater and soil. The accumulation of agricultural pollutants such as pesticides, herbicides and excess fertilizer present in the soil and heavy metals causes resources and agricultural productivity to decrease. Nanomaterials-based devices have great potential to be used as nanobiosensors in an eco-friendly manner. Nanosensors and nanobiosensors increasing the sensitivity by signal amplification are alternatives to classical quantification methods with their nanoscale sizes were reviewed in this chapter. This results in an improved electrochemical, optical and piezoelectric response with real-time sensing in the food industry and agriculture. Similarly, the determination of food pollutants such as antibiotics, toxins, preservatives, heavy metals, pathogens, and microbial load is essential, and so are traceability of temperature, moisture, gas, and aroma of foodstuffs. Nanobiosensor technology currently includes many innovative studies and is convenient for development. It is possible to design much more economical and portable devices with nanobiosensors combined with innovative materials that can result in increased sensitivity. Considering today's food demand, there is still a need for upgrades to facilitate fast, effective and *in situ* analyses, especially in agricultural lands. Nanobiosensor systems integrated with smart devices can be considered among the developments in terms of ease of use. More automated nanobiosensors with the use of new materials can meet future demands. In this section, strategies used by some authors to develop fast and sensitive optical, electrochemical and piezoelectric sensors were summarized. Therefore, a collaborative work of researchers in the fields of biosensor and agriculture will improve the detection mechanisms that result in protecting human health and preventing economic drawbacks caused by contaminated food. In this chapter where we deal mainly with the role of nanobiosensors in agriculture, it is evident that nanobiosensors promisingly have very superior properties such as improved specific and sensitive detection capabilities and compatibility for applications in different fields (food quality control, environmental and bioprocess control, biodefense, agriculture, and medical applications) compared to currently used biosensors.

Acknowledgments

We thank all the contributors for their assistance in designing, writing and also editing the chapter manuscript. The author responsibilities were as follows; Professor Dr Adil Denizli conducted the chapter writing and had primary responsibility for the final content of the manuscript. Merve Çalışır edited the final chapter language. All authors contributed to the writing, reading and editing of the article, and they read and approved the chapter. None of the authors had a conflict of interest, and there is also no funding support for chapter writing.

References

[1] Babinard J 2001 A short history of agricultural biotechnology *Genet. Mod. Org. Agri.* **2001** 271–4

[2] Spillane C 1999 Recent developments in biotechnology as they relate to plant genetic resources for food and agriculture *FAO Comm. Genet. Resour.* **9** 64

[3] Barrow C J 2014 Environmental problems associated with agriculture *Developing the Environment* (London: Routledge) pp 179–220

[4] Chen H and Yada R 2011 Nanotechnologies in agriculture: new tools for sustainable development *Trends Food Sci. Technol.* **2211** 585–94

[5] Salamanca-Buentello F and Salamanca-Gómez F 2005 Genetics and women in science *Am. J. Med. Genet Part* **136A4** 418

[6] Subramanian K S and Tarafdar J C 2011 Prospects of nanotechnology in Indian farming *Indian. J. Agric. Sci.* **81** 887–93

[7] Rakhimol K R, Thomas S, Kalarikkal N and Jayachandran K 2021 Nanotechnology in controlled-release fertilizers *Controlled Release Fertilizers for Sustainable Agriculture* (Amsterdam: Elsevier) pp 169–81

[8] Kah M, Kookana R S, Gogos A and Bucheli T D 2018 A critical evaluation of nanopesticides and nanofertilizers against their conventional analogues *Nat. Nanotechnol.* **13** 677–84
[9] Liu R and Lal R 2014 Synthetic apatite nanoparticles as a phosphorus fertilizer for soybean (Glycine Max) *Sci. Rep.* **4** 5686
[10] Giroto A S, Guimarães G G, Foschini M and Ribeiro C 2017 Role of slow-release nanocomposite fertilizers on nitrogen and phosphate availability in soil *Sci. Rep.* **7** 46032
[11] Resh V H and Cardé R T 2009 *Encyclopedia of Insects* (New York: Academic)
[12] Carriger J F, Rand G M, Gardinali P R, Perry W B, Tompkins M S and Fernandez A M 2006 Pesticides of potential ecological concern in sediment from South Florida canals: an ecological risk prioritization for aquatic arthropods *Soil. Sediment. Contam.: An. Int. J.* **15** 21–45
[13] Rai M and Ingle A 2012 Role of nanotechnology in agriculture with special reference to management of insect pests *Appl. Microbiol. Biotechnol.* **94** 287–93
[14] Chhipa H 2016 Nanofertilizers and nanopesticides for agriculture *Environ. Chem. Lett.* **15** 15–22
[15] Kumpiene J, Lagerkvist A and Maurice C 2008 Stabilization of as, CR, CU, PB and Zn in soil using amendments—a review *Waste Manag.* **28** 215–25
[16] Kah M and Hofmann T 2014 Nanopesticide research: current trends and future priorities *Environ. Int.* **63** 224–35
[17] Dadhore H, Kaur C, Shrivastava S and Choudhary R 2019 Synthesis and characterization of zero valent iron nanoparticles *J. Mater. Res. Technol.* **6** 178–83
[18] Scognamiglio V 2013 Nanotechnology in glucose monitoring: advances and challenges in the last 10 years *Biosens. Bioelectron.* **47** 12–25
[19] Sang S, Wang Y, Feng Q, Wei Y, Ji J and Zhang W 2015 Progress of new label-free techniques for biosensors: a review *Crit. Rev. Biotechnol.* 1–17
[20] Bawa R, Bawa S R, Maebius S B, Flynn T and Wei C 2005 Protecting new ideas and inventions in nanomedicine with patents *Nanomedicine* **1** 150–8
[21] Urban G A 2009 Micro and nanobiosensors state of the art and trends *Meas. Sci. Technol.* **20** 012001
[22] Patel P 2002 (Bio)sensors for measurement of analytes implicated in food safety: a review *TrAC. Trends Anal. Chem.* **21** 96–115
[23] Cheng N, Song Y, Fu Q, Du D, Luo Y, Wang Y, Xu W and Lin Y 2018 Aptasensor based on fluorophore-quencher nano-pair and smartphone spectrum reader for on-site quantification of multi-pesticides *Biosens. Bioelectron.* **117** 75–83
[24] Safran V, Göktürk I, Derazshamshir A, Yılmaz F, Sağlam N and Denizli A 2019 Rapid sensing of Cu^{+2} in water and biological samples by sensitive molecularly imprinted based plasmonic biosensor *Microchem. J.* **148** 141–50
[25] Nie Y, Teng Y, Li P, Liu W, Shi Q and Zhang Y 2018 Label-free aptamer-based sensor for specific detection of malathion residues by surface-enhanced Raman scattering *Spectrochim. Acta Part. A: Mol. Biomol. Spectrosc.* **191** 271–6
[26] Chawla P, Kaushik R, Shiva Swaraj V and Kumar N 2018 Organophosphorus pesticides residues in food and their colorimetric detection *Environ. Nanotechnol. Monit. Manag.* **10** 292–307
[27] Musarurwa H and Tawanda Tavengwa N 2021 Extraction and electrochemical sensing of pesticides in food and environmental samples by use of polydopamine-based materials *Chemosphere.* **266** 129222

[28] Facure M H, Mercante L, Mattoso L H and Correa D S 2017 Detection of trace levels of organophosphate pesticides using an electronic tongue based on graphene hybrid nanocomposites *Talanta* **167** 59–66
[29] Shang Z, Xu Y, Gu Y, Wang Y, Wei D and Zhan L 2011 A rapid detection of pesticide residue based on piezoelectric biosensor *Procedia Eng.* **15** 4480–5
[30] Çiçek Ç, Yılmaz F, Özgür E, Yavuz H and Denizli A 2016 Molecularly imprinted quartz crystal microbalance sensor (QCM) for bilirubin detection *Chemosens.* **4** 21
[31] Atay S, Pişkin K, Yılmaz F, Çakır C, Yavuz H and Denizli A 2016 Quartz crystal microbalance based biosensors for detecting highly metastatic breast cancer cells via their transferrin receptors *Anal. Methods* **8** 153–61
[32] Kuchmenko T A and Lvova L B 2019 A perspective on recent advances in piezoelectric chemical sensors for environmental monitoring and foodstuffs analysis *Chemosens.* **7** 39
[33] Srivastava A K, Dev A and Karmakar S 2017 Nanosensors and nanobiosensors in food and agriculture *Environ. Chem. Lett.* **16** 161–82
[34] Yan J, Guan H, Yu J and Chi D 2013 Acetylcholinesterase biosensor based on assembly of multiwall carbon nanotubes onto liposome bioreactors for detection of organophosphates pesticides *Pesticide Biochem. Physiol.* **105** 197–202
[35] Xu Y, Ding J, Chen H, Zhao Q, Hou J, Yan J, Wang H, Ding L and Ren N 2013 Fast determination of sulfonamides from egg samples using magnetic multiwalled carbon nanotubes as adsorbents followed by liquid chromatography–tandem mass spectrometry *Food Chem.* **140** 83–90
[36] Lin K, Hong C and Chen S 2013 Simultaneous determination for toxic ractopamine and salbutamol in pork sample using hybrid carbon nanotubes *Sens. Actuators B: Chem.* **177** 428–36
[37] Antiochia R, Vinci G and Gorton L 2013 Rapid and direct determination of fructose in food: a new osmium-polymer mediated biosensor *Food Chem.* **140** 742–7
[38] Nasirizadeh N, Hajihosseini S, Shekari Z and Ghaani M 2014 A novel electrochemical biosensor based on a modified gold electrode for hydrogen peroxide determination in different beverage samples *Food Anal. Methods* **8** 1546–55
[39] Nesakumar N, Sethuraman S, Krishnan U and Rayappan J B 2016 Electrochemical acetylcholinesterase biosensor based on ZnO nanocuboids modified platinum electrode for the detection of carbosulfan in rice *Biosens. Bioelectron.* **77** 1070–7
[40] Singh A, Choudhary M, Singh M, Verma H, Singh S P and Arora K 2015 DNA functionalized direct electro-deposited gold nanoaggregates for efficient detection of *Salmonella typhi Bioelectrochemistry* **105** 7–15
[41] Yavuz S, Erkal A, Kariper İ A, Solak A O, Jeon S, Mülazımoğlu İ E and Üstündağ Z 2015 Carbonaceous materials-12: a novel highly sensitive graphene oxide-based carbon electrode: preparation, characterization, and heavy metal analysis in food samples *Food Anal. Methods* **9** 322–31
[42] Koncki R 2000 Potentiometric determination of dialysate urea nitrogen *Talanta* **52** 13–7
[43] Christopher F C, Kumar P S, Christopher F J, Joshiba G J and Madhesh P 2020 Recent advancements in rapid analysis of pesticides using nano biosensors: a present and future perspective *J. Clean. Prod.* **269** 122356
[44] European Food Safety Authority (EFSA) 2017 Dietary reference values for nutrients summary report *EFSA Support. Publ.* **14**

[45] Akre C 2016 The use of pyrethroids, carbamates, organophosphates, and other pesticides in veterinary medicine *Chemical Analysis of Non-antimicrobial Veterinary Drug Residues in Food* (Hoboken, NJ: Wiley) pp 383–426

[46] Yadav I C and Devi N L 2017 Pesticides classification and its impact on environment *Environ. Sci. Eng.* **6** 140–58

[47] Health & safety executive 1999 *BMUS Bull.* **7** 50–1

[48] Sjerps R M, Kooij P J, Van Loon A and Van Wezel A P 2019 Occurrence of pesticides in Dutch drinking water sources *Chemosphere* **235** 510–8

[49] Arduini F, Cinti S, Caratelli V, Amendola L, Palleschi G and Moscone D 2019 Origami multiple paper-based electrochemical biosensors for pesticide detection *Biosens. Bioelectron.* **126** 346–54

[50] Zheng Y, Liu Z, Jing Y, Li J and Zhan H 2015 An acetylcholinesterase biosensor based on ionic liquid functionalized graphene–gelatin-modified electrode for sensitive detection of pesticides *Sens. Actuators B: Chem.* **210** 389–97

[51] Bhardwaj S K, Bhardwaj N, Mohanta G C, Kumar P, Sharma A L, Kim K and Deep A 2015 Immunosensing of atrazine with antibody-functionalized Cu-MOF conducting thin films *ACS Appl. Mater. Interfaces* **7** 26124–30

[52] Pollap A and Kochana J 2019 Electrochemical immunosensors for antibiotic detection *Biosensors* **9** 61

[53] Morawski F D, Winiarski J P, De Campos C E, Parize A L and Jost C L 2020 Sensitive simultaneous voltammetric determination of the herbicides diuron and isoproturon at a platinum/chitosan bio-based sensing platform *Ecotoxicol. Environ. Saf.* **206** 111181

[54] Caetano K D, Da Rosa D S, Pizzolato T M, Dos Santos P A, Hinrichs R, Benvenutti E V, Dias S L, Arenas L T and Costa T M 2020 MWCNT/zirconia porous composite applied as electrochemical sensor for determination of methyl parathion *Microporous Mesoporous Mater.* **309** 110583

[55] Roushani M and Sarabaegi M 2014 Electrochemical detection of butylated hydroxyanisole based on glassy carbon electrode modified by iridium oxide nanoparticles *J. Electroanal. Chem.* **717–8** 147–52

[56] Ilager D, Seo H, Shetti N P, Kalanur S S and Aminabhavi T M 2020 Electrocatalytic detection of herbicide, amitrole at $WO_3{\cdot}0.33H_2O$ modified carbon paste electrode for environmental applications *Sci. Total. Environ.* **743** 140691

[57] Li X, Gao X, Gai P, Liu X and Li F 2020 Degradable metal-organic framework/methylene blue composites-based homogeneous electrochemical strategy for pesticide assay *Sens. Actuators B: Chem.* **323** 128701

[58] Gençdağ Ş K, Muti M and Karagözler A E 2020 Highly selective molecularly imprinting polymer-based sensor for the electrochemical determination of metoxuron *Microchem. J.* **158** 105178

[59] Kaur R, Rana S, Lalit K, Singh P and Kaur K 2020 Electrochemical detection of methyl parathion via a novel biosensor tailored on highly biocompatible electrochemically reduced graphene oxide-chitosan-hemoglobin coatings *Biosens. Bioelectron.* **167** 112486

[60] He Y, Wu T, Wang J, Ye J, Xu C, Li F and Guo Q 2020 A sensitive pyrimethanil sensor based on porous $NiCo_2S_4$/graphitized carbon nanofiber film *Talanta* **219** 121277

[61] De Oliveira G A, Gevaerd A, Blaskievicz S F, Zarbin A J, Orth E S, Bergamini M F and Marcolino-Junior L H 2020 A simple enzymeless approach for paraoxon determination using imidazole-functionalized carbon nanotubes *Mater. Sci. Eng.: C.* **116** 111140

[62] Baksh H, Buledi J A, Khand N H, Solangi A R, Mallah A, Sherazi S T and Abro M I 2020 Ultra-selective determination of carbofuran by electrochemical sensor based on nickel oxide nanoparticles stabilized by ionic liquid *Monatsh. Chem.–Chem. Monthly* **151** 1689–96
[63] Yao J, Liu Z, Jin M, Zou Y, Chen J, Xie P, Wang X, Akinoglu E M, Zhou G and Shui L 2020 Uniform honeycomb CNT-microparticles prepared via droplet-microfluidics and sacrificial nanoparticles for electrochemical determination of methyl parathion *Sens. Actuators B: Chem.* **321** 128517
[64] Kumaravel A and Murugananthan M 2021 Electrochemical detection of fenitrothion usingnanosilver/dodecane modified glassy carbon electrode *Sens. Actuators B: Chem.* **331** 129467
[65] Tu X, Gao F, Ma X, Zou J, Yu Y, Li M, Qu F, Huang X and Lu L 2020 Mxene/carbon nanohorn/β-cyclodextrin-Metal-organic frameworks as high-performance electrochemical sensing platform for sensitive detection of carbendazim pesticide *J. Hazard. Mater.* **396** 122776
[66] Kumaravel A, Murugananthan M, Mangalam R and Jayakumar S 2020 A novel, biocompatible and electrocatalytic stearic acid/nanosilver modified glassy carbon electrode for the sensing of paraoxon pesticide in food samples and commercial formulations *Food Chem.* **323** 126814
[67] Nagabooshanam S, John A T, Wadhwa S, Mathur A, Krishnamurthy S and Bharadwaj L M 2020 Electro-deposited nano-webbed structures based on polyaniline/multi walled carbon nanotubes for enzymatic detection of organophosphates *Food Chem.* **323** 126784
[68] Yao Y *et al* 2020 Ultrasensitive aptamer-based biosensor for acetamiprid using tetrahedral DNA nanostructures *J. Mater. Sci.* **55** 15975–87
[69] Tcheumi H L, Kameni Wendji A P, Tonle I K and Ngameni E 2020 A low-cost layered double hydroxide (LDH) based amperometric sensor for the detection of isoproturon in water using carbon paste modified electrode *J. Anal. Methods Chem.* 1–10
[70] Amouzad F and Zarei K 2020 Electrochemical determination of dinitramine in water samples using a pencil graphite electrode modified with poLy-L-cystein-goLd nanoparticle *Chem. Pap.* **75** 493–501
[71] Shanmugam R, Manavalan S, Chen S, Keerthi M and Lin L 2020 Methyl parathion detection using SnS_2/N, s–co-doped reduced graphene oxide nanocomposite *ACS Sustain. Chem. Eng.* **8** 11194–203
[72] Sun Y, Xiong P, Tang J, Zeng Z and Tang D 2020 Ultrasensitive split-type electrochemical sensing platform for sensitive determination of organophosphorus pesticides based on MnO2 nanoflower-electron mediator as a signal transduction system *Anal. Bioanal. Chem.* **412** 6939–45
[73] Kumar T H, Raman Pillai S K, Chan-Park M B and Sundramoorthy A K 2020 Highly selective detection of an organophosphorus pesticide, methyl parathion, using ag–zno–SWCNT based field-effect transistors *J. Mater. Chem.* **8** 8864–75
[74] Zuo Y, Wang X, Fan L, Han Y, Zhang G and Guo Y 2020 The fabrication of an electrochemical sensing interface based on Pt nanoparticles/MoS2/graphene nanocomposite for sensitive detection of binaphthol *Int. J. Environ. Anal. Chem.* 1–12
[75] Kumar S, Vasylieva N, Singh V, Hammock B and Singh S G 2020 A facile, sensitive and rapid sensing platform based on CoZnO for detection of Fipronil; an environmental toxin *Electroanalysis* **32** 2056–64

[76] Liu C, Cao Y, Zhao T, Wang X, Fang G and Wang S 2020 A novel multi-purpose MIP for SPE-HPLC and QCM detection of carbaryl residues in foods *Food Anal. Methods* **14** 331–43

[77] Kamyabi M A and Moharramnezhad M 2020 Nickel foam decorated with ZnO nanocrystals using mesoporous silica templates for ultrasensitive electrogenerated chemiluminescence determination of diazinon *Microchem. J.* **154** 104540

[78] Thanalechumi P, Yusoff A R and Yusop Z 2020 Novel electrochemical sensor based on nylon 6,6-Modified graphite HB pencil electrode for chlorothalonil determination by differential pulsecathodic stripping voltammetry *Water, Air, Soil. Pollut.* **231**

[79] Xiao F, Li H, Yan X, Yan L, Zhang X, Wang M, Qian C and Wang Y 2020 Graphitic carbon nitride/graphene oxide(g-C3N4/GO) nanocomposites covalently linked with ferrocene containing dendrimer for ultrasensitive detection of pesticide *Anal. Chim. Acta* **1103** 84–96

[80] Li H, Shi Y and Gao Y 2020 A simple but highly sensitive electropolymerization of L-citrulline and β-cyclodextrin based voltammetric sensor for metribuzin *Int. J. Environ. Anal. Chem.* 1–9

[81] Kumar T V and Sundramoorthy A K 2019 Electrochemical biosensor for methyl parathion based on single-walled carbon nanotube/glutaraldehyde crosslinked acetylcholinesterase-wrapped bovine serum albumin nanocomposites *Anal. Chim. Acta* **1074** 131–41

[82] Gao N, He C, Ma M, Cai Z, Zhou Y, Chang G, Wang X and He Y 2019 Electrochemical co-deposition synthesis of AU-Zro_2-graphene nanocomposite for a nonenzymatic methyl parathion sensor *Anal. Chim. Acta* **1072** 25–34

[83] Jin R, Kong D, Zhao X, Li H, Yan X, Liu F, Sun P, Du D, Lin Y and Lu G 2019 Tandem catalysis driven by enzymes directed hybrid nanoflowers for on-site ultrasensitive detection of organophosphorus pesticide *Biosens. Bioelectron.* **141** 111473

[84] Brycht M, Leniart A, Zavašnik J, Nosal–Wiercińska A, Wasiński K, Półrolniczak P, Skrzypek S and Kalcher K 2018 Synthesis and characterization of the thermally reduced graphene oxide in argon atmosphere, and its application to construct graphene paste electrode as a naptalam electrochemical sensor *Anal. Chim. Acta* **1035** 22–31

[85] Wei P, Gan T and Wu K 2018 N-methyl-2-pyrrolidone exfoliated graphene as highly sensitive analytical platform for carbendazim *Sens. Actuators B: Chem.* **274** 551–9

[86] Roushani M, Nezhadali A and Jalilian Z 2018 An electrochemical chlorpyrifos aptasensor based on the use of a glassy carbon electrode modified with an electropolymerized aptamer-imprinted polymer and gold nanorods *Microchim. Acta* **185**

[87] Rahmani T, Bagheri H, Behbahani M, Hajian A and Afkhami A 2018 Modified 3D graphene-Au as a novel sensing layer for direct and sensitive electrochemical determination of carbaryl pesticide in fruit, vegetable, and water samples *Food Anal. Methods* **11** 3005–14

[88] Lang Q, Han L, Hou C, Wang F and Liu A 2016 A sensitive acetylcholinesterase biosensor based on gold nanorods modified electrode for detection of organophosphate pesticide *Talanta* **156–7** 34–41

[89] Arduini F, Forchielli M, Amine A, Neagu D, Cacciotti I, Nanni F, Moscone D and Palleschi G 2014 Screen-printed biosensor modified with carbon black nanoparticles for the determination of paraoxon based on the inhibition of butyrylcholinesterase *Microchim. Acta* **182** 643–51

[90] Bahadır E B and Sezgintürk M K 2015 Applications of commercial biosensors in clinical, food, environmental, and biothreat/biowarfare analyses *Anal. Biochem.* **478** 107–20

[91] Peng L, Dong S, Wei W, Yuan X and Huang T 2017 Synthesis of reticulated hollow spheres structure $NiCo_2S_4$ and its application in organophosphate pesticides biosensor *Biosens. Bioelectron.* **92** 563–9
[92] Deng Y, Liu K, Liu Y, Dong H and Li S 2016 An novel acetylcholinesterase biosensor based on nano-porous pseudo carbon paste electrode modified with gold nanoparticles for detection of methyl parathion *J. Nanosci. Nanotechnol.* **16** 9460–7
[93] Jiao Y, Hou W, Fu J, Guo Y, Sun X, Wang X and Zhao J 2017 A nanostructured electrochemical aptasensor for highly sensitive detection of chlorpyrifos *Sens. Actuators B: Chem.* **243** 1164–70
[94] Sundarmurugasan R, Gumpu M B, Ramachandra B L, Nesakumar N, Sethuraman S, Krishnan U M and Rayappan J B 2016 Simultaneous detection of monocrotophos and dichlorvos in orange samples using acetylcholinesterase–zinc oxide modified platinum electrode with linear regression calibration *Sens. Actuators B: Chem.* **230** 306–13
[95] Wei M and Wang J 2015 A novel acetylcholinesterase biosensor based on ionic liquids-AuNPs-porous carbon composite matrix for detection of organophosphate pesticides *Sens. Actuators B: Chem.* **211** 290–6
[96] Fei A, Liu Q, Huan J, Qian J, Dong X, Qiu B, Mao H and Wang K 2015 Label-free impedimetric aptasensor for detection of femtomole level acetamiprid using gold nanoparticles decorated multiwalled carbon nanotube-reduced graphene oxide nanoribbon composites *Biosens. Bioelectron.* **70** 122–9
[97] Jiang D, Du X, Liu Q, Zhou L, Dai L, Qian J and Wang K 2015 Silver nanoparticles anchored on nitrogen-doped graphene as a novel electrochemical biosensing platform with enhanced sensitivity for aptamer-based pesticide assay *Analyst* **140** 6404–11
[98] Madianos L, Tsekenis G, Skotadis E, Patsiouras L and Tsoukalas D 2018 A highly sensitive impedimetric aptasensor for the selective detection of acetamiprid and atrazine based on microwires formed by platinum nanoparticles *Biosens. Bioelectron.* **101** 268–74
[99] Belkhamssa N, Justino C I, Santos P S, Cardoso S, Lopes I, Duarte A C, Rocha-Santos T and Ksibi M 2016 Label-free disposable immunosensor for detection of atrazine *Talanta* **146** 430–4
[100] González-Techera A, Zon M A, Molina P G, Fernández H, González-Sapienza G and Arévalo F J 2015 Development of a highly sensitive noncompetitive electrochemical immunosensor for the detection of atrazine by phage anti-immunocomplex assay *Biosens. Bioelectron.* **64** 650–6
[101] Li Z, Qu S, Cui L and Zhang S 2017 Detection of carbofuran pesticide in seawater by using an enzyme biosensor *J. Coast. Res.* **80** 1–5
[102] Liu Q, Fei A, Huan J, Mao H and Wang K 2015 Effective amperometric biosensor for carbaryl detection based on covalent immobilization acetylcholinesterase on multiwall carbon nanotubes/graphene oxide nanoribbons nanostructure *J. Electroanal. Chem.* **740** 8–13
[103] Li Y, Shi L, Han G, Xiao Y and Zhou W 2017 Electrochemical biosensing of carbaryl based on acetylcholinesterase immobilized onto electrochemically inducing porous graphene oxide network *Sens. Actuators B: Chem.* **238** 945–53
[104] Nguyen V, Kwon Y S and Gu M B 2017 Aptamer-based environmental biosensors for small molecule contaminants *Curr. Opin. Biotechnol.* **45** 15–23
[105] Lin X H, Aik S X, Angkasa J, Le Q, Chooi K S and Li S F 2018 Selective and sensitive sensors based on molecularly imprinted poly(vinylidene fluoride) for determination of pesticides and chemical threat agent simulants *Sens. Actuators B: Chem.* **258** 228–37

[106] Gui R, Jin H, Guo H and Wang Z 2018 Recent advances and future prospects in molecularly imprinted polymers-based electrochemical biosensors *Biosens. Bioelectron.* **100** 56–70
[107] Dong J, Zhao H, Qiao F, Liu P, Wang X and Ai S 2013 Quantum dot immobilized acetylcholinesterase for the determination of organophosphate pesticides using graphene-chitosan nanocomposite modified electrode *Anal. Methods* **5** 2866
[108] Vinotha A and Mukherjee A A 2021 Review of recent developments on acetylcholinesterase inhibition based biosensors for organophosphorus pesticides detection *Microchem. J.* **161** 105779
[109] Liu Z, Xia X, Zhou G, Ge L and Li F 2020 Acetylcholinesterase-catalyzed silver deposition for ultrasensitive electrochemical biosensing of organophosphorus pesticides *Analyst* **145** 2339–44
[110] Vikas D, Anjum G, Neeraj D and Vikas H 2013 Acetylcholinesterase biosensors for electrochemical detection of organophosphorus compounds: a review *Biochem. Res. Int.* **731501** 18
[111] El-Maiss J, Cuccarese M, Maerten C, Lupattelli P, Chiummiento L, Funicello M, Schaaf P, Jierry L and Boulmedais F 2018 Mussel-inspired electro-cross-linking of enzymes for the development of biosensors *ACS Appl. Mater. Interfaces* **10** 18574–84
[112] Gui Q, Lawson T, Shan S, Yan L and Liu Y 2017 The application of whole cell-based biosensors for use in environmental analysis and in medical diagnostics *Sens.* **17** 1623
[113] Gao G, Fang D, Yu Y, Wu L, Wang Y and Zhi J 2017 A double-mediator based whole cell electrochemical biosensor for acute biotoxicity assessment of wastewater *Talanta* **167** 208–16
[114] López Rodriguez M L, Benimeli C, Madrid R E and Giacomelli C E 2015 A simple streptomyces spore-based impedimetric biosensor to detect lindane pesticide *Sens. Actuators B: Chem.* **207** 447–54
[115] Pabbi M and Mittal S K 2017 An electrochemical algal biosensor based on silica coated ZnO quantum dots for selective determination of acephate *Anal. Methods* **9** 1672–80
[116] Tucci M, Grattieri M, Schievano A, Cristiani P and Minteer S D 2019 Microbial amperometric biosensor for online herbicide detection: photocurrent inhibition of *Anabaena variabilis Electrochim. Acta* **302** 102–8
[117] Arora K, Chand S and Malhotra B 2006 Recent developments in bio-molecular electronics techniques for food pathogens *Analytica Chim. Acta* **568** 259–74
[118] Arora K, Chaubey A, Singhal R, Singh R P, Pandey M, Samanta S, Malhotra B and Chand S 2006 Application of electrochemically prepared polypyrrole–polyvinyl sulphonate films to DNA biosensor *Biosens. Bioelectron.* **21** 1777–83
[119] Izadi Z, Sheikh-Zeinoddin M, Ensafi A A and Soleimanian-Zad S 2016 Fabrication of an electrochemical DNA-based biosensor for bacillus cereus detection in milk and infant formula *Biosens. Bioelectron.* **80** 582–9
[120] Wan Y, Lin Z, Zhang D, Wang Y and Hou B 2011 Impedimetric immunosensor doped with reduced graphene sheets fabricated by controllable electrodepositi on for the non-labelled detection of bacteria *Biosens. Bioelectron.* **26** 1959–64
[121] Chen A and Shah B 2013 Electrochemical sensing and biosensing based on square wave voltammetry *Anal. Methods* **5** 2158
[122] Khan S, Ansari Z A, Alothman O Y, Fouad H and Ansari S G 2017 Application of amine and copper doped magnesium oxide nanoparticles in electrochemical immunosensors for detecting *Brucella abortus Nanosci. Nanotechnol. Lett.* **9** 1656–64

[123] Zhang W, Tang H, Geng P, Wang Q, Jin L and Wu Z 2007 Amperometric method for rapid detection of *escherichia coli* by flow injection analysis using a bismuth nano-film modified glassy carbon electrode *Electrochem. Commun.* **9** 833–38

[124] Vaisocherová-Lísalová H *et al* 2016 Copolymer brush-based ultralow-fouling biorecognition surface platform for food safety *Anal. Chem.* **88** 10533–9

[125] Kuan G C, Sheng L P, Rijiravanich P, Marimuthu K, Ravichandran M, Yin L S, Lertanantawong B and Surareungchai W 2013 Gold-nanoparticle based electrochemical DNA sensor for the detection of fish pathogen *Aphanomyces invadans Talanta* **117** 312–7

[126] Wang Z, Wei F, Liu S, Xu Q, Huang J, Dong X, Yu J, Yang Q, Zhao Y and Chen H 2010 Electrocatalytic oxidation of phytohormone salicylic acid at copper nanoparticles-modified gold electrode and its detection in oilseed rape infected with fungal pathogen *Sclerotinia sclerotiorum Talanta* **80** 1277–81

[127] Noori J S, Mortensen J and Geto A 2020 Recent development on the electrochemical detection of selected pesticides: a focused review *Sensors* **20** 2221

[128] Adachi T, Kitazumi Y, Shirai O and Kano K 2020 Development perspective of bioelectrocatalysis-based biosensors *Sensors* **20** 4826

[129] Ali Q *et al* 2021 Research advances and applications of biosensing technology for the diagnosis of pathogens in sustainable agriculture *Environ. Sci. Pollut. Res.* **28** 9002–19

[130] Xiang C, Li R, Adhikari B, She Z, Li Y and Kraatz H 2015 Sensitive electrochemical detection of *Salmonella* with chitosan–gold nanoparticles composite film *Talanta* **140** 122–27

[131] Fei J, Dou W and Zhao G 2015 A sandwich electrochemical immunosensor for *Salmonella pullorum* and *Salmonella gallinarum* based on a screen-printed carbon electrode modified with an ionic liquid and electrodeposited gold nanoparticles *Microchim. Acta* **182** 2267–75

[132] Afonso A S, Pérez-López B, Faria R C, Mattoso L H, Hernández-Herrero M, Roig-Sagués A X, Maltez-da Costa M and Merkoçi A 2013 Electrochemical detection of Salmonella using gold nanoparticles *Biosens. Bioelectron.* **40** 121–6

[133] Wang D, Wang Z, Chen J, Kinchla A J and Nugen S R 2016 Rapid detection of Salmonella using a redox cycling-based electrochemical method *Food Control.* **62** 81–8

[134] Punbusayakul N, Talapatra S, Ajayan P M and Surareungchai W 2013 Label-free as-grown double wall carbon nanotubes bundles for *Salmonella typhimuriumimmunoassay Chem. Cent. J.* **7** 102

[135] Chunyan Chai G L 2013 Rapid evaluation of *Salmonella pullorum* contamination in chicken based on a portable Amperometric sensor *J. Biosens. Bioelectron.* **4**

[136] Dou W, Tang W and Zhao G 2013 A disposable electrochemical immunosensor arrays using 4-channel screen-printed carbon electrode for simultaneous detection of *escherichia coli* O157:H7 and *Enterobacter sakazakii Electrochim. Acta* **97** 79–85

[137] Lau H Y and Botella J R 2017 Advanced DNA-based point-of-care diagnostic methods for plant diseases detection *Front. Plant. Sci.* **8** 2016

[138] Lin L, Ntambo M S, Rott P C, Wang Q, Lin Y, Fu H and Gao S 2018 Molecular detection and prevalence of *Xanthomonas albilineans*, the causal agent of sugarcane leaf scald, in China *Crop. Prot.* **109** 17–23

[139] Jarocka U, Wąsowicz M, Radecka H, Malinowski T, Michalczuk L and Radecki J 2011 Impedimetric immunosensor for detection of Plum Pox Virus in plant extract *Electroanalysis* **23** 2197–204

[140] Machini W and Oliveira-Brett A 2021 *In situ* electrochemical investigation of the interaction between bacteria *Xylella fastidiosa* DNA and copper(II) using DNA-electrochemical biosensors *Electrochem. Commun.* **125** 106975

[141] Alves R C, Barroso M F, González-García M B, Oliveira M B and Delerue-Matos C 2015 New trends in food allergens detection: toward biosensing strategies *Crit. Rev. Food Sci. Nutr.* **56** 2304–19

[142] Fang Y, Umasankar Y and Ramasamy R P 2014 Electrochemical detection of P-ethylguaiacol, a fungi infected fruit volatile using metal oxide nanoparticles *Analyst* **139** 3804–10

[143] Lau H Y, Wu H, Wee E J, Trau M, Wang Y and Botella J R 2017 Specific and sensitive isothermal electrochemical biosensor for plant pathogen DNA detection with colloidal gold nanoparticles as probes *Sci. Rep.* **7** 38896

[144] Tkac J, Whittaker J W and Ruzgas T 2007 The use of single walled carbon nanotubes dispersed in a chitosan matrix for preparation of a galactose biosensor *Biosens. Bioelectron.* **22** 1820–4

[145] Thandavan K, Gandhi S, Nesakumar N, Sethuraman S, Rayappan J B and Krishnan U M 2015 Hydrogen peroxide biosensor utilizing a hybrid nano-interface of iron oxide nanoparticles and carbon nanotubes to assess the quality of milk *Sens. Actuators B: Chem.* **215** 166–73

[146] Wu L, Gao B, Zhang F, Sun X, Zhang Y and Li Z 2013 A novel electrochemical immunosensor based on magnetosomes for detection of staphylococcal enterotoxin B in milk *Talanta* **106** 360–6

[147] Durán N and Marcato P D 2012 Nanobiotechnology perspectives. Role of nanotechnology in the food industry: a review *Int. J. Food Sci. Technol.* **48** 1127–34

[148] Yola M L, Eren T and Atar N 2015 A sensitive molecular imprinted electrochemical sensor based on gold nanoparticles decorated graphene oxide: application to selective determination of tyrosine in milk *Sens. Actuators B: Chem.* **210** 149–57

[149] Dong F 2021 Electrochemical non-enzymatic biosensor for tyramine detection in food based on silver-substituted ZnO nano-flower modified glassy carbon electrode *Int. J. Electrochem. Sci.* **210234**

[150] Crevillén A G, Ávila M, Pumera M, González M C and Escarpa A 2007 Food analysis on microfluidic devices using ultrasensitive carbon nanotubes detectors *Anal. Chem.* **79** 7408–15

[151] Crevillén A G, Pumera M, González M C and Escarpa A 2008 Carbon nanotube disposable detectors in microchip capillary electrophoresis for water-soluble vitamin determination: analytical possibilities in pharmaceutical quality control *Electrophoreses* **29** 2997–3004

[152] Ho J A, Hsu W, Liao W, Chiu J, Chen M, Chang H and Li C 2010 Ultrasensitive electrochemical detection of biotin using electrically addressable site-oriented antibody immobilization approach via aminophenyl boronic acid *Biosens. Bioelectron.* **26** 1021–7

[153] Wang X, Lu X and Chen J 2014 Development of biosensor technologies for analysis of environmental contaminants *Trends Environ. Anal. Chem.* **2** 25–32

[154] The European Parlament and the Council of the European Union Directives of 12 August 2013 Amending Directives 2000/60/EC and 2008/105/EC as Regards Priority Substances in the Field of Water Policy *Off. J. Eur. Union* 2013 1–17

[155] Brennan L M, Widder M W, McAleer M K, Mayo M W, Greis A P and Van der Schalie W H 2016 Preparation and testing of impedance-based fluidic biochips with RTgill-W1 cells for rapid evaluation of drinking water samples for toxicity *J. Visualized Exp.* **109** e53555
[156] Le V T, Vasseghian Y, Dragoi E-N, Moradi M and Mousavi Khaneghah A 2021 A review on graphene-based electrochemical sensor for mycotoxins detection *Food Chem. Toxicol.* **148** 111931
[157] Sharma A, Kumar A and Khan R 2017 Electrochemical immunosensor based on poly (3,4-ethylenedioxythiophene) modified with gold nanoparticle to detect aflatoxin B1 *Mater. Sci. Eng.: C.* **76** 802–9
[158] Kundu M, Rajesh Krishnan P and Gajjala S 2021 Comparative studies of screen-printed electrode based electrochemical biosensor with the optical biosensor for formaldehyde detection in corn *Food Bioprocess. Technol.* **14** 726–38
[159] Zhang Y, Wu B, Xu H, Liu H, Wang M, He Y and Pan B 2016 Nanomaterials-enabled water and wastewater treatment *NanoImpact* **3–4** 22–39
[160] Gan T, Li K and Wu K 2008 Multi-wall carbon nanotube-based electrochemical sensor for sensitive determination of Sudan I *Sens. Actuators B: Chem.* **132** 134–9
[161] Siregar A S, Sulistyo I and Prayogo N A 2020 Heavy metal contamination in water, sediments and *Planiliza Subviridis* tissue in the Donan River, Indonesia *J. Water Land. Dev.* **45** 157–64
[162] Obasi P N and Akudinobi B B 2020 Potential health risk and levels of heavy metals in water resources of lead–zinc mining communities of Abakaliki, southeast Nigeria *Appl. Water Sci.* **10**
[163] Kladsomboon S, Jaiyen C, Choprathumma C, Tusai T and Apilux A 2019 Heavy metals contamination in soil, surface water, crops, and resident blood in Uthai district, Phra Nakhon Si Ayutthaya, Thailand *Environ. Geochem. Health* **42** 545–61
[164] Kumar P, Kim K, Bansal V, Lazarides T and Kumar N 2017 Progress in the sensing techniques for heavy metal ions using nanomaterials *J. Ind. Eng. Chem.* **54** 30–43
[165] World Helath Organization Chemical fact sheets www.who.int/water_sanitation_health/dwq/GDW12rev1and2.pdfWorldHealth
[166] Xu Z, Liu J, Wang K, Yan B, Hu S, Ren X and Gao Z 2021 Facile synthesis of N-doped carbon dots for direct/indirect detection of heavy metal ions and cell imaging *Environ. Sci. Pollut. Res.* **28** 19878–89
[167] Smirnova S V, Ilin D V and Pletnev I V 2021 Extraction and ICP-OES determination of heavy metals using tetrabutylammonium bromide aqueous biphasic system and oleophilic collector *Talanta* **221** 121485
[168] Ghorbani Y A, Ghoreishi S M and Ghani M 2020 Derived N-doped carbon through core–shell structured metal-organic frameworks as a novel sorbent for dispersive solid phase extraction of Cr(III) and Pb(II) from water samples followed by quantitation through flame atomic absorption spectrometry *Microchem. J.* **155** 104786
[169] Do J and Lin K 2016 Kinetics of urease inhibition-based amperometric biosensors for mercury and lead ions detection *J. Taiwan. Inst. Chem. Eng.* **63** 25–32
[170] Liu L, Chen C, Chen C, Kang X, Zhang H, Tao Y, Xie Q and Yao S 2019 Poly (noradrenalin) based Bi-enzyme biosensor for ultrasensitive multi-analyte determination *Talanta* **194** 343–9

[171] Da Silva W, Ghica M E and Brett C M 2020 Biotoxic trace metal ion detection by enzymatic inhibition of a glucose biosensor based on a poly(brilliant green)–deep eutectic solvent/carbon nanotube modified electrode *Talanta* **208** 120427

[172] Dalkıran B 2020 Amperometric determination of heavy metal using an HRP inhibition biosensor based on ITO nanoparticles-ruthenium (III) hexamine trichloride composite: central composite design optimization *Bioelectrochemistry* **135** 107569

[173] Elsebai B, Ghica M E, Abbas M N and Brett C M 2017 Catalase based hydrogen peroxide biosensor for mercury determination by inhibition measurements *J. Hazard. Mater.* **340** 344–50

[174] Mukherjee S *et al* 2021 Sensory development for heavy metal detection: a review on translation from conventional analysis to field-portable sensor *Trends Food Sci. Technol.* **109** 674–89

[175] Gao G, Qian J, Fang D, Yu Y and Zhi J 2016 Development of a mediated whole cell-based electrochemical biosensor for joint toxicity assessment of multi-pollutants using a mixed microbial consortium *Anal. Chim. Acta* **924** 21–8

[176] Khan A *et al* 2020 A novel biosensor for zinc detection based on microbial fuel cell system *Biosens. Bioelectron.* **147** 111763

[177] Yu D, Bai L, Zhai J, Wang Y and Dong S 2017 Toxicity detection in water containing heavy metal ions with a self-powered microbial fuel cell-based biosensor *Talanta* **168** 210–6

[178] Gupta N, Renugopalakrishnan V, Liepmann D, Paulmurugan R and Malhotra B D 2019 Cell-based biosensors: recent trends, challenges and future perspectives *Biosens. Bioelectron.* **141** 111435

[179] Fang D, Gao G, Shen J, Yu Y and Zhi J 2016 A reagentless electrochemical biosensor based on thionine wrapped *E. coli* and chitosan-entrapped carbon nanodots film modified glassy carbon electrode for wastewater toxicity assessment *Electrochim. Acta* **222** 303–11

[180] Adekunle A, Raghavan V and Tartakovsky B 2019 On-line monitoring of heavy metals-related toxicity with a microbial fuel cell biosensor *Biosens. Bioelectron.* **132** 382–90

[181] Bae J, Lim J and Kim T 2018 Reusable and storable whole-cell microbial biosensors with a microchemostat platform for *in situ* on-demand heavy metal detection *Sens. Actuators B: Chem.* **264** 372–81

[182] Liu Y, Guo M, Du R, Chi J, He X, Xie Z, Huang K, Luo Y and Xu W 2020 A gas reporting whole-cell microbial biosensor system for rapid on-site detection of mercury contamination in soils *Biosens. Bioelectron.* **170** 112660

[183] Diculescu V C, Chiorcea-Paquim A and Oliveira-Brett A M 2016 Applications of a DNA-electrochemical biosensor *TrAC. Trends Anal. Chem.* **79** 23–36

[184] Chauhan N, Jain U and Soni S 2019 Sensors for food quality monitoring *Nanoscience for Sustainable Agriculture* (Cham: Springer) pp 601–26

[185] Meshram B D, Agrawal A K, Adil S, Ranvir S and Sande K K 2018 Biosensor and its application infood and dairy industry: a review *Int. J. Curr. Microbiol. Appl. Sci.* **7** 3305–24

[186] Srivastava A K, Dev A and Karmakar S 2017 Nanosensors for food and agriculture *Nanoscience in Food and Agriculture 5Sustainable Agriculture Reviews* (Cham: Springer) pp 41–79

[187] Hu L, Deng L, Alsaiari S, Zhang D and Khashab N M 2014 'Light-on' sensing of antioxidants using gold nanoclusters *Anal. Chem.* **86** 4989–94

[188] Chen C, Yuan Z Q, Chang H T, Lu F N, Li Z H and Lu C 2016 Silver nanoclusters as fluorescent nanosensors for selective and sensitive nitrite detection *Anal. Methods* **8** 2628–33

[189] Bera D, Qian L, Tseng T-K and Holloway P H 2010 Quantum dots and their multimodal applications: a review *Materials* **3** 2260–345
[190] Liu H, Zhou K, Wu D, Wang J and Sun B 2016 A novel quantum dots-labeled on the surface of molecularly imprinted polymer for turn-off optosensing of dicyandiamide in dairy products *Biosens. Bioelectron.* **77** 512–7
[191] Kalita H, Palaparthy V S, Baghini M S and Aslam M 2016 Graphene quantum dot soil moisture sensor *Sens. Actuators B: Chem.* **233** 582–90
[192] Uzek R, Sari E and Denizli A 2021 Detection of melamine by using fluorescent nanocomposites with specific recognition sites *ChemistrySelect* **6** 2149
[193] Kretschmann E 1971 The determination of the optical constants of metals by excitation of surface plasmons *Z. Phys.* **241** 313–24
[194] Jatschka J, Dathe A and Csáki A 2016 Propagating and localized surface plasmon resonance sensing - A critical comparison based on measurements and theory *Sens. Bio-Sens. Res.* **7** 62–70
[195] Zeng S W, Baillargeat D, Ho H P and Yong K T 2014 Nanomaterials enhanced surface plasmon resonance for biological and chemical sensing applications *Chem. Soc. Rev.* **43** 3426–52
[196] Saylan Y, Yılmaz F, Özgür E, Derazshamshir A and Denizli A 2020 Plasmonic smart nanosensors for the determination of environmental pollutants *Emerging Carbon-Based Nanocomposites for Environmental Applications* ed A K Mishra, C M Hussain and S B Mishra (Hoboken, NJ: Wiley)
[197] Akgönüllü S, Yavuz H and Denizli A 2020 SPR nanosensor based on molecularly imprinted polymer film with gold nanoparticles for sensitive detection of aflatoxin B1 *Talanta* **219** 121219
[198] Saylan Y and Denizli A 2021 Fundamentals and applications of molecularly imprinted systems *Molecular Imprinting for Nanosensors and Other Sensing Applications* ed A Denizli (Amsterdam: Elsevier) pp 1–17
[199] Akgönüllü S, Yavuz H and Denizli A 2021 Molecularly imprinted based sensors for detection of allergens *Molecular Imprinting for Nanosensors and Other Sensing Applications)* ed A Denizli (Amsterdam: Elsevier) pp 309–34
[200] Saylan Y, Akgönüllü S, Çimen D, Derazshamshir A, Bereli N, Yılmaz F and Denizli A 2017 Development of surface plasmon resonance sensors based on molecularly imprinted nanofilms for sensitive and selective detection of pesticides *Sens. Actuators, B Chem.* **241** 446–54
[201] Yılmaz E, Özgür E, Bereli N, Türkmen D and Denizli A 2017 Plastic antibody-based surface plasmon resonance nanosensors for selective atrazine detection *Mater. Sci. Eng. C.* **73** 603–10
[202] He S, Xie W, Zhang W, Zhang L, Wang Y, Liu X, Liu Y and Du C 2015 Multivariate qualitative analysis of banned additives in food safety using surface enhanced Raman scattering spectroscopy *Spectrochim. Acta Part. A: Mol. Biomol. Spectrosc.* **137** 1092–9
[203] Çimen D, Bereli N, Yavuz H and Denizli A 2021 Sensors for the detection of food contaminants *Nanosensors for Environment, Food and Agriculture* 1 (Cham: Springer) pp 169–82
[204] Qureshi T, Türkmen D and Denizli A 2021 Current trends of plasmonic nanosensors use in agriculture *Biosensors in Agriculture: Recent Trends and Future Perspectives* (Cham: Springer) pp 97–113

[205] Wu S, Liu L, Duan N, Wang W, Yu Q and Wang Z 2018 A test strip for Ochratoxin A based on the use of aptamer-modified fluorescence upconversion nanoparticles *Microchim. Acta* **185** 497

[206] Divsar F, Habibzadeh K, Shariati S and Shahriarinour M 2015 Aptamer conjugated silver nanoparticles for the colorimetric detection of arsenic ions using response surface methodology *Anal. Methods* **7** 4568–76

[207] Ameri M, Shabaninejad Z, Movahedpour A, Sahebkar S, Mohammadi S, Hosseindoost S, Ebrahimi M S, Savardashtaki A, Karimipour M and Mirzaei H 2020 Biosensors for detection of Tau protein as an Alzheimer's disease marker *Int. J. Biol. Macromol.* **162** 1100–8

[208] Wu Y, Li X, Mao E, Du Y and Yang F 2020 Design and development of monitoring device for corn grain cleaning loss based on piezoelectric effect *Comput. Electron. Agri.* **179** 105793

[209] Debabhuti N *et al* 2021 Development of QCM sensor to detect α-terpinyl acetate in cardamom *Sens. Actuators A: Phys.* **319** 112521

[210] Bakhshpour M, Göktürk I, Çakır O, Yılmaz F, Baysal Z and Denizli A 2020 Detection of pesticide via nanoparticle based quartz crystal microbalance sensor *Hacet. J. Biol. Chem.* **48** 75–82

[211] Cervera-Chiner L, Juan-Borrás M, March C, Arnau A, Escriche I, Montoya Á and Jiménez Y 2018 High fundamental frequency quartz crystal microbalance (HFF-QCM) immunosensor for pesticide detection in honey *Food Control.* **92** 1–6

IOP Publishing

Bionanomaterials for Environmental and Agricultural Applications

Ravindra Pratap Singh and Kshitij RB Singh

Chapter 5

Role of biopesticides derived from bionanomaterials for enhanced food security and sustainable agriculture

Charles Oluwaseun Adetunji, Olugbemi T Olaniyan, Kshitij RB Singh, Abel Inobeme, Vanya Nayak, Jay Singh and Ravindra Pratap Singh

Sustainable agriculture is study utilizing green chemistry and innovative technologies to develop various biological products for a supportable agricultural system. Bionanomaterials utilization is gaining serious attention in environmental and agriculture sectors as it helps in the diagnosis, management and treatment of pests and other pathogenic organisms, which restores the soil quality, environmental resources, and generates bio-renewable resources like biofuels, biofertilizers, bio-herbicides, bio-stimulants, etc. These innovative sustainable green chemistry principles-based approaches are renewable, eco-friendly, non-toxic and can develop highly valued agricultural bioproducts like enzymes, polysaccharides, and oils. Hence, the aim of the chapter is to give an overview of biopesticides derived from bionanomaterials.

5.1 Introduction

Today, clear-cut signals concerning the toxic role of chemicals on health and the environment are routinely utilized in conventional agricultural processes. Sustainable agriculture is a robust approach that uses green chemistry innovative technologies to develop bioproducts and biomaterials of sustainable farm systems. Nanotechnology is gaining serious attention across diverse sectors such as environmental, agriculture, and medicine by providing effective bio-nanoparticles and green alternatives for the diagnosis, management, and treatment of pests and other pathogenic organisms. Moreover, many reviews and book chapters have started to focus on the green synthesis of various bionanomaterials and their utilities in

doi:10.1088/978-0-7503-3863-9ch5

different domains like medical, agriculture, environment, animal husbandry, sensor development, etc (1–10).

Several authors have reported the role of biopesticides derived from bionanomaterials such as insect pheromones, plants extract and oils and other beneficial microorganisms for enhanced food security and sustainable agriculture. Many of these biopesticides promote plant growth, attack pathogenic pests, and enhance the soil and water microenvironment for the plant, and for animal health in sustainable agricultural processes. The field of biopesticides has gained tremendous attention due to diverse, sustainable green chemistry principles-based approaches; this involves protecting soil quality, environmental resources, water, wildlife, and the generation of renewable resources like biofuels, biofertilizers, bioherbicides, biostimulants, and bioinsecticides. Innovative sustainable green chemistry principles-based approaches for renewable bioresources and feedstocks apply eco-friendly, non-toxic and sustainable technology to develop green bio-based materials, chemicals and agricultural high-valued bio-products like enzymes, polysaccharides and oils. These bio-based derived bio-products are less toxic, biodegradable, specific in action, and environmentally friendly than conventional pesticides, synthetic chemicals, or materials. The application of biopesticides as a tool for helping farmers transit from the very toxic synthetic pesticides into the season of sustainable agriculture is the key focus. The field of biopesticides is very broad, thus a large amount of research is needed to elucidate many green chemistry solutions and opportunities in a multi-disciplinary approach. Various disciplines like entomology, chemistry, mycology, microbiology, physiology, biochemistry, and entrepreneurship are essential to develop sustainable active ingredients for bioproducts. Also, the standard test for the efficacy, toxicity and health implications of many of these bioproducts utilized for biopesticides will be required to be elucidated by environmental physiologists. Insect pheromones are chemicals released by some insects during cell communications, which contain fragrances and flavors. The pheromones semiochemicals are equally derived from plants and some animals released as messenger substances and may function in various physiological regulatory ways like attracting the opposite sex, sending an alarm, and attacking pests. Many of these chemicals like plant extracts and oils are useful in pest management owing to high specificity, eco-friendly nature and reduced toxicity. Some of the pheromones are 8 Dodecen-1-yl acetate, 8 Dodecen-1-yl acetates,—8 Dodecen-1-ol, pyrethrum, nicotine, rotenone, lemongrass oil, thymol, enzyme extract, plant growth regulators like gibberellins, auxins, cytokines, S-Methoprene and chitin, and they have powerful regulatory and inhibitory activities. Microbial-based biopesticides involve numerous microorganisms such as bacteria, baculoviruses, nematode-associated bacteria, and fungi that can inhibit pests and other pathogens in agro-ecosystems. Thus, this chapter provides an update to the current knowledge on microencapsulation and nanoformulations based biopesticides technologies and other availability active substances derived from plants in integrated pest management, crop protection, and sustainable agricultural practices to enhance crop growth and production, eradicate the utilization of synthetic pesticides and increase food security, as shown in figure 5.1.

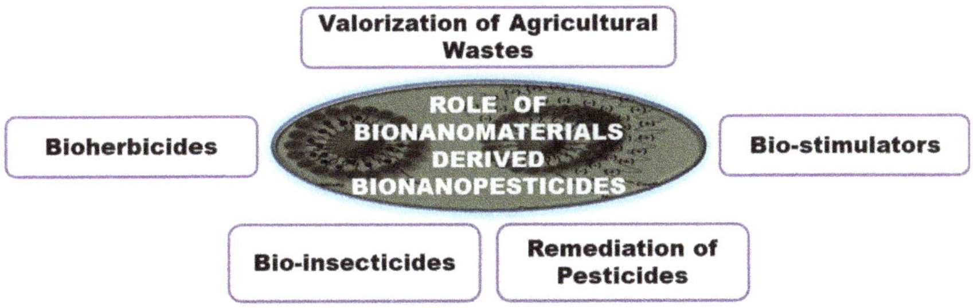

Figure 5.1. Diagrammatic illustration of the discussed details in this chapter.

5.2 Utilization of bionanomaterials for valorization of several agricultural wastes

It has been reported that to promote a sustainable lifestyle, waste valorization for novel technology and an improved environment seems to be the new concept for many circular economies. The potential of bioconversion of many plants' active ingredients from waste into novel bionanomaterials is a current trend among biomedical scientists. Studies have shown that the generation of agricultural waste and its disposal in the environment generate a serious concern. Thus, there is a greater need for novel strategies to address agricultural waste treatment and recycling. Biological treatment of agricultural waste using nanomaterials derived from plants is one of the most leveraging research studies focused on ecological quality, environmental monitoring, and agricultural wastewater abatement through bioremediation (11). Moreover, active molecules from plants are utilized to produce bionanomaterials used for phytoremediation of textile dyes, contaminant wastewater, and volatile agricultural pollutants. Figure 5.2 shows the application strategies for various nanopesticides/biopesticides in food security and sustainable agriculture (12).

It was observed that agricultural waste represents the most abundant pool of untapped bioresources for economic gain and environmental protection. Using plant-based nanoparticles, these bioresources can be converted into bio-based products, bioenergy and biomolecules for environmental and food industries. Even though these agricultural waste chains are difficult to assess due to so many reasons like seasonality, region and complexity, many approaches can be adopted to harness residual agricultural resources to produce high-value chemicals for industrial use (13). Moreover, it was also reported that agricultural wastes are valuable bioresources that can produce a high valued chain of bioproducts and nanoparticles with potential applications in nanoelectronics, renewable solar energy, biomedical science, thermal insulation, and drug-gene delivery through eco-friendly, environmentally renewable strategy. The authors discussed some of the agricultural wastes such as sugar cane, bagasse and leaves, egg shell, bamboo leaf, walnut shell, wheat straw, banana peels, and coconut shells (14).

José *et al* explained that many of the sea plants and marine waste through biotechnology could be utilized to produce peptones for diverse industrial

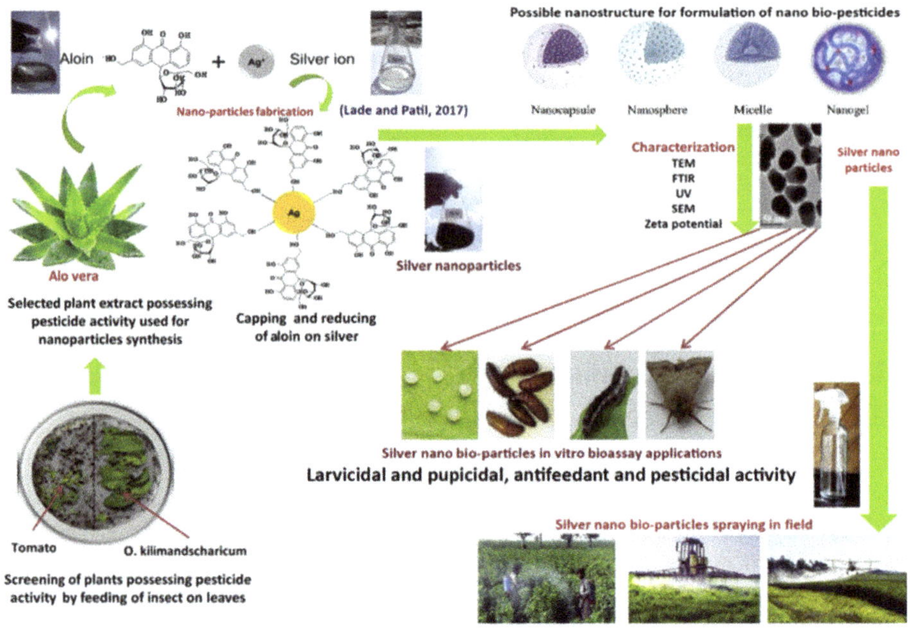

Figure 5.2. Various application strategies that are used for nanopesticides/biopesticides in food security and sustainable agriculture. Reproduced from (12), copyright (2019) with permission from Elsevier.

applications (15). It was reported that the current global environmental and energy problem could be tackled using nanomaterials generated from byproducts materials like palm oil. This approach is inexpensive and can be adopted to transform agricultural waste into highly valued sustainable products for diverse industrial sectors (16). The present world is facing serious economic challenges due to pandemics and over-reliance on fossil sources of energy which are not sustainable. Bio-waste from agricultural sources is rich in sources of biomolecules for several industrial applications. Further, it was reported that tomato residues could be upgraded and recovered by applying novel natural halloysite nanotubes and green nanocomposites based on tomato peels combined with carvacrol, which is a potent natural antimicrobial agent. They showed that the materials are a good candidate for industrial utilization in food packaging (17). Similarly, it was shown that waste derived from agriculture could be recycled through nanotechnology. The authors noted that plant species are part of the agricultural residues that source lignocellulosic biomass to produce biofuels, chemicals, and materials for industrial applications like automotive, construction, and packaging (18). Also, it was noted that many value-added bionanoproducts such as fine chemicals, industrial enzymes, microbial polysaccharides, organic acids and biofertilizers could be generated from lignocellulosic agricultural residues for economic benefits.

The rapid increase in the global population has caused a growing awareness of sustainability in the circular economy based on recycling resources and waste reduction processes. Waste from agricultural sectors like plants is currently being explored for innovative opportunities like cellulose nanofibers, Rice husk-derived Si nanomaterials and graphene (19). Further, it was reported that silver and gold nanoparticles through green synthesis from plant extracts could be utilized for diverse agricultural applications, particularly environmentally-friendly antimicrobial agents (20). Furthermore, it was reported that agricultural lignocellulosic residues of corn stalk, soy stalk, wheat straw, and perennial grasses, such as Miscanthus switchgrass are currently being utilized for the production of nano-value-added biobased composite materials for different industrial applications (21). Cristiane *et al* also described the production of nanocellulose from sugarcane lignocellulosic biomass by utilizing enzymatic hydrolysis conversion into various products such as high-value materials biofuels, biochemicals for diverse industrial applications like biorefineries (22).

5.3 Utilization of bionanomaterials as bioherbicides

Currently, plants possess chitin, lignin and other active materials that can be utilized to produce nanoparticles for bioherbicides applications against fungi, viruses and other pathogenic organisms (23). Bioherbicides are being used for weed management and control of pest invasion. Further, it was also reported that new bioherbicides for controlling weeds and pathogens derived from natural products like extracts of natural materials, corn gluten meal, dried distiller's grains with soluble, essential oil extracts and mustard seed meals had shown promising effects. These natural plants derived bionanoherbicides have been indicated to be cheap, effective, and have limited effects on the environment and humans. The authors demonstrated that these bioherbicides have greater advantages such as increasing crop yields, herbicide resistance, and efficiency (24). Hoewever, not much work has been done on the bionanomaterials which can be utilized as bioherbicides and researchers need to focus on this domain.

5.4 Application of bionanomaterials as bio-stimulators

It has been reported that developing nanomaterials from naturally derived polymer for agriculture and food-based industries will be a sustainable approach for both human and animal safety. The application of microbial extracellular enzymes on polymeric materials for nano-delivery systems for agri-food purposes is witnessing huge transformation and development among many scientists. Further, it was also observed that these natural polymers are derivative of animals, food waste, plants, and bio-fabricated biological products using microorganisms or non-enzymatic mechanisms for generating renewable nanomaterials. Many of these materials are used to produce synthetic polymers like polycaprolactone, polyvinyl alcohol, polyethylene, acrylate-based polymers. Further, it was also observed that many extracellular polymeric nanomaterials are derived from plants' active molecules like cellulose, protein, k-carrageenan, polysaccharides, chitosan, lignin, alginate, which can be utilized as nutraceuticals, food and agro-businesses. The modern way of farming involves using agrochemicals composed of bionanomaterials to enhance the quality and quantity of crops, and equally to reduce the negative impact of these chemicals on humans and the environment. Due to excessive use of several synthetic bio-stimulants, numerous side effects on soil, air, water, and plant have developed. Therefore, many scientists have now started to develop bionanomaterials from plants' active molecules that can serve as an active bio-stimulant with greater advantages over the synthetic ones. These bionanomaterials are found to be biodegradable, can serve as bio-stimulants, and improve the nutritional utilization, decrease the use of non-toxic products, are efficient and very biocompatible (25). Moreover, it was also reported that algal and other marine plant extracts are utilized for the generation of nanoparticles currently used for bio-stimulants in agricultural industries. These plants are known to be the most important sustainable resources with potential usage due to a large amount of active molecules such as fucoidan, galactan, alginate, laminarin, lectins, chlorophylls, phycobiliproteins, carotenoids, phenolic acids, cinnamic acid, flavonoids, isoflavones, lignans, benzoic acid, quercetin, cytokinins, gibberellins, auxins, and abscisic acid (26).

Owing to the growth potentials and activities of these bionanostimulants, many of them are currently being utilized for crop production and increased yield, therefore, a study reports the renewed interest of many scientists for the use of plant bio-stimulants derived from waste and several other sources for boosting vegetable crop production. It was also further explained that these plant-based derived bionanomaterial based bio-stimulants could increase crop performance, quality and growth. Moreover, it was also observed that brown macro-algae and protein hydrolysates for increasing sustainable vegetable production, are cost-effective and can form an innovative approach in the food and agro sector. These bionanostimulants can promote plant growth by interacting with the surface membrane charges of plant cells and adherence to the corona (27). Further, bionanofertilizers as bio-stimulants were reported to be developed from banana peel. The study demonstrated that banana peel contains certain biomolecules and nutrients that could be utilized through recycling for diverse purposes, especially for bio-stimulants in the agricultural sector. The extracted bionanofertilizer from banana peel was tested on fenugreek and tomato plants, and it was discovered that the percentage of germination significantly improved with increased use of extracted bionanofertilizer from banana peel. These bionanostimulants can reduce soil toxicity, improve the nutritional efficiency of plants, and avert negative impacts due to overdose (28).

5.5 Application of bionanomaterials as bioinsecticides

Generally, nanoparticles are utilized for the management of several agricultural and environmental veterinary control of pest and invading pathogens. Studies have reported several economic benefits of arthropods particularly phytophagus mites and insects, but many are known to cause devastating effects on agricultural crops and products (29). In a similar pattern, *Tribolium castaneum* popularly known as red flour beetle has been reported to cause destruction to stored agricultural products across the globe, thus reducing the quality, weight, nutritive value, economic value and fertility or germination properties (30).

The use of several inorganic nanomaterials such as nano-TiO_2, nano-Ag, and CdS to mitigate the negative effects of *Spodoptera litura* Fab have been reported (31). The role of pungamoil-based metallic nanoparticles and commercially available neem pesticides were tested against the *Pericallia ricini* larval survival, growth, gravimetric and reproduction. From the analysis, it was discovered that the pungamoil-based metallic nanoparticles produced a significant effect against *Pericallia ricini* larva (32).

Several metallic nano-insecticidal products derived from *Aloe vera* extract in combination with chitosan are currently being utilized as bionanopesticides (12). These bionanopesticides were tested against lepidopteran pests and it was reported that these bionanopesticides produced significant effects by suppressing the feeding pattern, weight and development of the lepidopteran pest's larva (33). Studies have shown that chin-chitosan derived nanoinsecticide produces more significant results in the elimination of pathogens and pests, particularly the plants based pests (34–36).

In another study, DNA tagged metallic nanoparticles were designed to mitigate against *S. litura larva.* From their study, it was discovered that the nanoparticles produced a significant result by causing maximum mortality. Also similarly, ecdysteroid analogs nanoparticles were tested against *Corcyra cephalonica* and they showed that these nanoparticles caused maximum toxicity just like the commercially available pesticides (37, 38).

Bionanoparticles produced from plants or microorganisms such as *Aristolochia indica* utilizing their metabolites or enzymes were tested against the larva of *H. armigera* and reported to give maximum cytotoxic effects against the pest (39, 40). It was revealed that these bionanoparticles caused destruction in embryonic development, suppression of egg laying ability, reduced oviposition and increased biopesticidal activity against several pests and pathogens like *T. castaneum* and *Tribolium confusum* (36, 41–43).

Polyethylene glycol-coated nanoparticles in combination with garlic essential oil and *Euphorbia prostrata* were tested against *Tribolium castaneum* pathogen. From the results, it was demonstrated that these nanoparticles produced 100% mortality within a short period of time against the tested pathogen (44, 45).

5.6 Application of bionanomaterials from plants in remediation of pesticides

Scientists globally have developed green technology that can produce clean and safe food products for healthy living and well-being. Studies have shown that nanoparticles and their application in several agricultural activities are safe and nontoxic, thus their application can be effective in mitigating against food insecurity, increasing sustainable agricultural practices like pest control and curbing global food crisis (32). Thus, in line with these recommendations, several countries are beginning to implement reliable green technological approaches like metallic nanoparticles in agricultural practices, particularly in pest control (46–50). Though, many countries have adopted these nanoparticles for pest control green technology in agriculture, more is needed to be done in this regard. The utilization of microbial metabolites and plants' bioactive molecules can further be synthesized and developed into bionanosensors for the detection of pollutants, pest and environmental toxins which are cost effective, eco-friendly and safe (51). Studies have revealed that diverse techniques are available to reduce the pesticidal contents of agrochemicals like biological degradation, surface adsorption, and membrane filtration. Many of these techniques are reported to be less sensitive, fast and specific, thus the use of bionanotechnology has yielded greater advantages compared to the other techniques. Bionanotechnology is very useful in the degradation, detection, remediation, sensing, and elimination of pesticides. Bionanomaterials have excellent physiochemical properties for the elimination of pesticides in the environment (52). The use of carbon nanotubes containing graphitic carbons is novel. Unique macromolecules nanoparticles have several physiochemical properties such as thermal stability, one-dimensional structure, and dynamic chemical nature with tremendous ability to serve as potential pesticides removing agents due to high adsorption capacity, surface functional groups. This adsorption is done through π-π interactions, hydrophobic effect, covalent bonding, electrostatic interactions and hydrogen bonding (53–56).

Studies have revealed that the chemical and physical characteristics of these nanoparticles and bionanocomposites are important factors in the membrane selection and removal of pesticides due to relative molecular weight, porosity, desalting degree, retention properties, pH and membrane potentials (57, 58). The optimal conditions such as pH, agitating time, initial amount of the pesticide, adsorbent dose, of the bionanocomposites must be considered when searching for suitable bionanocomposites like chitosan-based metal particles composites to be utilized in the elimination of pollutants (59–63). Schematic illustration represents the application of bionanomaterials in pesticide delivery figure 5.3 (64). Microbial enzymes, proteins and metabolites could also be generated to serve as bionanocomposites that can be developed into agents with high level of application in the elimination of pollutants in the environment causing degradation of organic pollutants, phenol and aniline into catechol through β-ketoadipate pathway (65, 66).

5.7 Conclusion and future recommendations

Biopesticides have great potential in revolutionizing agricultural practices through innovative strategies and with the onset of nanotechnology, these biopesticides have been modified and are considered as safe, non-toxic and sustainable eco-friendly agriculture approaches to be adopted in developing many of these bionanopesticides. Currently, the knowledge and utilization of bionanopesticides in pest management are replacing synthetic pesticides that are known to pose serious environmental risks and health challenges to plants, animals and humans. Relevant stakeholders like governments should provide serious regulatory policies and more research output, scientists and industrialists on reducing the production and development of synthetic chemicals and the replacement with bionanopesticides to improve pest management, crop protection and sustainable agricultural practices to achieve enhanced crop production, suppress the utilization of synthetic insecticides and

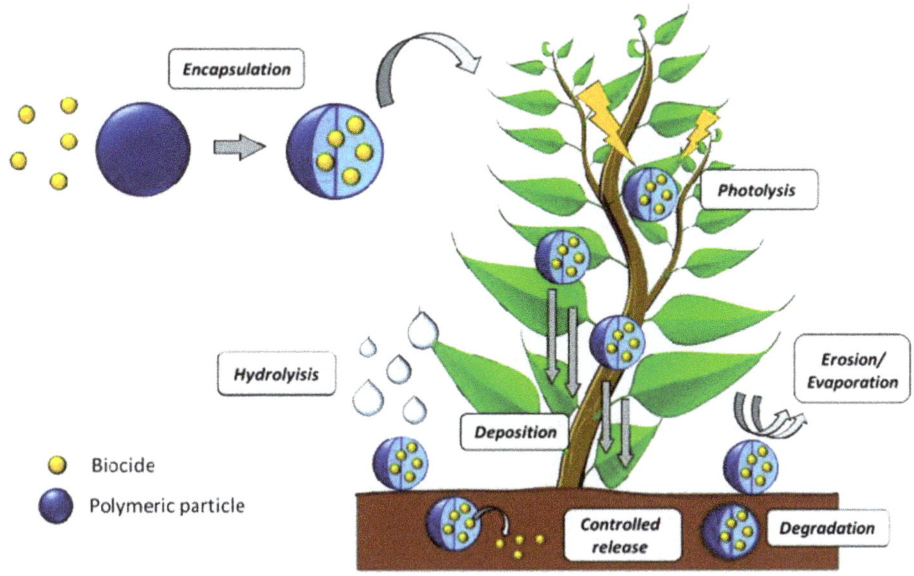

Figure 5.3. Application of bionanomaterials in pesticide delivery. Reproduced from (64), with permission from MDPI [CC BY 4.0] .

increase food security. However, much research like safety, toxicity, biocompatibility, degradability, etc is still left to be studied and demands immediate attention as its commercial large-scale utilization can increase the agriculture market and will also help in developing sustainable agricultural practices.

Acknowledgment

COA, OTO, and AI are thankful to their respective universities for helping them to prepare this chapter. KRBS would like to thank Professor A K Singh for giving constant encouragement throughout the work. JS expresses their gratitude for the DST-INSPIRE faculty Fellowship, BHU (IoE grant), and UGC New Delhi to provide financial assistance. RPS and VN are thankful to Vice-Chancellor, IGNTU, Amarkantak, India, for providing financial assistance and for motivating good science.

References

[1] Singh K R, Nayak V, Sarkar T and Singh R P 2020 Cerium oxide nanoparticles: properties, biosynthesis and biomedical application *RSC Adv.* **10** 27194–214

[2] Nayak V, Singh K R, Singh A K and Singh R P 2021 Potentialities of selenium nanoparticles in biomedical science *New J. Chem.* **45** 2849–78

[3] Ukhurebor K E, Singh K R R, Nayak V and UK-Eghonghon G 2021 Influence of SARS-CoV-2 pandemic: a review from the climate change perspective *Environ. Sci. Process Impacts* **23** 1060–78

[4] Singh P, Singh K R, Singh J, Das S N and Singh R P 2021 Tunable electrochemistry and efficient antibacterial activity of plant-mediated copper oxide nanoparticles synthesized by *Annona squamosa* seed extract for agricultural utility *RSC Adv.* **11** 18050–60

[5] Fernandes M, Singh K R, Sarkar T, Singh P and Singh R P 2020 Recent applications of magnesium oxide (MgO) nanoparticles in various domains *Adv. Mater. Lett.* **11** 20081543

[6] Singh R P 2017 Application of nanomaterials toward development of nanobiosensors and their utility in agriculture *Nanotechnology* (Singapore: Springer) pp 293–303

[7] Singh K R and Singh R P 2021 Utility of nanobiosensors in environmental analysis and monitoring *Nanosensors for Environment, Food and Agriculture* vol 1 *Environmental Chemistry for a Sustainable World* vol 60 (Cham: Springer) pp 229–46 https://doi.org/10.1007/978-3-030-63245-8_11.

[8] Singh R P and Singh K R B 2021 Nanobiotechnology in animal production and health *Advances in Animal Genomics* ed S Mondal and R L Singh (Amsterdam: Elsevier) pp 185–98

[9] Singh R P 2011 Prospects of nanobiomaterials for biosensing *Int. J. Electrochem.* **2011** 1–30

[10] Singh R P 2021 Recent trends, prospects, and challenges of nanobiosensors in agriculture *Biosensors in Agriculture: Recent Trends and Future Perspectives* (Cham: Springer) pp 3–13

[11] Xu C, Nasrollahzadeh M, Selva M, Issaabadi Z and Luque R 2019 Waste-to-wealth: biowaste valorization into valuable bio(nano)materials *Chem. Soc. Rev.* **48** 4791–822

[12] Lade B D, Gogle D P, Lade D B, moon G M, Nandeshwar S B and Kumbhare S D 2019 Nanobiopesticide formulations: application strategies today and future perspectives *Nano-Biopesticides Today and Future Perspectives* (Amsterdam: Elsevier) pp 179–206

[13] Gontard N, Sonesson U, Birkved M, Majone M, Bolzonella D and Celli A *et al* 2018 Mar A research challenge vision regarding management of agricultural waste in a circular bio-based economy *Crit. Rev. Environ. Sci. Technol.* **48** 614–54

[14] Zamani A, Marjani A P and Mousavi Z 2019 Agricultural waste biomass-assisted nano-structures: synthesis and application *Green Process Synth.* **8** 421–9

[15] Vázquez J A, Durán A I, Menduíña A and Nogueira M 2020 Biotechnological valorization of food marine wastes: microbial productions on peptones obtained from aquaculture by-products *Biomolecules* **10** 1184

[16] Kaniapan S, Hassan S, Ya H, Patma Nesan K and Azeem M 2021 Mar The utilisation of palm oil and oil palm residues and the related challenges as a sustainable alternative in biofuel, bioenergy, and transportation sector: a review *Sustainability* **13** 3110

[17] Bugatti V, Brachi P, Viscusi G and Gorrasi G 2019 Valorization of tomato processing residues through the production of active bio-composites for packaging applications *Front. Mater.* **6** 34

[18] Baudrit J V 2017 Recycling and elimination of wastes obtained from agriculture by using nanotechnology: nanosensors *Int.* J **3** 368–75 . Biosens. Bioelectron.

[19] Soni S K 2019 Value added products from lignocellulosic agricultural residues: an overview *Int.* J **9** 101–5 Available from http://ndpublisher.in/admin/issues/IJFFTv9n2g.pdf. Food Ferment. Technol.

[20] Castillo-Henríquez L, Alfaro-Aguilar K, Ugalde-Álvarez J, Vega-Fernández L, Montes de Oca-Vásquez G and Vega-Baudrit J R 2020 Green synthesis of gold and silver nanoparticles from plant extracts and their possible applications as antimicrobial agents in the agricultural area *Nanomaterials* **10** 1763

[21] Nagarajan V, Mohanty A K and Misra M 2013 Sustainable green composites: value addition to agricultural residues and perennial grasses *ACS Sustain. Chem. Eng.* **1** 325–33

[22] Farinas C S, Marconcini J M and Mattoso L H C 2018 Mar Enzymatic conversion of sugarcane lignocellulosic biomass as a platform for the production of ethanol, enzymes and nanocellulose *J. Renew. Mater.* **6** 203–16

[23] Morganti P and Stoller M 2017 Chitin and lignin: natural ingredients from waste materials to make innovative and healthy products for humans and plant *Chem. Eng. Trans.* **60** 319–24

[24] Cai X and Gu M 2016 Bioherbicides in organic horticulture *Horticulturae* **2** 3

[25] Sampathkumar K, Tan K X and Loo S C J 2020 Developing nano-delivery systems for agriculture and food applications with nature-derived polymers *iScience* **23** 101055

[26] Szymańska E and Winnicka K 2015 Stability of chitosan—a challenge for pharmaceutical and biomedical applications *Mar. Drugs.* **13** 1819–46

[27] Petropoulos S A 2020 Practical applications of plant biostimulants in greenhouse vegetable crop production *Agronomy* **10** 1569

[28] Hussein H S, Shaarawy H H, Hussien N H and Hawash S I 2019 Preparation of nano-fertilizer blend from banana peels *Bull. Natl. Res. Cent.* **43** 26

[29] Yang Q, Kučerová Z, Li Z, Kalinović I, Stejskal V and Opit G *et al* 2012 Jan Diagnosis of *Liposcelis entomophila* (Insecta: Psocodea: Liposcelididae) based on morphological characteristics and DNA barcodes *J. Stored Prod. Res.* **48** 120–5

[30] Adarkwah C, Obeng-Ofori D, Buttner C, Reichmuth C and Schöller M 2010 *Bio-rational Control of Red Flour Beetle Tribolium castaneum (Herbst) in Stored Wheat with Calneem®oil Derived from Neem Seeds* (Berlin: Springer)

[31] Chakravarthy A K, Chandrashekharaiah, Kandakoor S B, Bhattacharya A, Dhanabala K G and Ramesh P 2012 Bio efficacy of inorganic nanoparticles CdS, Nano-Ag and Nano-TiO_2 against *Spodoptera litura* (Fabricius) (Lepidoptera: Noctuidae) *Curr. Biot.* **3** 271–81

[32] Kitherian S 2016 Nano and bio-nanoparticles for insect control *Res. J. Nanosci. Nanotechnol.* **7** 1–9

[33] Yasur J and Usha Rani P 2015 Lepidopteran insect susceptibility to silver nanoparticles and measurement of changes in their growth, development and physiology *Chemosphere* **124** 92–102

[34] Zheng L, Hong F, Lu S and Liu C 2005 Effect of nano-TiO_2 on strength of naturally aged seeds and growth of spinach *Biol. Trace Elem. Res.* 104 83–92

[35] Rabea E I, Badawy M E I, Rogge T M, Stevens C V, Höfte M and Steurbaut W *et al* 2005 Oct Insecticidal and fungicidal activity of new synthesized chitosan derivatives *Pest Manag. Sci.* 61 951–60

[36] Chandra J H, Raj L F A A, Namasivayam S K R and Bharani R S A 2013 RSA. Improved pesticidal activity of fungal metabolite from nomureae rileyi with chitosan nanoparticles Int. Conf. on Advanced Nanomaterials & Emerging Engineering Technologies (Piscataway, NJ: IEEE) 387–90

[37] Chandrashekharaiah M, Kandakoor S B, Basana Gowda G, Kammar V and Chakravarthy A K 2015 Nanomaterials: a review of their action and application in pest management and evaluation of DNA-tagged particles *New Horizons in Insect Science: Towards Sustainable Pest Management* (New Delhi: Springer India) 113–26

[38] Chakravarthy A K 2012 DNA-tagged nano gold: a new tool for the control of the armyworm, *Spodoptera litura* Fab. (Lepidoptera: Noctuidae) *African* J **11** 883 . Biotechnol.

[39] Sahayaraj K, Roobadevi M, Rajesh S and Azizi S 2015 *Vernonia cinerea* (L.) Less. silver nanocomposite and its antibacterial activity against a cotton pathogen *Res. Chem. Intermed.* 41 5495–507

[40] Siva C and Kumar S 2015 Pesticidal activity of eco-friendly synthesized silver nanoparticles using Aristolochia indica extract against *Helicoverpa armigera* Hubner (Lepidoptera: Noctuidae) *Int. J. Adv. Sci. Tech. Res.* **2**

[41] Christofoli M, Costa E C C, Bicalho K U, de Cássia Domingues V, Peixoto M F and Alves C C F *et al* 2015 Insecticidal effect of nanoencapsulated essential oils from *Zanthoxylum rhoifolium* (Rutaceae) in *Bemisia tabaci* populations *Ind. Crops Prod.* **70** 301–8

[42] Sahab A F, Waly A I, Sabbour M M and Nawar S L 2015 Synthesis, antifungal and insecticidal potential of Chitosan (CS)-g-poly (acrylic acid) (PAA) nanoparticles against some seed borne fungi and insects of soybean *Chem. Tech.* **8** 589–98

[43] Sabbour M M A 2020 Efficacy of nano-formulated certain essential oils on the red flour beetle Tribolium castaneum and confused flour beetle, *Tribolium confusum* (Coleoptera: Tenebrionidae) under laboratory and storage conditions *Bull. Natl. Res. Cent.* **44** 111

[44] Yang F-L, Li X-G, Zhu F and Lei C-L 2009 Structural characterization of nanoparticles loaded with garlic essential oil and their insecticidal activity against *Tribolium castaneum* (herbst) (coleoptera: tenebrionidae) *J. Agric. Food Chem.* **57** 10156–62

[45] Abduz A, Zahir A, Bagavan C, Kamaraj G and Elango A A R 2012 Efficacy of plant-mediated synthesized silver nanoparticles against *Sitophilus oryzae JBioPest.* **5** 95–102

[46] Naik B R, Gowreeswari G S, Singh Y, Satyavathi R, Daravath S S and Reddy P R 2014 Bio-Synthesis of silver nanoparticles from leaf extract of pongamia pinnata as an effective larvicide on dengue vector aedes albopictus(skuse) (diptera: culicidae) *Adv. Entomol.* **2** 93–101

[47] Raut R W, Kolekar N S, Lakkakula J R, Mendhulkar V D and Kashid S B 2010 Extracellular synthesis of silver nanoparticles using dried leaves of pongamia pinnata (L) pierre *Nano-Micro. Lett.* 2 106–13

[48] Velusamy P, Das J, Pachaiappan R, Vaseeharan B and Pandian K 2015 Greener approach for synthesis of antibacterial silver nanoparticles using aqueous solution of neem gum (*Azadirachta indica* L.) *Ind. Crops Prod.* **66** 103–9

[49] Khan Z, Hussain J I and Hashmi A A 2012 Jun Shape-directing role of cetyltrimethylammonium bromide in the green synthesis of Ag-nanoparticles using Neem (*Azadirachta indica*) leaf extract *Colloids Surf. B Biointerfaces* **95** 229–34

[50] Vivek R, Thangam R, Muthuchelian K, Gunasekaran P, Kaveri K and Kannan S 2012 Green biosynthesis of silver nanoparticles from *Annona squamosa* leaf extract and its *in vitro* cytotoxic effect on MCF-7 cells *Process. Biochem.* **47** 2405–10

[51] Das S, Chakraborty J, Chatterjee S and Kumar H 2018 Prospects of biosynthesized nanomaterials for the remediation of organic and inorganic environmental contaminants *Environ. Sci. Nano* **5** 2784–808

[52] JEONG G-H 2009 Surface functionalization of single-walled carbon nanotubes using metal nanoparticles *Trans. Nonferrous Met. Soc. China* **19** 1009–12

[53] Ren X, Chen C, Nagatsu M and Wang X 2011 Jun Carbon nanotubes as adsorbents in environmental pollution management: a review *Chem. Eng. J.* **170** 395–410

[54] Yang K, Wu W, Jing Q and Zhu L 2008 Nov Aqueous adsorption of aniline, phenol, and their substitutes by multi-walled carbon nanotubes *Environ. Sci. Technol.* **42** 7931–6

[55] Cho H-H, Smith B A, Wnuk J D, Fairbrother D H and Ball W P 2008 Influence of surface oxides on the adsorption of naphthalene onto multiwalled carbon nanotubes *Environ. Sci. Technol.* **42** 2899–905

[56] Fryxell G E and Cao G 2007 *Environmental Applications of Manomaterials* (London: Imperial College Press)
[57] Nghiem L D and Schafer A 2004 Trace contaminant removal with nanofiltration *Nanofiltration—Principles and Applications* (Amsterdam: Elsevier) pp 479–524
[58] Causserand C, Aimar P, Cravedi J P and Singlande E 2005 Dichloroaniline retention by nanofiltration membranes *Water Res.* **39** 1594–600
[59] Moradi Dehaghi S, Rahmanifar B, Moradi A M and Azar P A 2014 Removal of permethrin pesticide from water by chitosan–zinc oxide nanoparticles composite as an adsorbent *J. Saudi Chem. Soc.* **18** 348–55
[60] Zainal Z, Hui L K, Hussein M Z, Abdullah A H, Hamadneh I and Moh'd K R 2009 May Characterization of TiO_2–chitosan/glass photocatalyst for the removal of a monoazo dye via photodegradation–adsorption process *J. Hazard Mater.* **164** 138–45
[61] Saifuddin N, Nian C Y, Zhan L W and Ning K X 2011 Chitosan-silver nanoparticles composite as point-of-use drinking water filtration system for household to remove pesticides in water *Asian J. Biochem.* **6** 142–59
[62] Rahmanifar B and Moradi Dehaghi S 2014 Removal of organochlorine pesticides by chitosan loaded with silver oxide nanoparticles from water *Clean Technol. Environ. Policy* 16 1781–6
[63] Kumari B and Singh D P 2016 A review on multifaceted application of nanoparticles in the field of bioremediation of petroleum hydrocarbons *Ecol. Eng.* **97** 98–105
[64] Álvarez-Paino M, Muñoz-Bonilla A and Fernández-García M 2017 Antimicrobial polymers in the nano-world *Nanomaterials* **7** 48
[65] Hou J, Liu F, Wu N, Ju J and Yu B 2016 Efficient biodegradation of chlorophenols in aqueous phase by magnetically immobilized aniline-degrading *Rhodococcus rhodochrous* strain *J. Nanobiotechnol.* **14** 5
[66] Li J, Liu G, Zhou J, Wang A, Wang J and Jin R 2016 Redox activity of lignite and its accelerating effects on the chemical reduction of azo dye by sulfide *RSC Adv.* **6** 66930–7

IOP Publishing

Bionanomaterials for Environmental and Agricultural Applications

Ravindra Pratap Singh and Kshitij RB Singh

Chapter 6

Bionanoformulations: special emphasis on agricultural crop protection and growth

Gunjan Vyas and Umesh Kumar

Agriculture is one of the most important sectors across the world for food, feed, and fibres production. The major challenge today's world is facing is decline in culturable land, poor soil fertility and loss of crop yield. To overcome this problem, the use of chemicals during the green revolution compromises target specificity, environmental safety concerns and could create eco-system imbalance issues in the future. Nanotechnology seems promising to overcome these problems by replacing traditional practices with nanopesticides, nanoherbicides, nanofertilizers, nanosensors etc. Bionanoformulation is a green alternative to conventional methods for boosting soil fertility and plant growth, enhancing yield, and controlling plant disease.

6.1 Introduction

Agriculture is one of the most pivotal sectors across various countries for food, feed, and fibres production. This stable and important sector is amajor contributor to the economy of the world because of production and provision of raw materials for food and feed industries. Issues with global natural resource management (production land, soil, water resources etc) as well an exponential increment in world population has raised serious concerns for production and productivity in agriculture. Such limitations critically demand agriculture development to be viable, efficient as well as environmentally friendly from a safety, futuristic and economic viewpoint [1]. As per the US Census Bureau of The International Data Base [2] world population has doubled in 40 years from only 3 billion in 1959 to 6 billion in 1999. It is expected that this number will be 9 billion by the year 2044,

doi:10.1088/978-0-7503-3863-9ch6

which is almost a 50% increase within a 45 years span. The main challenge with this is sensible determination of our planet's capacity to sustain this huge growing population number and to fulfil their food, nutrition, fibre, raw materials and subsistence requirements. For example, 0.5 hectare of productive land per person will be required if we consider only food subsistence for people with diverse diet habits. At present the available productive land is 0.27 hectare per capita only, which will drastically reduce to 0.14 hectare per capita within the coming 40 years because of increasing populations, urbanization, pollution challenges, desertification, deforestation and reduction of non-renewable resources [3]. Such challenges may increase shortage of food, malnutrition, constraints with agricultural raw materials, feed limitation, and ultimately affect the world economy. Currently, each available developed technological approach is working in the direction of the agricultural area to meet the needs of increases in crop production and productivity. There is a need for a 70% increase in agricultural productivity by 2050 to combat the lag generated in the agricultural area [4]. Among the technologies contributing to the agricultural field, biotechnology and genetic engineering are providing input to overcome challenges at ground level. The field has successfully achieved a breakthrough with the development of genetically modified (GM) products. Insect resistant *Bacillus thuringiensis* (Bt) cotton, insect resistant maize, herbicide tolerant soybean etc are successful technology-based GM crop products. GM products use includes benefits with decreased use of pesticides and herbicides, improved nutrition quality, increased yield, development of resistant varieties, decreased labour costs, stress tolerant varieties, productivity and production improvement per area of the fertile land [5]. Such technology missions for improving the quality and quantity of crops have increased both production and productivity. The mission includes modern inputs such as adoption of improved hybrids, advancement in production protocols, integrated water management, seed improvement, technologies for pest management, technologies for nutrient management, post-harvest technology etc in agriculture. Enhanced crop productivity target can be achieved by using knowledge of GM varieties and hybrids, integrated nutrient management, integrated pests management, insecticide resistance management, modules of integrated disease management and high density planting systems [6]. Usage of GM hybrids with upgraded single or two traits has tremendously improved farming while on the other hand all remaining dimensions with insecticide dosage, usage against sucking pests, chemical use against fungi, bacteria and viruses are still there. So, the overall challenges pertaining to the excessive use of chemicals in the form of pesticides/insecticides/herbicides or fertilizers in the agricultural field are still there to be addressed [7]. Presently, the whole world is undergoing revolutionary changes with breakthrough nanotechnology applications in all dimensions and fields. Even the agriculture sector is revolutionizing with tremendous potential of nanotechnology-based applications. Agriculture and the environment work together by supporting each other to rectify various human population problems. Hence, this chapter comprises nanotechnology, its advancement and inputs for the improvement in agriculture.

6.1.1 Global scenario and challenges in agriculture

Despite increasing population, urbanization, technological and industrial intrusion, increasing land scarcity, pollution challenges and rising land values, the developing world has evidenced an extraordinary era of crop and food productivity in the past five or more decades as a result of green revolution impacts. Leaving population jump behind, the green revolution has witnessed dramatic increment in crop productivity by 30% in cultivated land areas. In this revolution time there has been major investment in research and development for agriculture from infrastructure and innovations to marketing. The green revolution strategy for crop productivity was based on the central part of breeding crops as well wide usage of chemical-based fertilizers, pesticides and herbicides to combat challenges the world was facing. The first green revolutionary phase between 1966–1985 widely used improved crop germplasm varieties for wheat, maize and rice with positive impact in production, but this early success soon faced a challenge with sustaining productivity gains, adapting to climate change and enhancing smallholder competitiveness across production systems in the late green revolutionary phase [8]. Apart from this, the green revolution has also resulted in major drawbacks associated with constancy of technological outcomes, sustainability of productivity in global geological variations and rectitude to the urgent need of novel implementation for agricultural research results and progress [9]. Post green revolution dependency on chemicals has generated serious issues with health hazards with negative impacts on the environment and ecosystems. Thus, the idea developed for replacement of chemicals with biobased fertilizers and pesticides as an innovative environment friendly approach. But the idea was compromised by issues with *in situ* stability, poor shelf life, high dose requirement due to nonspecific action and sustainability in fluctuating environmental and climatic conditions [10]. By 2050 it is estimated that the planet's requirement for food and feed will double, which will generate staple food of only 62% from the total 80% of practiced rain fed globally. Non-agricultural usage of water has increased demand and there is a scarcity of water area used under irrigation. However, temperature is one of the most challenging factors as by the mid-21st century it is likely to increase by about 1–3 °C, and by the late 21st century by 2–5 °C on the basis of regional and pollution scenarios. Historical data calculates each 'degree day' spent above 30 °C reduces the agricultural product yield by 1% under normal conditions, which rises up to 1.7% under water stress conditions [11]. Biotechnology, genetic engineering, and genome-assisted breeding are giving their helpful inputs to overcome such problems of crop loss. Genetically engineered varieties of cotton, maize, soybean, and canola were widely planted in North America and Asia in 2005 while there was minimal use of such varieties in Africa and Europe. The majority of transgenic conferred varieties are with a single trait in crops with improvements to multiple trait varieties were farmed of 20% in 2005 of total transgenic crops. After the era of the green revolution, the global use of transgenic crops expanded from 1 million hectares in 1996 to 90 million hectares in 2005 which rose to 191.7 million hectares by 2018. Genetic modification was commercialized to four major crops: cotton, soybean, maize and oilseed rape.

Global GM cotton cultivation was 37 million with 68% among which 25.1 million hectares covered GM products. With regard to maize, the total cultivation area used is 184 million hectares, where GM product covers 55.2 million hectares, or 30% of GM maize. Globally, soybean farming covers 111 million hectares, among which 82%, or around 90.7 million hectares, is GM soybean cultivation. Of 36 million hectare areas of canola grown, 9 million, or 25% of canola, is of the GM category. France along with USA has begun field trials of GM herbicide resistant tobacco plants. With 90% of the adaption rate of GM crops presently, the USA is the leading producer of biotech crops in a 73.1 million hectares area which covers 40% of the global agricultural area. Apart from this, USA is in the advance of GM insect resistant and herbicide resistant papaya, drought tolerant maize, less acrylamide-innate potato, and less lignin-containing alfalfa [12]. Canada increased GM crop cultivation from 1.7 million hectares to 11.6 million hectares in 2014. About 85% of total the GM canola is covered by Canada. Apart from these, other major GM crops grown are soybean, sugar beet and maize. Brazil is the second largest grower of GM after USA with the adoption of GM soybean of 93.2% plantation in a 29.1 million hectare area, 12.5 million hectares plantation for GM maize and 65.1% adoption rate for GM cotton, insect resistant sugarcane for sugar and ethanol production, and rice, citrus and wheats are also in the pipeline [12]. In Europe 86 GM crops are approved by The European Union, among which GM maize, cotton, soybean, sugar beet and oil seed rape are authorised to be registered for food and feed [12]. In Africa, Sudan, South Africa and Burkina Faso have approved commercialization of GM maize, sugarcane, cassava, potato and cotton for improved traits such as drought tolerance, herbicide tolerance, insect resistance and starch enhancement [12]. The Asia-pacific region including India, China, Australia, Indonesia, Philippines, Japan, Thailand and Bangladesh have widely accepted GM products. Presently, India is the country with the most cotton hectares globally, with 46% biotech cotton planted. Around 96% of the total area is cultivated with GM cotton in India, which has increased 230-fold in commercial production of cotton. India's market share for global cotton production doubled from 12% to 25% from 2002 to 2014. The Genetic Engineering Appraisal Committee of India has permitted field trials and then commercialization of biotech brinjal, chick pea, mustard and nitrogen use efficiency rice [13]. Such engineered crops resulted in advantages of enhanced production and quality, and disease resistance varieties by developing antimicrobial compounds and resistance genes, intensifying abiotic stress tolerance, plants with anti-herbivory effect and replacement of chemical fertilizers and herbicides, and decreased input costs to farmers, independent of weather conditions.

Despite the world widely moving towards commercialization of GM crops there are knowledge drawbacks that agriculture is facing. One of the most challenging is development of antibiotic resistance in pathogens as a result of higher mutation rates in microorganisms as pathogens evolving faster. There the challenges of multi- drug resistant super-pathogens as the resistance gene may horizontally transfer to pathogenic microorganisms. The introduction of viral promotors and selectable markers from reproductively incompatible species to economically important crops as a pest management strategy was not found to be as effective because of sudden and unfamiliar changes with altered plant characteristics and its association with non-target organisms [14]. In other ways biodiversity and ecosystems are also challenged because of biotech proteins which are expressed in an active form from different parts of the plant and at different growth stages. This results in exposure of different toxins to different untargeted microflora and ultimately ecosystems and ecology in the long term. Food chains are also closely dependent on weeds, thus limitless control over weeds may disturb predators and parasitoids adversely. Extra-capabilities of plants to survive in adverse conditions may also convert crops to weeds due to limitless growth. Genetically altered crops may also produce toxic compounds harmful to the human body. Also, GM traits may produce misfolded proteins which may become disease causing and carcinogenic to humans. The alterations may create allergens to humans [15]. Including humans in the environmental balance cannot be compromised with such gaps in biotechnological innovation. The sixth most revolutionary technology in the modern era is overtaking all breakthrough applications in multiple disciplines. Agriculture being the most important covered sector for economic needs as well as food is not exceptional in benefitting from nanotechnological applications. Recently, agriculture has been emerging with nanotechnology as 'Nano Agro Technology' for future hope and success.

6.1.2 Nanotechnological interventions in agriculture

Human life is impacted with novel frontiers in industrial applications with the development of nanotechnology. The technology has profoundly covered multiple applications in the fields of medicine, cosmetics, textiles, engineering, computing, information technology food science, textiles, energy sectors and agriculture. The market value of the nanotechnology industry was calculated to touch US$ 75.8 billion by 2020 because of worldwide popularity and expansion of commercialization [16]. Strong applications in wide industrial sectors is the contribution of nanotechnology-based research, which has remarkably impacted global growth. The fact is, nanotechnology has tremendous potential to benefit society by transforming the agriculture sector. The innovative technology has impacted on the agri business sector with annual growth rate of 25%, which is estimated to grow from 1.08 billion US$ to 3.4 trillion US$ by 2020 [17]. Among all preceding revolutions in the world after the green revolution this sixth nano technological revolution has immensely influenced the agriculture field. From chemical usage under the green revolution to usage of bio-based fertilizers and pesticides created issues on environmental impacts as well on field stability and shelf life of introducing materials. Currently, nanoparticles based formulations are markedly showing success over these classical bioformulations [18]. Thus, nanoformulations in modern agriculture are beginning to overcome challenges such as production and productivity, climate change, crop sustainability and crop protection and security. Metal and metal oxide based nanoparticles as well as carbon based nanodimension materials are being widely commercialized in the agricultural field for various applications [19]. Nanomaterials are being explored as nanofertilizers, nanopesticides, nanoherbicides, nano-based delivery carriers for controlled release of agrochemicals; with nanoparticles increasing stress tolerance in plants and much more. The potentiality of the ideas has improved productivity with low input cost and energy [20, 21].

Nanotechnology based devices are also widely accepted as sensing devices, for plant breeding and genetic engineering inputs, crop productivity and disease management. Several nanotechnology based studies are focusing on pesticide encapsulation and its sustained release for plant protection, crop improvement based on nanoparticles based genetic material delivery, rain-fed crops in which there is carbon nanotube-based seed germination, for enhanced nutritional availability and productivity applications of nanofertilizers, nanoherbicides for weed management and nanopesticides for plant disease management, nanosensors for soil monitoring and microflora management. On the other side, probable accumulation of metal and metal oxide based nanomaterials and increase of co-contaminants are a challenge the world has to face [22]. Nanoparticles such as silver, zinc, aluminum, cerium oxide, zinc oxide, titanium oxide and titanium dioxide show antimicrobial activity against a wide range of microorganisms infecting various crops. Many applications of nanoparticles can be attained with their unique properties. Besides a number of positive effects, adverse effects such as phytotoxicity and reduced plant growth can also be major problems in agriculture. Manufactured non- biocompatible nanoparticles show adverse effects such as impact on plant growth, impact on physiological and biochemical traits of plants, impact at genetic level and impact on soil microbial community.

6.1.2.1 Impact of non-biocompatible nanoparticles on plants

6.1.2.1.1 Impact on plant growth inhibition

A lot of interest has been generated in studying the potential benefits of nanotechnology in the agriculture sector as it can enhance agricultural productivity with low input of cost and energy. Importantly, nanoparticles have offered enormous potential in the agriculture sector including nanopesticides, nanoherbicides, nanofertilizers, and metal and/or metal oxide nanoparticles based smart delivery systems for controlled release of agrochemicals to different crops [23]. But on the dark side of this, these nanoformulations exhibit damage to agricultural products and soil on bioaccumulation. Several researchers have studied the adverse effects of such metal nanoparticles on the growth and development of various plant types. The studies include the effect of nanoparticles on germination, root growth, and shoot growth of various plants. Over 90% of land plants roots are in mutualistic symbiosis with arbuscular mycorrhizal fungi (AMF). Silver nanoparticles (AgNPS) and iron oxide nanoparticles (FeONPs) show significantly reduced mycorrhizal clover (*Trifolium repens*) biomass along a concentration gradient of each. FeO NPs at concentration of 3.2 mg kg^{-1} show reduction by 34% and AgNPs at 0.01 mg kg^{-1} concentration inhibit mycorrhizal clover growth due to agglomeration of nanoparticles and its elevated stress [24]. Also, reports indicate that plant growth is greatly inhibited by nanoparticles that adversely affect plant roots. The negative effects include, alteration of root apical meristem, varied metaxylem frequency of root cells, decreased rate of root growth and also inhibition of root development. Citrate-coated silver nanoparticles (citrate-AgNPs) and zinc oxide nanoparticles (ZnO NPs) affect the roots of two agriculturally important plants; maize (*Zea mays* L.) and cabbage (*Brassica oleracea* var. *capitata* L.), while copper oxide nanoparticles (CuO NPs) affect radish (*Raphanus sativus*), perennial ryegrass (*Lolium perenne*), and annual ryegrass (*Lolium rigidum*) due to significant accumulation of oxidatively modified mutagenic DNA lesions in root cells [25, 26]. Reduction of zucchini (*Cucurbita pepo*) plant biomass because of toxic effects of various nanoparticles–cell contacts, uptake and subsequent peroxidation of cellular membrane have been reported. Plant biomass reduction by 77%, 60% and 75% had been observed due to adverse effects of copper nanoparticles (CuNPs), multiwalled carbon nanotubes (MWCNTs) and AgNPs, respectively [27]. Another report on the effect of metal oxide nanoparticles viz. aluminum oxide (Al_2O_3) nanoparticles, silicon dioxide (SiO_2) nanoparticles, magnetite (Fe_3O_4) nanoparticles and zinc oxide (ZnO) nanoparticles on the development of mouse-ear cress (*Arabidopsis thaliana*) show noticeable retardation in plant growth and development by inhibiting seed germination, number of leaves and root elongation mechanism. Among these, ZnO nanoparticles show complete retardation of plant growth at 400 mg L^{-1} concentration [28]. Also, ZnO nanoparticles show decreased root growth in soybean (*Glycine max*) by mutating plant DNA at its seedling stage which may lead to overall hormonal imbalance of growth hormones [29]. Bioavailability studies of water insoluble CuNPs to important crops such as mung bean (*Phaseolus radiatus*) and wheat (*Triticum aestivum*) show concentration dependent bioaccumulation within cells which results in phytotoxicity. This affects seedlings of both plants along with root growth and shoot growth in wheat and mung bean, respectively [28].

6.1.2.1.2 Impact on physiological and biochemical traits of plants

Accumulation of nanoparticles within plant cells leads to changes in biochemical compositions and thus physical traits such as cell morphology and integrity. Various impacts have been reported by researchers to date. Among them, major biochemical alterations include increased level of reactive oxygen species (ROS), which are further scavenged by important biomolecules of the cell and lead to cell destruction and toxic impacts. Reports emphasize single walled carbon nanotubes (SWCNTs) induced programmed cell death in Arabidopsis and rice (*Oryza sativa*). Apoptosis is because of generation and accumulation of ROS at high concentration, chromatin condensation, nuclear DNA damage and ultra-structural changes in organelles at cellular level [30]. Also, ROS productions in red spinach on exposure to MWCNTs results in high response type necrotic lesions of leaf cells/tissue. Morphology of root and leaf changes, and resultant increased sensitivity to pathogens leads to cell necrosis and death by protein, lipid and DNA oxidation [31]. Decreased transpiration rate in *Cucurbita pepo* has been observed because of induced stomata closing or reduced stomata opening on exposure to AgNPs and CuNPs [32]. Another major effect was observed when Ag colloidal NPs affect *Allium cepa* cell growth by penetrating the plant system and impairing various stages of cell division; because of generation of chromatin bridge, stickiness of chromosomes, disturbed metaphase due to disturbance in the spindle apparatus, as well as multiple chromosomal breaks and disintegration of cells [33]. On the other hand, ZnO NPs show toxicity in green peas (*Pisum sativum* L.) by increasing hydrogen peroxide content in leaves and by decreasing chlorophyll content, which leads to reduction in shoots and leaves of the plant [34].

6.1.2.1.3 Impact on plants at genetic level

Reports on soybean (Glycine max) show the biotransformation of ZnO NPs and cerium dioxide nanoparticles (CeO_2 NPs) in plant seedlings, which further mutates host cell DNA and causes genotoxic effect on the plant cells [29]. Studies on maize (*Zea mays* L.) and rice (*Oryza sativa* L.) show significant phytotoxic effect due to ZnO NPs and CuO NPs at molecular level. Damage to the molecular mechanism is seen as a result of oxidation of DNA nitrogenous bases to 7,8-dihydro-8-oxoguanine, 2,6-diamino-4-hydroxy-5-fromamidopyrimidine, and 4,6-diamino-5-formamidopyrimidine. Such mutagenic lesions of genes lead to genetic, metabolic and phenotypic level changes in plants [35, 26]. Titanium dioxide nanoparticles (TiO_2 NPs) show genotoxicity to *Nicotiana tobacum* and to *Allium cepa* by resulting in chromosomal aberration and formation of bridges in structures of anaphase and telophase chromatids in plants. These lead to the reduction in root growth. Also, DNA damage of the plant cells is due to lipid peroxidation in the plant because of bioaccumulation and toxicity of nanoparticles [36].

6.1.2.1.4 Impact on soil microbial community of plant

In agricultural soil, TiO_2 nanoparticles and ZnO nanoparticles negatively affect wheat growth and soil enzyme activities. Both nanoparticles reduce biomass of wheat among which, TiO_2 nanoparticles are retained in the soil for a longer period of time and adhere to the cell wall of wheat cells while ZnO nanoparticles dissolve in the soil leading to enhanced uptake of toxic Zn by wheat cells. These NPs also show significant inhibition of soil enzymes activities, such as protease, catalase, peroxidase which are bioindicators of soil quality and health [37]. Ag nanoparticles and CuO nanoparticles shows major impact of increasing toxicity towards agriculturally important soil microbes such as *Pseudomonas chlororaphis*, *Rhizobiales*, *Sphingobacteriaceae* [38, 39]. Thus, the futuristic approach is to develop nanocarriers or nanoformulations as biocompatible and biodegradable which is promising, environmentally friendly and long lasting to balance the ecosystem.

6.1.3 What are bionanoformulations

Bio-origin materials are non-harmful, biocompatible as well as nature acceptable. The approach is developed to synthesize nanoparticles from biomaterials, with the main vision and concern of biodegradability and environmental friendliness. The bionanoformulation approach is development of bio-origin nanoparticles and use of them as a carrier for agricultural material delivery for crop improvement, protection as well productivity increment.

6.1.4 Significance of bionanoformulations

In the current scenario, the agricultural sector is excessively using fertilizer to increase production, which not only affects the soil quality, but also affects the environment. Some of the most significant uses of bionanoformulation are listed and summarized in figure 6.1.

6.2 Different types/materials used for bionanoformulations synthesis

In agriculture, the emerging bionanoformulations based controlled release technology is a versatile tool to reduce severe damage to the environment and humans. Controlled release is a target specific permeation regulated transfer from a reservoir of an active ingredient at predetermined concentration level for a specified period of time. Conventional applications require high initial dose with less effective concentration of an active ingredient. Controlled release formulations may maintain low effective concentration for longer time in a controllable manner. Macromolecular biodegradable polymeric matrices such as guargum, pectin, alginate, carboxymethyl cellulose-based controlled release have been applied in agriculture where crop protection is mandatory for various outcomes such as environment protection and integrity maintenance, protection of soil microflora etc. Some advantages are ease in handling, constant level delivery of an active ingredient for desired time, decrease in environmental pollution, least phytotoxicity, reduction in evaporation losses [40]. Main objectives are enhanced safety and reliability, more effective treatment, prolonged efficacy and reduced side effects. An ideal bionanoformulation would maintain an adequate level of active ingredient for pest control, leaving minimum crop and environmental residue. Nanotechnology exploits the observation that materials at a scale of few nanometers varying from 10 to 100 nm have very different properties than at micro or macro levels. The active ingredient is dissolved, attached,

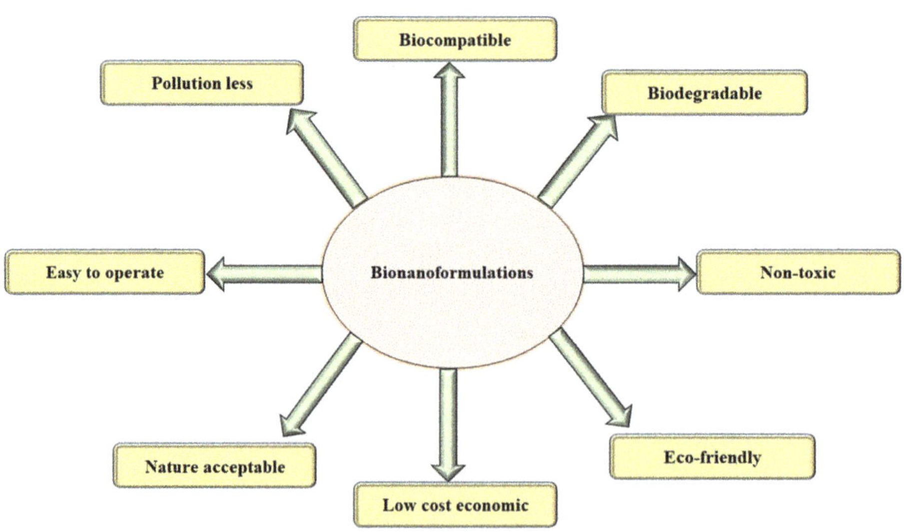

Figure 6.1. Significance of bionanoformulations.

entrapped or encapsulated to nanoparticles for controlled release. The most commonly used biodegradable natural polymers are polysaccharides, proteins and essential oils. Polysaccharides include cellulose, dextran, agarose, carrageenan, alginates, starch, chitosan and proteins such as gelatin, zein and albumin.

6.2.1 Cellulose nanoparticles and their derivatives

The linear biopolymer cellulose consists of alternating amorphous and crystalline domains of D-anhydroglucopyranose units linked together by $\beta 1 \rightarrow 4$-glycosidic bonds. It is the most predominant basic structural component of all plant fibers and a most important organic and renewable biopolymer produced worldwide. It is biodegradable and a nontoxic biopolymer. It is a value added versatile starting material for various cellulose based fibers and films at domestic and industrial levels. Cellulose derivatives are generally formed by interfering with its hydrogen bonding in orderly crystal units, which results in their increased property of water solubility. Methylcellulose, ethylcellulose, cellulose acetate, carboxymethylcellulose and native cellulose have been used in various native and domestic applications such as lubricants, thickeners, emulsifiers, optical and biomedical devices, resins and filters, glue and binders, pharmaceutical raw materials. Cellulose is important because of its favourable properties of low cost, toughness, transparency, natural feel, and softness [41]. Cellulose, being abundantly available and renewable, can be isolated from various natural resources viz. wood, cotton, weed straws, coconut husk, sesame husk, sugarcane bagasse and banana rachis. Cellulose nanoparticles for drug delivery application have been studied and reported very effectively using methylene blue (MB) as a model hydrophilic drug. Nanoparticles of cellulose were fabricated by nanoprecipitation method from cotton fibres. Cellulose nanoparticles between size ranges from 70 to 365 nm were investigated, and the highest loading efficiency of MB at 89% had been achieved. Drug release study kinetics showed its stability and drug releasing efficiency at physiological conditions of pH around 7.4 and 30 °C–37 °C with complete sustained release maximally from 56 to 96 h [42].

6.2.2 Dextran nanoparticles and their derivatives

Polysaccharide dextran synthesized by bacteria has wide applications in effective delivery systems. Dextran is a biocompatible and biodegradable carbohydrate. Dextran has been extensively used in the medical field as a drug delivery carrier [43]. Various dextran formulations as drug delivery carriers have been widely reported. Reports show that water insoluble boronate dextran nanoparticles in water can be used as drug delivery carrier for hydrophobic anticancer drug doxorubicin. These nanoparticles are acid responsive polymer (B-Dex). Under a mild acidic microenvironment of pH around 5.0 at 37 °C, nanocarriers show three-fold increase in their degradation rate and four-fold increase in drug release compared to pH 7.0 at 37 °C. B-Dex nanoparticles of size 273 nm are stable at pH 7.4 and have 25.6% drug encapsulation efficiency with 90% drug release efficiency from 20 to 96 h [44]. Also, linear amphiphilic Dex-b-PLA nanoparticles composed of poly (D, L-lactide) and dextran, formulated via self-assembly of block copolymers have also been reported for drug delivery. Precisely tuned sized nanoparticles from 15 to 70 nm show 21% of doxorubicin drug payload with 100% drug release efficiency at 37 °C temperature and 7.4 pH for 96 h [45].

6.2.3 Chitosan and carrageenan nanoparticles

Chitosan is a natural origin polymer composed of repeating units of N-acetylglucosamine and D-glucosamine and is obtained from deacetylation of chitin, the main component of a crustacean's exoskeleton. Carrageenan is a polysaccharide extracted from red seaweed composed of galactose and anhydro galactose units linked by glycosidic bonds. Depending on the algal source and methods of carrageenan extraction, three main types of carrageenan are obtained viz. kappa, iota and lambda, which differ structurally in the degree of substitution of sulfate groups. Depending on the ionic nature, all three types of carrageenan have a tendency of reversible gel formation. Studies based on application of chitosan/carrageenan nanoparticles for ovalbumin as a model protein delivery have been reported. Reports show loading capacity of ovalbumin ranging from 4% to 17%, with excellent controlled releasing property up to three weeks *in vitro* [46]. Chitosan exhibits properties of low toxicity, antibacterial property, biodegradability, and biocompatibility. Chitosan nanoparticles of 44 nm size have been reported as a drug delivery system for perindopril erbumine. Nanoparticles showed 94.1% loading efficiency of the drug. The drug can be incorporated into the carrier system by chemical conjugation or by passive absorption. The sustained release of the drug was obtained up to 99.8% over a period of 2 h at physiological pH and temperature [47].

6.2.4 Starch nanoparticles

Starch is an abundantly available source of polysaccharides in nature. Its non-toxicity, biocompatibility, renewability, biodegradability and low cost make starch a promising precursor for nanoparticle formation and use at industrial and domestic level. Starch nanoparticles with controllable size can be synthesized via precipitation in water-in-oil micro emulsion approach with advantages of ultra-low interfacial tension, large interfacial area and thermodynamic stability which produce stable monodispersed nanoparticles. The mean particle size of starch nanoparticles depends on synthesis parameters such as ratio of oil/cosurfactants, stirring rates, oil phases, co-surfactants, and ratio of water/oil. By using this method, nanoparticle size varying from 83 to 109 nm can be achieved [48]. Starch nanoparticles produced by method of precipitation in water-in-oil microemulsion for drug delivery application have been reported. The preparation technique offers advantages as it does not require extreme pressure and no special and expensive instruments. Also, preparation method affords monodispersed particle size distribution of nanoparticles. Starch nanoparticles produced within the size range between 50 and 80 nm were shown to have a curcumin loading efficiency of 78%. Effective controlled and sustained release from stable drug loaded nanoparticles was observed, where the nanocarrier exhibited 93% of curcumin releasing efficiency at physiological pH 7.4 over 10 days period [48].

6.2.5 Gelatin nanoparticles

Gelatin is a natural macromolecule obtained from collagen. It is a polyampholyte with both cationic and anionic groups along with hydrophobic groups. The biopolymer contains a polypeptide with—(Gly-X-Pro) n where X is amino acid mostly lysine, arginine, methionine and valine. The gelatin molecule contains approximately 13% of positively charged (lysine and arginine), 12% negatively charged (glutamic acid and aspartic acid) while 11% of the chain is hydrophobic in nature (leucine, isoleucine, methionine and valine). Glycine, proline and hydroxyproline covers the rest of the chain. The positive:negative:hydrophobic segments are at approximate ration of 1:1:1. [49] Also, gelatin is a biodegradable, biocompatible, non-toxic and low cost molecule with biological origin. It is easy to cross link and chemically modified gelatin molecules gives it an immense potential to be used as drug delivery system. Gelatin nanoparticles can be prepared by various methods such as, water-in-oil emulsion, nanoencapsulation, desolvation, coacervation-phase separation etc. Research has been reported of using gelatin nanoparticles as carrier system for bovine serum albumin. Gelatin nanoparticles prepared with mean diameter of 200–300 nm showed 87.4% loading efficiency of bovine serum albumin. An 80% release efficiency of bovine serum albumin was observed at 37 °C over a period of 7 days [50].

These, various biocompatible nanocarriers show very high drug loading and drug releasing efficiency in the mammalian system. Thus, they can also be considered for use in plant systems.

6.3 Different approaches and methods for bionanoformulations synthesis

Nanotechnology has emerged as a potential field of science. Nanoparticles are the key factor of the technology which are in size ranging from 1 to 100 nm. Mainly nanoparticles are categorised under inorganic nanoparticles, organic nanoparticles, carbon-based nanoparticles, and ceramic nanomaterials. The metal based and metal oxide based ones are categorised under inorganic nanomaterials while carbon nanotubes, fullerene, graphene, carbon nanofiber etc are carbon-based nanomaterials. Each type of nanomaterial is classified under zero-, one-, two- and three-dimensional nanoparticles. There are mainly two approaches of nanoparticles synthesis. Nanoparticle synthesis approaches are listed in figure 6.2.

6.3.1 Top-down approach

This is a method where a larger molecule or bulk material is decomposed into smaller particles followed by nanoparticles synthesis. Mechanical/ball milling, thermal decomposition, laser ablation are examples of the destructive top-down approach of synthesis.

6.3.2 Bottom-up approach

Bottom up is a constructive approach of synthesis. In this method nanoparticles are synthesized from monomer units or simpler substances. Chemical vapour

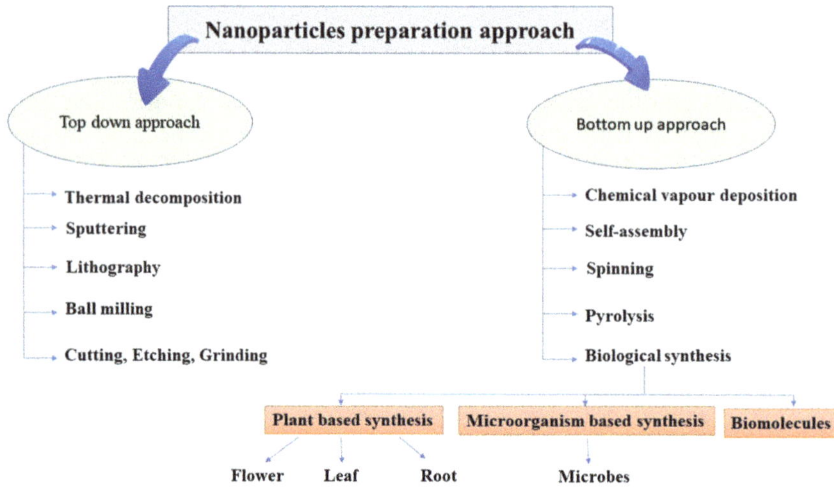

Figure 6.2. Approaches of nanoparticles preparation.

deposition, sol-gel, biological synthesis from raw materials are examples of bottom-up approach.

6.3.3 Methods of bionanoformulations synthesis

Synthesis of polymeric nanoparticles can be achieved by dispersing polymers using different methods. The methods such as nanoprecipitation, solvent evaporation, dialysis, salting-out, super critical fluid technology are widely accepted and performed among all. The fundamental property of these methods involves two components: nanoparticles containing organic phase and stabilizer containing water phase [51, 52].

6.3.3.1 Solvent evaporation

Most biodegradable polymeric nanoparticles are synthesized by solvent evaporation, the most common and widely used method for polymeric nanocarrier synthesis for delivery applications. The method involves dissolution of polymer and drug in organic solvent and aqueous phase containing surfactant being used to formulate emulsion. There are reports showing usage of dichloromethane, ethyl acetate and chloroform used for dissolution of the hydrophobic drug. Two main strategies on which the solvent evaporation works is single emulsion (oil-in-water) another is double emulsion (water-in-oil-in-water). For the synthesis method high speed homogenizer or ultra-sonication is used to prepare emulsion. The organic solvent phase is evaporated after the synthesis protocol and particles are collected with ultracentrifugation separation technique. Water wash can be applied to remove additional surfactant. Lyophilisation can be used to obtain particles in powdered form. As the method requires external energy input within the process, the particle size and morphology vary depending on such external factors as sonication or homogenizer speed and time of reaction and temperature inputs during synthesis.

6.3.3.2 Salting out

Salting out effect is being used in the method for separation of water miscible organic solvents in the synthesis process. In the method, polymer is dissolved in water miscible organic solvents such as ethanol, acetone, tetrahydrofuran and this polymer dissolved organic phase is added to aqueous phase containing water along with stabilizer and salting-out agent. Electrolytes such as magnesium chloride, calcium chloride or sucrose as non-electrolyte are used as salting out agents. This is followed by addition of a sufficient amount of water and resulting nanoparticles are separated via filtration method. The method is used for stress sensitive proteins and also encapsulation. In this method, as heat is not a factorial part of synthesis, heat sensitive polymeric nanoparticles are developed. That frequent washing is crucial in synthesis by this method is considered a disadvantage. Mainly lipophilic drugs are used for encapsulation via this procedure.

6.3.3.3 Nanoprecipitation

Nanoprecipitation method is quite simple and a fast reproducible approach to prepare nanospheres and nanocapsules. This is a method of solvent displacement which is based on the principle of substitutional polymer collapse from water miscible splitting solvent in lipophilic solution. Natural, synthetic or semisynthetic polymers, organic solvents of the polymer (non-solvent) and water are the three main basic components of the system. Here mostly water miscible organic solvents are used which can be easily evaporated with heat. Such a non-solvent system is used for precipitation of polymer.

6.3.3.4 Dialysis

This is an effective method for synthesis of small polymeric nanoparticles. An organic solvent is used to solubilize the polymer and fills the dialysis bag with coordinating molecular weight. During the process the solvent loses its solubility by displacement, which progressively synthesizes aggregation of homogenous suspension of nanoparticles that can be collected.

6.3.3.5 Supercritical fluid technology

Supercritical fluids are ones that do not change in phase despite the change of pressure. For supercritical technology mostly CO_2 is used because it is non-flammable, nontoxic, inexpensive and compatible with the critical temperature ($T_c = 31.1$ °C, Pc = 73.8 bar). This technique is effective in particle synthesis with more advantages over traditional methods as precipitated products do not contain solvents. Rapid expansion of supercritical solvent is a type of method in which the substance dissolved in organic solvent is released to supercritical fluid where the organic phase rapidly dissolves in supercritical fluid and the nanosize particles are separated out. Another type is supercritical anti-solvent method in which the polymer and active component are both directly dissolved into supercritical solvent directly at high pressure and particles in nanosize are synthesized.

6.3.3.6 Mini emulsion

This is a typical mixture of water, monomer, initiator, co-stabilizer and surfactant which is used for low molecular co-stabilizer compounds. These are critically balanced and require high shear stress of homogenization for synthesis.

6.4 Examples and current status of bionanoformulations for crop protection and growth

Different types of non-polymeric nanoparticles such as metal, metal oxide, carbon based materials are used in agriculture as carrier systems for delivering fertilizers, insecticides, herbicides, weed controllers, and plant growth regulators which cause multiple damage to the environment and show multiple disadvantages [23]. Thus, as a solution and from the advantages point of view polymeric nanoparticles as delivery carrier applications are emerging and successful across the field in all aspects. Moreover, their fragility, simplicity and softness of material make them biocompatible, harmless and biodegradable. These carrier systems encapsulate agrochemicals efficiently and control their release in the environment [53]. This green nanotechnology has exhibited a successful role in the environment as waste-water treatment, desalination treatment, combat drug-resistance pathogens, sustainable chemical synthesis route together with agriculture, alternative green energy generation and food production. In agriculture it provides advantages of controlled and sustained release of encapsulated chemicals, reduction of exposure of chemicals to the environment, reduced pollution to soil and water bodies, and it controls toxicity to normal flora of soil. As a result, this approach escapes hazardous chemicals and organic solvents and uses high energy inputs. The synthesis and availability produces economic, biodegradable, biocompatible and non-toxic nanoparticles. Chitosan and pectin are widely successful nanocarrier materials [23]. Chitosan is the N-deacetylated form of chitin, which is a natural polysaccharide from exoskeletons of crustaceans, fungi and insects. It is among the most popular polymers on account of being biocompatible, biodegradable, showing properties of antimicrobial as well insecticidal activities, muco-adhesive, with a high active compound loading capacity and high permeability. It shows controlled delivery in agriculture with the ability to chelate inorganic compounds, make plant stress tolerance against biotic as well abiotic stress, can penetrate the hydrophobic surface of plants, and its being antimicrobial reduces plant pathogen infections [53, 54]. Chitosan-tripolyphosphate nanomaterial is reported to deliver paraquat herbicide with reduced genotoxicity that shows post-emergent control over herbs of *Brassicai* sp. and *Zeya mays* and reduced cytotoxicity in mammalian cells [55]. Caboxymethyl chitosan nanoparticles are also successfully applicable to deliver diuron herbicide, which results in increments in pre-emergent herbicidal activity against *Echinochloa crusgalli* at doses of 3 kg per hectare [56]. Chitosan nanoparticles showed fungicidal action against *Fusarium graminearum* in doses in the range 100–5000 ppm [57]. Reports [58] show hexaconazole fungicide encapsulation in chitosan nanocarrier with lower toxicity to non-target cells and increased antifungal activity against *Rhizoctonia solani* at doze range to 0.1–1 ppm. Insecticide methomyl was also encapsulated in chitosan nanocapsules, showing increased insecticidal activity against armyworms at 50 to 100 ppm concentration [59]. Neem essential oil was encapsulated into chitosan nanoparticles which show insecticidal activity against *Helicoverpa armigera* with decreased larval activity, feeding and pupal activity at 0.1%–0.3% working concentration [60]. Plant growth regulator S-Nitroso-mercaptosuccinic acid was encapsulated in chitosan-tripolyphosphate nanoparticles and tested for improved slat stress tolerance in Zea mays at 50 to 100 uM concentration [61]. Gibberellic acid was encapsulated in chitosan based composites as alginate/chitosan and chitosan/tripolyphosphate, which show increased leaf area as well as photosynthesis pigments contents of *Phaseolus vulgaris* with increased plant growth [55] gamma-Polyglutamic acid/chitosan nanoparticles encapsulating gibberellic acid show enhancement of germination in leaf area and root development of *Phaseolus vulgaris* [62]. As with chitin, pectin is also a widely acceptable and successful implementation of polysaccharide in the field of food industries, biomedicine and agriculture. Being a constituent of the plant cell wall as 1,4-linked 2-D-galactosyluronic acid residues, pectin is most suitable for agriculture. Synthetic polymers such as poly (lactide-co-glycolides) (PLGA), poly(epsilon-caprolactone) (PCL), polyglutamic acid or polyglycolides (PGA), poly (vinyl alcohol) (PVA), poly(citric acid) (PCA), and polyethylene glycol (PEG) are widely used [63, 64]. PCL as nanocapsules is reported to encapsulate herbicides such as ametryn and atrazine, which showed reduction in genotoxicity in *Allium cepa* chromosome aberration at 100 mg ml^{-1} concentration [55] PCL as nanocapsule and nanosphere was reported to encapsulate atrazine herbicide with reduced genotoxicity in *Allium cepa* and pre-emergent control of *Brassica* sp [65]. PCL and PLGA nanoparticles are reported to encapsulate metolachlor herbicide which shoe internalization in *Oryza sativa* and *Digitaria sanguinalis* at concentration 10–400 mg l^{-1} [64]. Fungicide azomethine [66] was reported with encapsulation in PCL nanoparticles showing effective antifungal activity against *Sclerotium rolfsii, Rhizoctonia bataticola* and *Rhizoctonia solani* at concentration range of 3.9–250 ug ml^{-1}. Garlic essential oil as insecticide was reported to be encapsulated in PEG nanoparticles, which at a concentration of 2000–8000 mg kg^{-1} show insecticidal activity against *Tribolium castaneum* [63]. PCL nanocapsules were reported [67] as encapsulating carrier for neem essential oil as insecticidal agent against *Plutella xylostella* at a concentration range of 4000–5000 mg kg^{-1}. PEG nanoparticles [68] were reported for acephate insecticide carrier against increased insecticidal activity against *Sitophilus oryzae* at 180–300 ppm concentration. Imidacloprid insecticide was reported [69] encapsulating inside PEG and PCA nanoparticles, which acted with increased mortality rate of *Glyphodes pyloalis* at 25–300 ppm. Zein nanoparticles are reported for biodistribution in soybean as well sugarcane [70]. Sodium alginate was reported to encapsulate gibberellic acid plant hormone to increase efficiency of the active ingredient [71]. Such bionanoformulations have proved themselves for promoting a sustained site-specific release of active ingredient with improved activity. Also, bionanoformulations reduce excess run-off, reduce toxic effect, increase thermal as well as phytochemical stability or formulation and ingredient, thus they are expected to reduce required dosage to gain desired activity. As a result, these

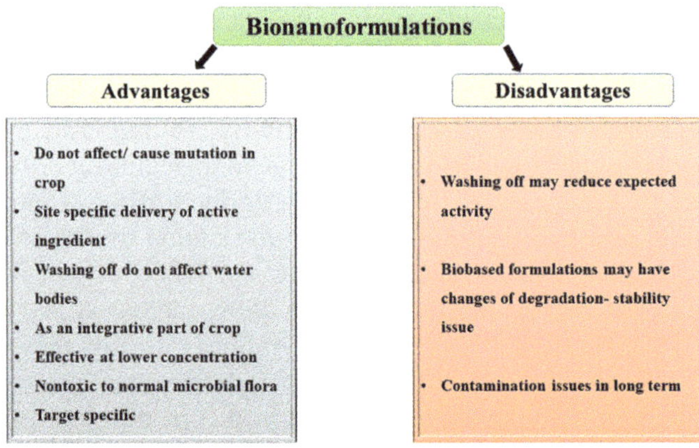

Figure 6.3. Listed advantages and disadvantages of bionanoformulation.

bionanoformulations are expected to maximize agriculture output, crop yield, and production with minimizing inputs as herbicides, fertilizers and pesticides. Reduction of nanotoxicity to the environment is achieved.

6.5 Advantages and disadvantages of bionanoformulations

Some of the utmost advantages and disadvantages of bionanoformulations as reported by several researchers are enlisted and summarized in figure 6.3.

6.6 Gaps in agriculture crop protection

Though agriculture approaches with bionanoformulations are successfully emerging with the modern era that takes various factors into account, there are still a few gaps to fill. For example, target specific delivery, slow release nanocarriers may further reduce the working concentration of bionanoformulations. Various approaches that explore research towards developing hydrophobic nanocarriers that reduce the amount of chemicals that are washed off during rain or watering or irrigation should be promoted. Development of bio-based formulation can concentrate on application of biodegradable nanocarriers in agriculture, which will reduce the problem of bio-accumulation and pose no threat to normal flora and fauna.

6.7 Future prospects of bionanoformulations in agriculture

Bionanoformulations seem promising in providing viable alternatives in future to turn mankind, research as well science towards environment sensitivity. The future approach may look forward to replacement of encapsulated materials to complete bioactive component formulations. The research must be promoted for commercialization of bionanoformulations and its widespread awareness and usage in the field of agriculture. Boosting agricultural production by intervention of nanotechnology is essential to meet peoples' needs, but its impact on the environment needs to be assessed holistically, therefore, novel approaches need to be considered. Promising application of nanotechnology in the form of nanofertilizers, nanobiosensors, nanopesticides or as environmental remediation agents has already shown its promising results in the agriculture sector.

6.8 Conclusion

Bionanoformulation is one of the most emerging areas in bionanotechnology. Selective successful experiments are reported for bionanoformulations applications like *in vitro* germination, field-based experimentation for increased crop productivity and their commercial usage. This is a very novel conceptual area for environmentally friendly and sustainable agriculture. The approach is different in terms of eco-friendliness from conventional nanotechnology. Bionanoformulation is a highly innovative area which requires concerted efforts from scientists across the globe.

Acknowledgment

The authors would like to thank CUG library for providing support with the various resources and plagiarism software.

References

[1] Oluwaseun A C, Phazang P and Sarin N B 2018 *Biosensors: A Fast-Growing Technology for Pathogen Detection in Agriculture and Food Sector Detection in Agriculture and Food Sector* (Rijeka: InTech)

[2] UN Department of Economic and Social affairs 2013 *World Population Prospects: The 2012 Revision* https://www.un.org/en/development/desa/publications/world-population-prospects-the-2012-revision.html

[3] Pimentel D and Pimentel M 2006 Global environmental resources versus world population growth *Ecol. Econ.* **59** 195–8

[4] Delaney B 2015 Safety assessment of foods from genetically modified crops in countries with developing economies *Food Chem. Toxicol.* **86** 132–43

[5] Klümper W and Qaim M 2014 A meta-analysis of the impacts of genetically modified crops *PLoS One.* **9** e111629

[6] Sabesh M, Prakash A H and Bhaskaran G 2014 Shift in Indian cotton scenario due to shift in cotton production technology *Indian Soc. Cotton Improve.* **6** 75–82

[7] Gutierrez A P, Ponti L and Baumgärtner J 2017 A critique on the paper 'Agricultural biotechnology and crop productivity: macro-level evidences on contribution of Bt cotton in India' *Curr. Sci.* **112** 690–3

[8] Pingali P L 2012 Green revolution: impacts, limits, andthe path ahead *Proc. Natl. Acad. Sci.* **109** 12302–8

[9] Conway G R and Barbie E B 1988 After the green revolution. Sustainable and equitable agricultural development *Futures* **20** 651–70

[10] Arora N K 2015 Plant microbes symbiosis: applied facets *Plant Microbes Symbiosis: Applies Facets* (Cham: Springer) pp 1–381

[11] Lobell D B, Bänziger M, Magorokosho C and Vivek B 2011 Nonlinear heat effects on African maize as evidenced by historical yield trials *Nat. Clim. Chang.* **1** 42–5

[12] James C 2014 Global status of commercialized biotech/GM Crops: 2010 global area of biotech crops million hectares (1996–2010) *ISAAA Brief 42-2010: Executive Summary* pp 16–27 https://www.isaaa.org/resources/publications/briefs/42/executivesummary/

[13] James C 2006 Global Status of Commercialized Biotech/GM Crops: 2006 *Applications ISAAA Briefs BRIEF 35* pp 1–107

[14] Antoniou M 2021 Genetically engineered food – panacea or pandora's box *Nutr. Ther. Today* **6** 8–11

[15] Graf L, Hayder H and Mueller U 2014 Endogenous allergens in the regulatory assessment of genetically engineered crops *Food Chem. Toxicol.* **73** 17–20

[16] PR News Wire 2015 Global nanotechnology market outlook 2015–20—industry will grow to reach US$ 75.8 billion. Párr.1. http://prnewswire.com/news-releases/global-nanotechnology-market-outlook-2015-2020---industry-will-grow-to-reach-us-758-billion-507155671.html

[17] Sodano V and Verneau F 2014 Competition policy and food sector in the European Union *J. Int. Food Agribus Mark.* **26** 155–72

[18] Auffan M, Rose J, Bottero J Y, Lowry G V, Jolivet J P and Wiesner M 2009 Towards a definition of inorganic nanoparticles from an environmental, health and safety perspective *Nat. Nanotechnol.* **4** 634–41

[19] Nair R, Varghese S H, Nair B G, Maekawa T, Yoshida Y and Kumar D S 2010 Nanoparticulate material delivery to plants *Plant Sci.* **179** 154–63

[20] Campos E V R, de Oliveira J L and Fraceto L F 2014 Applications of controlled release systems for fungicides, herbicides, acaricides, nutrients, and plant growth hormones: a review *Adv. Sci. Eng. Med.* **6** 373–87

[21] Grillo R, Abhilash P C and Fraceto L F 2016 Nanotechnology applied to bio-encapsulation of pesticides *J. Nanosci. Nanotechnol.* **16** 1231–4

[22] Servin A D and White J C 2016 Nanotechnology in agriculture: next steps for understanding engineered nanoparticle exposure and risk *NanoImpact.* **1** 9–12

[23] Mishra S, Keswani C, Abhilash P C, Fraceto L F and Singh H B 2017 Integrated approach of agri-nanotechnology: challenges and future trends *Front. Plant Sci.* **8** 1–12

[24] Feng Y, Cui X, He S, Dong G, Chen M and Wang J *et al* 2013 The role of metal nanoparticles in influencing arbuscular mycorrhizal fungi effects on plant growth *Environ. Sci. Technol.* **47** 9496–504

[25] Pokhrel L R and Dubey B 2013 Evaluation of developmental responses of two crop plants exposed to silver and zinc oxide nanoparticles *Sci. Total Environ.* **452–3** 321–32

[26] Atha D H, Wang H, Petersen E J, Cleveland D, Holbrook R D and Jaruga P *et al* 2012 Copper oxide nanoparticle mediated DNA damage in terrestrial plant models *Environ. Sci. Technol.* **46** 1819–27

[27] Stampoulis D, Sinha S K and White J C 2009 Assay-dependent phytotoxicity of nanoparticles to plants *Environ. Sci. Technol.* **43** 9473–9

[28] Lee C W, Mahendra S, Zodrow K, Li D, Tsai Y C and Braam J *et al* 2010 Developmental phytotoxicity of metal oxide nanoparticles to *Arabidopsis thaliana Environ. Toxicol. Chem.* **9** 669–75

[29] Lu C, Zhang C, Wen J, Wu G and Tao M 2002 Research of the effect of nanometer materials on germination and growth enhancement of Glycine max and its mechanism *Soybean Sci.* **21** 168–71

[30] Shen C X, Zhang Q F, Li J, Bi F C and Yao N 2010 Induction of programmed cell death in Arabidopsis and rice by single-wall carbon nanotubes *Am. J. Bot.* **97** 1602–9

[31] Begum P and Fugetsu B 2012 Phytotoxicity of multi-walled carbon nanotubes on red spinach (*Amaranthus tricolor* L) and the role of ascorbic acid as an antioxidant *J. Hazard. Mater.* **243** 212–22

[32] Sundukov Y N 2006 First record of the ground beetle *Trechoblemus postilenatus* (Coleoptera, Carabidae) in Primorskii krai *Far. East Entomol.* **65** 16
[33] Kumari M, Mukherjee A and Chandrasekaran N 2009 Genotoxicity of silver nanoparticles in *Allium cepa Sci. Total Environ.* **407** 5243–6
[34] Mukherjee A, Peralta-Videa J R, Bandyopadhyay S, Rico C M, Zhao L and Gardea-Torresdey J L 2014 Physiological effects of nanoparticulate ZnO in green peas (*Pisum sativum* L.) cultivated in soil *Metallomics.* **6** 132–8
[35] Castiglione M R, Giorgetti L, Geri C and Cremonini R 2011 The effects of nano-TiO_2 on seed germination, development and mitosis of root tip cells of *Vicia narbonensis* L. and *Zea mays* L *J. Nanoparticle Res.* **13** 2443–9
[36] Ghosh M, Bandyopadhyay M and Mukherjee A 2010 Genotoxicity of titanium dioxide (TiO_2) nanoparticles at two trophic levels: plant and human lymphocytes *Chemosphere* **81** 1253–62
[37] Du W, Sun Y, Ji R, Zhu J, Wu J and Guo H 2011 TiO_2 and ZnO nanoparticles negatively affect wheat growth and soil enzyme activities in agricultural soil *J. Environ. Monit.* **13** 822–8
[38] Calder A J, Dimkpa C O, McLean J E, Britt D W, Johnson W and Anderson A J 2012 Soil components mitigate the antimicrobial effects of silver nanoparticles towards a beneficial soil bacterium, *Pseudomonas chlororaphis* O6 *Sci. Total Environ.* **429** 215–22
[39] Frenk S, Ben-Moshe T, Dror I, Berkowitz B and Minz D 2013 Effect of metal oxide nanoparticles on microbial community structure and function in two different soil types *PLoS One.* **8** 1–12
[40] Roy A, Singh S K, Bajpai J and Bajpai A K 2014 Controlled pesticide release from biodegradable polymers *Cent. Eur. J. Chem.* **12** 453–69
[41] Azeh Y 2017 Synthesis and characterization of cellulose nanoparticles and its derivatives using a combination of spectro-analytical techniques *Int. J. Nanotechnol. Med. Eng.* **2** 65–94
[42] Chin S F, Jimmy F B and Pang S C 2018 Size controlled fabrication of cellulose nanoparticles for drug delivery applications *J. Drug Deliv. Sci. Technol.* **43** 262–6
[43] Wasiak I, Kulikowska A, Janczewska M, Michalak M, Cymerman I A and Nagalski A *et al* 2016 Dextran nanoparticle synthesis and properties *PLoS One.* **11** 1–17
[44] Li L, Bai Z and Levkin P A 2013 Boronate-dextran: an acid-responsive biodegradable polymer for drug delivery *Biomaterials* **34** 8504–10
[45] Verma M S, Liu S, Chen Y Y, Meerasa A and Gu F X 2012 Size-tunable nanoparticles composed of dextran-b-poly(D,L-lactide) for drug delivery applications *Nano Res.* **5** 49–61
[46] Grenha A, Gomes M E, Rodrigues M, Santo V E, Mano J F and Neves N M *et al* 2010 Development of new chitosan/carrageenan nanoparticles for drug delivery applications *J. Biomed. Mater. Res. – Part* **92** 1265–72
[47] Barkoula N M, Alcock B, Cabrera N O and Peijs T 2008 Flame-retardancy properties of intumescent ammonium poly(phosphate) and mineral filler magnesium hydroxide in combination with graphene *Polym. Polym. Compos.* **16** 101–13
[48] Chin S F, Mohd Yazid S N A and Pang S C 2014 Preparation and characterization of starch nanoparticles for controlled release of curcumin *Int. J* **2014** 340121. Polym. Sci.
[49] Mohanty B, Aswal V K, Kohlbrecher J and Bohidar H B 2005 Synthesis of gelatin nanoparticles via simple coacervation *J. Surf. Sci. Technol.* **21** 149–60
[50] Azimi B, Nourpanah P, Rabiee M and Arbab S 2013 Producing gelatin nanoparticles as delivery system for bovine serum albumin *Iran. Biomed. J.* **18** 34–40

[51] Ijaz I, Gilani E, Nazir A and Bukhari A 2020 Detail review on chemical, physical and green synthesis, classification, characterizations and applications of nanoparticles *Green Chem. Lett. Rev.* **13** 59–81

[52] Katmıs A, Fide S, Karaismailoglu S and Derman S 2018 (Online First) Synthesis and characterization methods of polymeric nanoparticles *Charact. Appl. Nanomater.* **1** 1–9

[53] Kashyap P L, Xiang X and Heiden P 2015 Chitosan nanoparticle based delivery systems for sustainable agriculture *Int. J. Biol. Macromol.* **77** 36–51

[54] Malerba M and Cerana R 2016 Chitosan effects on plant systems *Int. J. Mol. Sci.* **17** 1–15

[55] Grillo R, Pereira A E S, Nishisaka C S, De Lima R, Oehlke K and Greiner R *et al* 2014 Chitosan/tripolyphosphate nanoparticles loaded with paraquat herbicide: an environmentally safer alternative for weed control *J. Hazard. Mater.* **278** 163–71

[56] Yu Z, Sun X, Song H, Wang W, Ye Z and Shi L *et al* 2015 Glutathione-responsive carboxymethyl chitosan nanoparticles for controlled release of herbicides *Mater. Sci. Appl.* **06** 591–604

[57] Kheiri A, Moosawi Jorf S A, Mallihipour A, Saremi H and Nikkhah M 2016 Application of chitosan and chitosan nanoparticles for the control of Fusarium head blight of wheat (*Fusarium graminearum*) *in vitro* and greenhouse *Int. J. Biol. Macromol.* **93** 1261–72

[58] Chauhan N, Dilbaghi N, Gopal M, Kumar R, Kim K H and Kumar S 2017 Development of chitosan nanocapsules for the controlled release of hexaconazole *Int. J. Biol. Macromol.* **97** 616–24

[59] Sun C, Shu K, Wang W, Ye Z, Liu T and Gao Y *et al* 2014 Encapsulation and controlled release of hydrophilic pesticide in shell cross-linked nanocapsules containing aqueous core *Int. J. Pharm.* **463** 108–14

[60] Gabriel Paulraj M, Ignacimuthu S, Gandhi M R, Shajahan A, Ganesan P and Packiam S M *et al* 2017 Comparative studies of tripolyphosphate and glutaraldehyde cross-linked chitosan-botanical pesticide nanoparticles and their agricultural applications *Int. J. Biol. Macromol.* **104** 1813–9

[61] Pereira A E S, Silva P M, Oliveira J L, Oliveira H C and Fraceto L F 2017 Chitosan nanoparticles as carrier systems for the plant growth hormone gibberellic acid *Colloids Surf.,* B 150 141–52

[62] Pereira A E S, Sandoval-Herrera I E, Zavala-Betancourt S A, Oliveira H C, Ledezma-Pérez A S and Romero J *et al* 2017 γ-Polyglutamic acid/chitosan nanoparticles for the plant growth regulator gibberellic acid: characterization and evaluation of biological activity *Carbohydr. Polym.* **157** 1862–73

[63] Yang F L, Li X G, Zhu F and Lei C L 2009 Structural characterization of nanoparticles loaded with garlic essential oil and their insecticidal activity against *Tribolium castaneum* (Herbst) (Coleoptera: Tenebrionidae) *J. Agric. Food Chem.* **57** 10156–62

[64] Tong Y, Wu Y, Zhao C, Xu Y, Lu J and Xiang S *et al* 2017 Polymeric nanoparticles as a metolachlor carrier: water-based formulation for hydrophobic pesticides and absorption by plants *J. Agric. Food Chem.* **65** 7371–8

[65] Pereira A E S, Grillo R, Mello N F S, Rosa A H and Fraceto L F 2014 Application of poly (epsilon-caprolactone) nanoparticles containing atrazine herbicide as an alternative technique to control weeds and reduce damage to the environment *J. Hazard. Mater.* **268** 207–15

[66] Mondal P, Kumar R and Gogoi R 2017 Azomethine based nano-chemicals: development, *in vitro* and *in vivo* fungicidal evaluation against *Sclerotium rolfsii*, *Rhizoctonia bataticola* and *Rhizoctonia solani Bioorg. Chem.* **70** 153–62

[67] Forim M R, Costa E S, Da Silva MFDGF, Fernandes J B, Mondego J M and Boiça Junior A L 2013 Development of a new method to prepare nano-/microparticles loaded with extracts of azadirachta indica, their characterization and use in controlling plutella xylostella *J. Agric. Food Chem.* **61** 9131–9

[68] Pradhan S, Roy I, Lodh G, Patra P, Choudhury S R and Samanta A *et al* 2013 Entomotoxicity and biosafety assessment of PEGylated acephate nanoparticles: a biologically safe alternative to neurotoxic pesticides *J. Environ. Sci. Heal. – Part* **48** 559–69

[69] Memarizadeh N, Ghadamyari M, Adeli M and Talebi K 2014 Preparation, characterization and efficiency of nanoencapsulated imidacloprid under laboratory conditions *Ecotoxicol. Environ. Saf.* **107** 77–83

[70] Prasad A, Astete C E, Bodoki A E, Windham M, Bodoki E and Sabliov C M 2018 Zein nanoparticles uptake and translocation in hydroponically grown sugar cane plants *J. Agric. Food Chem.* **66** 6544–51

[71] Pereira A, do E S, Oliveira H C and Fraceto L F 2019 Polymeric nanoparticles as an alternative for application of gibberellic acid in sustainable agriculture: a field study *Sci. Rep.* **9** 1–10

IOP Publishing

Bionanomaterials for Environmental and Agricultural Applications

Ravindra Pratap Singh and Kshitij RB Singh

Chapter 7

Utility of metal oxide-based bionanocomposites for wastewater treatment

Tanvi Vats and Pratibha Pandey

Rapid industrial growth coupled with other human activities is making the Earth 'water-stressed'. Water remediation has been an issue for the research community for quite a long time. In the present work, discuss how an environmentally benign biopolymer matrix can be successfully reinforced with nanomaterials to remove pollutants like heavy metals, dyes, and other secondary toxic pollutants. Moreover, the sustainability and practical applicability of nano biocomposite systems is also discussed in detail.

7.1 Introduction

The Earth is rightly called the blue planet, as 70% of the Earth is covered with water. Water is responsible for sustaining life on earth. It is imperative to note that only 3% of total water is freshwater, and hence we can say that as a whole the living planet is facing water scarcity! To exacerbate this scarcity, freshwater systems are jeopardized by various human activities. Figure 7.1 gives the classification of major water pollutants along with their origin and effects [1–3]. Further, rapid urbanization, heavy industrialization and various schemes to maximize the accessibility of water (water reservoirs and several irrigation schemes) have led to economic output and greater water availability but at the same time cause unprecedented damage to the ecosystem and biodiversity.

In recent years, massive discussions on the effect of anthropological activities on the ecosystem have generated great awareness of human–Nature co-existence. In the wake of this, water purification methods have become a focal point of research. With the advent of nanotechnology, scientists have come up with several approaches which combine conventional substances with nanomaterials and

doi:10.1088/978-0-7503-3863-9ch7

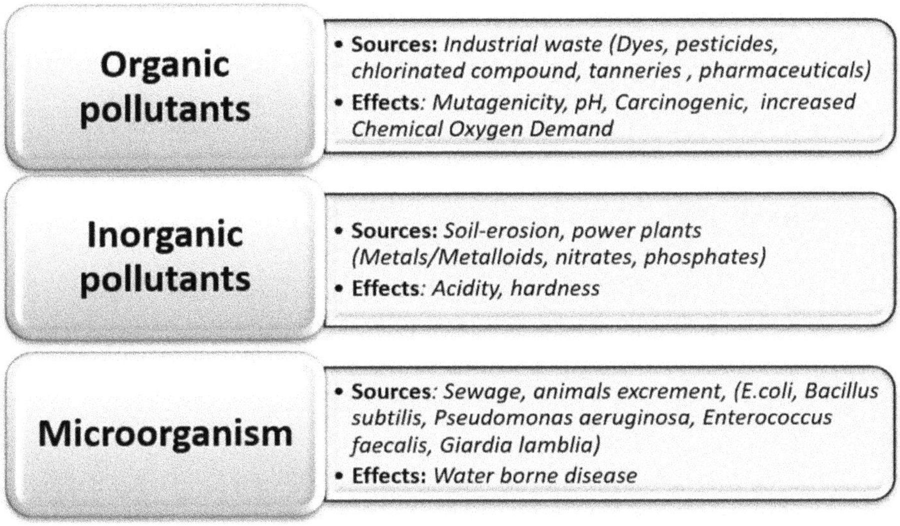

Figure 7.1. Water pollutants: classified.

have developed nanocomposites. One of the key issues which seems to be underemphasized is the potential toxicity of nanocomposites used for remediation. These nanomaterials are difficult to synthesize and are seemingly non biodegradable. Nanocomposites comprising biopolymers, i.e., naturally occurring polymers in permutation with an inorganic nanostructure, are classified into a separate materials group called bionanocomposites. Bionanocomposites were first named in 2004 and are also referred to as nanobiocomposites, green composites or at times biohybrides. Bionanocomposite materials can be considered as a beautiful interdisciplinary amalgamation of life sciences, science of materials and nanotechnology [4–7]. Marked enhancement in gas barrier properties, mechanical strength and optical properties is observed in these materials, which is attributed to the better blending of nanoparticles in the polymer matrix [8, 9]. The naturally occurring components are abundantly available, reduce the hassles of synthesis and are Nature compatable. These nanocomposites are relatively safe and also can be reused. The development of an interfacial bond between the biofibers and the other components in a composite system leads to a better comapatability and also better sorption of the pollutants [10–13]. Most commonly they are composited with metal nanoparticles, metal oxides, and carbon nanotubes for efficient enhancement in the properties. Figure 7.2 summarizes the properties, binding action and pollutants removed using bionanocomposites. In this chapter we will explore the potential application of non-toxic and biodegradable nanocomposites on efficient water remediation. Though several materials have been studied for potential application, we will focus on four natural components, namely cellulose, chitosan (CS), gum and clay based bionanocomposites and work done on them in the quest of greening the blue.

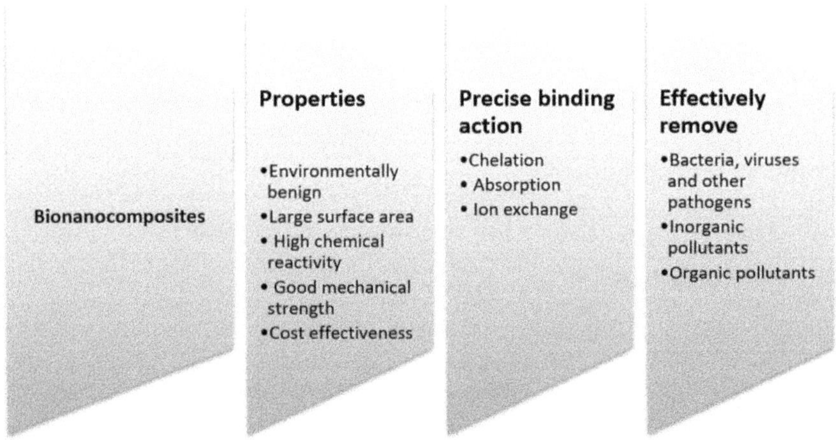

Figure 7.2. Bionanocomposites: properties, binding mechanism and pollutants removed.

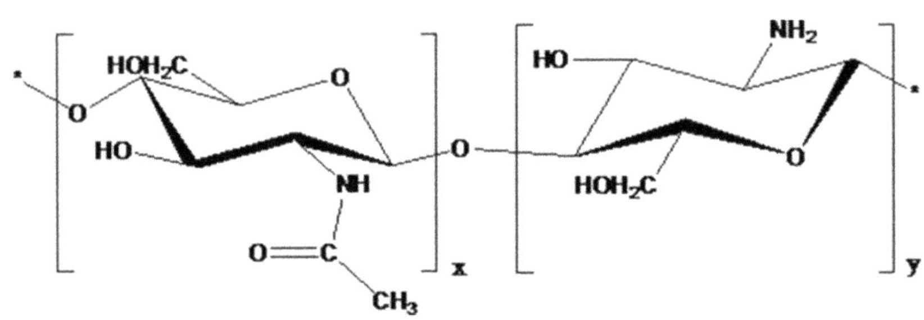

Figure 7.3. Structure of chitosan.

7.2 Chitosan based metal oxide bionanocomposites

CS can be termed as a semi-synthetic polymer obtained by the partial deacetylation of chitin using an alkaline substance like sodium hydroxide. It is a linear polysaccharide composed of β-(1→4)-linked D-glucosamine (deacetylated unit) and N-acetyl-D-glucosamine (acetylated unit), which are randomly distributed throughout the matrix (figure 7.3) [14, 15]. CS is one of the most copious biopolymers and is found in crab and shrimp shells, insects, crustaceans, annelids, molluscs, coelenterate and fungal mycelia [16]. This polycationicpolymer is non-toxic, biodegradable and biocompatible [17].

Through the reactive amino and hydroxyl functional groups CS undergoes various reactions like etherification, esterification and reductive amination and forms multiple covalant bonds [18, 19]. It is this versatility and abundant availability coupled with the nanodimensional aspect of a CS nanostructure (like a nanofibril) that make CS nanocomposites an ideal material for a variety of applications, including water purification. The two advantages of CS nanostructures are that their structural morphology and size can be controlled *in situ* and they can be easily modified with particular functional groups in order to remove specific pollutants [20, 21]. The added advantage of chitosan over other natural polymers is its polyfunctional nature. The functional groups of chitosan help it to form composites with appropriate groups which in turn enhances its capacity to adsorb water pollutants. Chitosan also holds the merit of being a single cationic biopolymer and can therefore flocculate anionic colloidal particles, thereby lowering the levels of chemical oxygen demand and turbidity in water [22, 23].

Several studies report that CS can easily bind with carbon nanotubes, silver, iron and gold nanoparticles. An important feature of these nanocomposites is that they are uniformly distributed and free from agglomeration [24, 25]. The physical, chemical and biological properties of CS can be modified to make it compatible for applications in water treatment. Metal chelation can be brought about by the amino and hydroxyl groups present in the structure. The very same groups interact with the thiol groups present in the bacterial cell wall and cell membranes disrupting the respiratory system of bacteria [26–28].

CS nanocomposites can be used in water disinfection in several manifestations like an adsorbent in beads and sponges; in water storage tanks they act as an excellent coating material or as an antimicrobial agent in membranes. The properties like bioactivity, biocompatibility, high antimicrobial activity and low toxicity to humans and animals make CS-based materials first among equals in their field of applications [29–31]. The antimicrobial activities of polycationic CS-based composites can be broadly studied under the following mechanistic pathways:

- One explaination infers that the microbial growth is restrained due to osmotic imbalance which in turn is brought about by the attraction forces between the amine groups of the CS (positively charged) and the microbial cell membrane (negatively) charged [32, 33].
- Another working mechanism states that the CS induces the peptidoglycans of the microbial cell wall to break down, which leads to the seepage of electrolytes present in the cell, which consist of potassium ions, proteins, nucleic acids, glucose and lactate dehydrogenase [34, 35].
- Yet another manifestation of antimicrobial behavior of CS is reticence of synthesis of mRNA by bonding with DNA of the bacteria. This takes place by dissemination of CS in the nucleus of the microbes [36, 37].
- Chelation of metals leading to containment of spores and binding to essential nutrients resulting in the death of cells is the fourth mechanistic explanation of CS action [38].

The major limitations encountered while using CS nanomaterials are its water solubility, thermal stability and working pH [39]. They also exhibit swelling behavior and low mechanical strength. In order to overcome these cross-linkers like epichlorohydrin and glutaraldehyde, inorganic materials such as alumina, silica, and titania are utilized to enhance their chemical constancy [40]. Another very popular reinforcing material being used these days are Multiwalled carbon nanotubes (MWCNTs) [41]. Researchers have also synthesized CS nanotubes, hollow fibers and porous nanostructures with the sole aim to improve efficiency [42].

Nthunya and co-workers [42] prepared a polymer blend of CS and other polymerspolyacrylamide and polyethylene glycol. These polymers when added in optimized amounts supplement the ability to electrospun CS. To this blend they added polyisopropene and functionalized MWCNTs which in turn resulted in better morphology, improved mechanical strength, thermal properties and reduced swelling behavior by twofold! The addition of functionalized MWCNTs was reported to advance the tensile strength of the nanofibres up to two times when compared to nanofibres without them. The thermal degradation of the nanofibres was improved by a magnitude of 50 °C. The group also came up with a possible disinfection application of the nanocomposites by embedding silver and iron nanoparticles as they have antibacterial properties.

Monaheng Masheane and co-workers [29] did an extensive study on antimicrobial CS-alumina/functionalized-MWCNT. The group devised a simple phase inversion method of nanoparticle synthesis. They used oxalic acid as cross-linker and employed scanning electron microscopy and Fourier transform infrared spectroscopy to establish nanocomposites formation. The group worked on twelve strains of bacteria namely: *E. faecalis, K. pneumonia, K. pneumoniae, P. aeruginosa, E. coli, S. sonnei, E. coli, P. mirabilis, K. oxytoca, B. cereus, S. boydii, E. cloaca*. They ascribed the antimicrobial activity of positively charged active sites namely amino and the hydroxyl groups in the CS which are attracted towards the bacterial cell wall (negative in charge) causing bacterial lysis.

The major outcome of the study indicated that though the CS-alumina/functionalized-MWCNTs exhibited similar antibacterial properties to those of pristine CS, they exhibited better stability at higher temperatures along with lower solubility. These properties rendered them useful at high temperature and lower pH without any deterioration.

In a recent study Alba *et al* [43] prepared CS based bionanocomposites with layered mica as inorganic component. The study revealed that controlled surface charge coupled with high cation exchange capacity and swelling behavior of sodium–manganese micas (n = 2 and 4) exhibit improved adsorption properties when compared to other inorganic solids. The group employed ion exchange reaction between CS and high charge mica sample. The composites were characterized using x-ray diffraction, Fourier transform infrared spectroscopy, thermal-analysis, MAS-NMR spectroscopy and zeta-potential. The results revealed that the CS is effectively adsorbed although partly retained in the outer surface as it is hydrogen-bonded to the tetrahedral sheet of the silicate. These composites exhibit higher thermal stability than CS, rendering them capable of functioning at higher temperatures.

Dehaghi *et al* [44] used sonication method to synthesise CS–zinc oxide nanocomposites. The nanocomposites were manifested in the form of beads and dissolution and swelling behavior was studied for the sample. The group tested and compared the adsorption performance of pristine CS and CS–zinc oxide nanocomposites by studying the adsorption from aqueous solution of permethrin 25% on a column of respective beads. The results indicated almost two-fold increase in removal efficiency from 49% of CS to 99% of CS–zinc oxide nanocomposites.

The nanocomposites could also be efficiently reused and it is claimed that 56% of the material is regenerated post three cycles of running the column. This could pave the way for benign methodology for pesticide removal from wastewater.

An extensive amount of work has been done on biocomposites of CS with iron oxides. Sureshkumar *et al* [45] employed co-precipitation method to prepare magnetite nanoparticles. They studied the effect of addition of magnetite nanoparticles on the adsorption affinity for chromium. The group used potassium dichromate solution as the source of Cr(VI) ions and the adsorption was carried out on the strips of CS as well as nanocomposites. The results revealed that the efficiency of CS strip in removing chromium was 29.39% while CS-magnetite nanocomposite strip exhibited a great improvement up to 92.33%.

CS-magnetite nanocomposites were developed as a potential adsorbent, chelating agent or bactericidal agent by Abd-Elhakeem *et al* [46] They tested a variety of substances and found that the adsorption capacities of CS magnetite nanocomposites for petroleum impurities were highest and were lowest for heavy metals. They also studied the effect on the multiplication of bacteria which was partly inhihibited at 0.1 g mL^{-1} of CS-magnetite nanocomposites and complete inhibition was reported at the concentration of 0.5 g/10 mL.

CS-magnetite nanocomposites were also studied for their adsorption of nickel and cobalt from water by Hritcu *et al* [47]. In their results they reported that the material has a maximum adsorption capacityof 588.24 mg g^{-1} for cobalt ions and, respectively, 833.34 mg g^{-1} for nickel ions. Also, they reported that these samples can be used three times exhibiting almost similar, repeatable efficiency each time.

Saifuddin and Dimara [48] employed CS-magnetite nanocomposites as a primary coagulant and flocculant for pre treating palm oil mill effluent. They performed extensive experiments in which dosage and pH were kept as the determiners and the evaluation of the process was carried out by analyzing the effect on turbidity, the amount of total suspended solids and decrease in chemical oxygen demand. At the most favorable pH and concentration, 98.8% reduction of turbidity, 97.6% of suspended solids and 62.5% for chemical oxygen demands was reported. Not only did the composites exhibit better efficiency for adsorption and removal of palm oil than pristine chitosan, but also it was a benign technique as compared to the use of alum and alum polychloride composites. The latter created hazardous residual waste.

Siddeeg and co-workers [49] added another dimension to the study of chitosan magnetite nanocomposites by immobilizing manganese peroxidase enzyme on its surface. In recent years, manganese peroxidase has emegerd as a potential material for its effectiveness as a degradant for organic pollutants. The nanocomposites provided a large surface area enzyme immobilization. The exhaustive study consisted of a detailed report on the effect of several environmental factors, namely temperature, pH, as well as storage duration on the activity of the extracted enzyme. The samples were characterized using Fourier transmission infrared spectroscopy, scanning electron microscope, x-ray diffraction and transmission electron microscope. The sample dyes used for the study were reactive orange 16 and methylene blue. The samples were shown to be effectively reused up to five times and allowed the removal of about 96% methylene blue and 98% reactive orange 16.

Morsi and co-workers [50] prepared multifunctional citosan nanocomposites i.e., more than one component was added to the chitosan polymer in the composites. Silver nanoparticles, copper nanoparticles and carbon nanotubes either as bi- or multifunctional nanocomposites were prepared. An important aspect of the study involved the observation of the overall morphology of the synthesized nanocomposites. In the case of twin component composites the nanoparticles were observed to be distributed homogenously while in the case of multifunctional composites different dimensional shapes were observed. The study also reported higher antimicrobial acivity for shorter contact timings against Gram negative bacteria (*E. coli*), Gram positive bacteria (*Staphylococcus aureus*) and fungal strain (*Aspergillus flavus*).

7.3 Cellulose based metal oxide nanobiocomposite

Cellulose is a biodegradable, non-toxic polysaccharide and it is made up of a linear chain of several thousands of β-linked D-glucose units. Natural resources and agricultural wastes including nut waste, banana and orange peel, coconut bunch waste, rice and wheat straw are a good source of cellulose [51]. However, the inflexibility of molecular chains and the strong interaction of intramolecular and intermolecular hydrogen bonds formed by a large number of hydroxyl groups in molecular chains of cellulose limit its applicability [52]. Henceforth, utilization value of cellulose has been improved either through chemical modification or by using layer-by-layer technique, mixing etc, so as to extend the application prospects of cellulose and its composites in various fields [55]. To explore the properties of cellulose, researchers have synthesized new functional composites by using natural polymer cellulose as the matrix and combining cellulose with metal oxides. Improved optical, thermal and mechanical properties have been shown by these nanocomposites and they are used in water treatment. Not only material properties of nanocomposites but also antibacterial, antioxidant and photocatalytic properties have been enriched by the incorporation of metal oxides or inorganic materials into cellulose. Different metal oxides like iron oxide, zinc oxide and titanium oxide nanocomposite with cellulose have been prepared and applied in removal of pollutants like heavy metal ions, dyes in water via adsorption, photodegradation etc techniques [53–55].

Luo *et al* have prepared magnetic cellulose-based beads with micro and nanopore structure via an optimal extrusion dropping technology from NaOH/Urea aq. solution. The composite beads are incorporated with carboxyl decorated magnetite nanoparticles and nitric acid modified activated carbon. These composites were highly effective in removal of Cu^{2+}, Pb^{2+} and Zn^{2+}using magnetic response technique [56]. Graphene oxide (GO) has also been incorporated into cellulose nanofibres to prepare biohybrids for water purification. The graphene oxide cellulose nanofibre biohybrid purifies water via adsorption technique. It has great adsorption capacity towards Cu(II) due to TOCNF (TOCNF is (2,2,6,6 tetramethyl piperidine-1-oxylradical) (TEMPO) mediated oxidized cellulose nanofibres) and GO phase [54]. Derivative of cellulose has also been utilized by researchers to prepare biocomposites. For example, Mahmoud *et al* [55] have prepared a biocomposite of titanium oxide nanoparticles and caboxymethyl cellulose. They have reported that nanotitanium oxide worked as a platform to immobilize carboxymethyl cellulose (CMC) using glutaraldehyde (glu) as an efficient cross linking reagent (nano-TiO_2-glu-CMC). Heavy metals like Cd (II), Pb (II) and Hg (II) etc could be easily adsorb using the prepared bionanocomposite

Iron oxide-bead cellulose nanocomposite has been prepared using iron oxy-hydroxide as precursor for *in situ* preparation of iron oxide nanoparticles and it has been utilized in the removal of arsenic ions from aqueous solutions via ion exchange or chelation process [56]. Cellulose based ZnO nanocomposite film from cellulose carbamate–NaOH/ZnO solutions was prepared through one-step coagulation in Na_2SO_4 aq. Solution. Cellulose based films were able to destroy viable bacteria within 6 h and antibacterial properties of ZnO have been utilized in purification of water and cellulose providing a biocompatible environment to the matrix [57].

Oyewo *et al* [58] have reported that when metal oxide blends with cellulose the hydrophobic and photocatalytic properties of metal oxide have been improved resulting in better adsorption property of the matrix. This property of bionano-composite has been utilized by the same group in removal of methylene blue (MB) dye using adsorption technique. They have prepared sawdust derived cellulose nanocrystal incorporated with ZnO for the removal of MB [59]. MB dye removal has been also reported by Kamal *et al* [60] using carboxymethyl cellulose grafted acrylamide (CMC-g-AM) embedded with porous carbon and citric acid modified magnetite. Recently, nanocellulose and copper nanoparticle was used to modify graphene oxide and it was used to remove malachite green (MG). This nano-composite was capable of adsorbing the toxic dyes. Fe_3O_4 nanoparticles were also incorporated into carboxymethyl cellulose using impinging stream-rotating packed bed technique.[61] These nanocomposites are able to remove Pb(II) and heavy metal ions. Fe_3O_4 nanorods were also synthesized using cellulose nanocrystals as template. This matrix is useful in adsorption of As (III) and As (V) and it could be used as an adsorbent in treatment of water [62]. TiO_2 nanocrystal has been synthesised in the vicinity of cellulose nanocrystal (CNC) using *in situ* hydrolysis and cellulose nanocrystal improved the photocalytic property of TiO_2 nanocrystals. Cr(VI) hazardous metal ion has been removed using this nanocomposite in the presence of sunlight only. They have also reported that crystal growth of TiO_2 has not only been enhanced by cellulose nanocrystal but its light harvesting and charge transfer peoperties are also modified [63]. Organic pollutants like MB were degraded photocatalyticaly using $H_3PW_{12}O_{40}/TiO_2$ (HPW/TiO_2) nanotubular composite. Layer-by-layer technique has been used to assemble nanocomposite on cellulose substance and form HPW/TiO_2/cellulose composite sheet. These nanocomposites worked as phtocatalytic membrane and degraded organic pollutants in direct sunlight [64]. Researchers reported that modified cellulose like sulfated carboxy-methyl cellulose (S-CMC) was usd as a template for preparation of TiO_2 nano-particles composite with cinnamon bark (CIN) (CIN/S-CMC/TiO_2). 4-nitrophenol,

(4 NP) and azo dyes such as cango red (CR), methyl orange were reduced using this bionanocomposite loaded with mono and bimetallic metal ion in zero oxidation state [65]. A magnetic composite was prepared using Fe_3O_4 particles and graphene oxide–carboxymethyl cellulose (FCG) and utilized in removal of dye molecules like congo red (CR), acid red 1 (AR1) and reactive red 2 (RR2). They have reported that adsorption of CR, AR1 and RR2 on graphene oxide–carboxymethyl cellulose composite followed the electrostatic attraction, hydrogen bonding, π–π stacking, and surface complexation mechanisms. 65% of FGC composite was regenerated over after five consecutive cycles using an aqueous 0.1 M NaOH solution and the composite could be used as a potential adsorbent for the removal of dye molecules from contaminated water [66].

7.4 Guar gum based metal oxide bionanocomposite

Another abundant polysaccharide which has a great potential to be used as a matrix in bionanocomposites is uar gum. It is a non-ionic polysaccharide which is derived from the seeds of *cyamopis tetragonolobus*, belonging to the Leguminosae family. The main areas of industrial application include food, textiles, paper, cosmetics and pharmaceuticals. Apart from being biodegradable, biocompatible and sustainable it conatins several functional groups in its polymer backbone and these various functional groups can be easily modified in order to increase the effective surface area along with the adsorption capacity [67, 68]. A rod-like structure of guar gum has been observed due to polysaccharides of galactomannans which are a linear chain of (1 → 4)-linked β-D-mannopyranosyl units with (1 → 6)-linked α-D-galacto-pyranosyl residues as side chains. Several properties of guar gum make it an ideal component of composite materials with majestic properties. These include long chain length, multiple cis–OH groups, steric factors, degree of polymerization and existence of several substituents. Guar gum is extremely soluble in cold water but is insoluble in organic solvents like esters, fats, alcohols, hydrocarbons, ketones. Solubility in cold water is depend on high galactose to mannose ratio. It is known to yield solutions of high viscocity even at low concentration [71–76].

The mechanism of colloid formation by guar gum is very interesting. The formation of highly viscous colloid can be attributed to the presence of multiple hydroxyl groups which in turn leads to extensive hydrogen bonding with water. Apart from this, the galactose side chains interact with water forming a network of highly entangled intermolecular chains and with incearse in gum concentration gel formation takes place. The large number of increased 'exposed hydroxyl groups' arising due to galactose–mannose ratio leads to strong interactions with aquous colloids. This peculiar mechanism makes guar gum an excellent dispersant for hydroxyl- and carboxyl-containg pollutants and a good coagulant for inorganic materials like clay.

In the field of water remediation guar gum based nanobiocomposites are used as hydrogels having an enormous network that is both porous and has a high water content. Both these properties make them an ideal material for solute diffusion and adsorption [69]. Some of the key features that give an edge to guar gum based hydrogels are low cost, diversification of materials to meet specific requirements, good mechanical properties and toughness towards strong stirring (so that the adsorbant can be reused). Heavy metal ions such as Pb(II), Cd(II), Cr(VI), Zn(II) can be easily adsorbed using these bionanocomposites and they have selective adsorption tendency towards ionic dye molecules which could be cationic, anionic or both. Several guar gum and metal/metal-oxide based bionanocomposites successfully degrade dye molecules by photochemical pathways [70].

Mittal and his co-workers have reported nanocomposite based on gum and iron oxide. They have utilized adsorption properties of gum polymer for the remediation of MB contaminated water [71]. They have also prepared xanthum gum based novel hydrogel nanocomposite via chemical cross-linking with grapheme oxide for highly efficient adsorption of crystal violet dye from contaminated waste water [72]. Another group of researchers has synthesized silver oxide guar gum for reduction of nitrobenzene and oxidation of benzyl alcohol [73]. Modification of guar gum has been done to improve the properties of the polymer. 2-hydroxymethyl methacrylate cross-linked gum tragacanth hydrogel (Gum-T-Cl-HEMA) was synthesized by simple microwave assisted green polymerization technique. TiO_2 nanoparticles were immobilized in the synthesized gum and these nanocomposites were utilized in removal of malachite [74]. Magnetic hydrogel beads were prepared using modified biopolymer gum traganth (GT), polyvinyl alcohol (PVA) and graphene oxide using gelatin method. Not only could cationic (crystal violet), anionic (cango red) dyes be removed easily but Pb(II) and Cu(II) heavy metal ions were also removed efficiently from aqueous solution using these nanocomposites [75].

7.5 Clay based bionanocomposites

Clay minerals are yet another economical, plentiful, characterized by high sorption properties and easily regeneratable material that can be used mainly as potential ion-exchange materials in water remediation [76, 77]. Structurally, nanoclays can be thought of as nanoparticles of stackable layered mineral silicates which form complex crystallites by stacking these layers in the form of octahedral and tetrahedral sheets [78, 79]. In the case of octahedral silicates, the sheets consist of aluminum or magnesium in a six-fold coordination with oxygen from a tetrahedral sheet and with hydroxyl group. In the case of tetrahedral sheets, three out of four silicon–oxygen sheets are connected to the adjoining tetrahedral sheets while the remaining one sheet is joined through covalent bond to an octahedral sheet [80, 81]. The properties of nanoclays are determined by stoichiometric composition of the arrangement of sheets which in turn determines various applications of these versatile materials [82].

The space between various layers of clay can be impregnated by biopolymers like cellulose, hydrogels etc to give clay based nanobiocomposites (figure 7.4). The various synthetic routes are for the synthesis of clay based composites include photopolymerization [83], click chemistry [84], radical-mediated polymerization [85, 86], tandem preparation [87], and mini-emulsion [88].

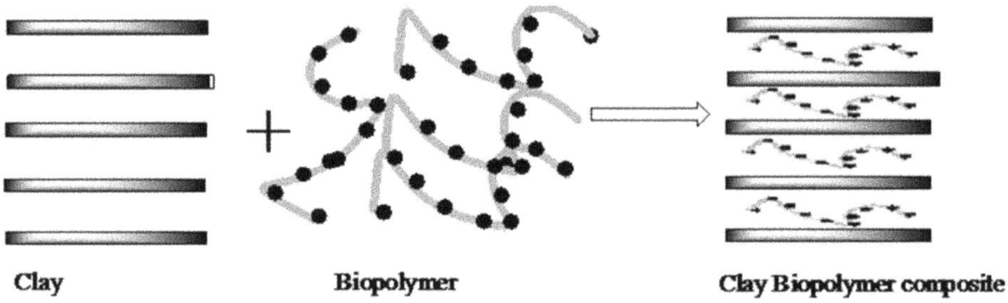

Figure 7.4. Schematic representation of clay biopolymer nanocomposites.

Clay based bionanocomposites are expected to be an ideal material for water treatment because of their large adsorption capacity and reusability. This is attributed to their efficient cation exchange, large surface area, easy processability, economic viability and low toxicity [89–97]. The main area of their application is dye removal from waste water.

Auta and Hameed worked on chitosan/modified ball clay composite in the context of dye adsorption [98]. They inferred that an increase in the initial concentration, solution pH and temperature of the process enhances the adsorption of MB on the bionanocomposites.

Another biopolymer that has attained focus in this area is carrageenan, which is a colloidal extract obtained from carrageen seaweed and other red algae. Reza-Mahdavinia *et al* used poly(acrylamide)/k-carrageenin/sodium alginate sodium montmorillonite (MMT) clay hydrogel nanocomposites to remove crystal violet dye [99]. The study revealed that dye adsorption capacity of the synthesized hydrogels depended on the clay content, the weight of biopolymer and the study also indicated that the lower pH favoured the dye adsorption capacity.

The same group also studied the effect of incorporation of magnetite nanoparticles on the composite efficiency [100]. They synthesized poly(*kappa*-carrageenin-*co*-PVA)/Fe3O4 magnetic nanocomposite hydrogels with the aim to remove crystal violet dye. The study inferred that there was decline in the magnitude of dye removal on inclusion of magnetic particles but a marked improvement was shown on increasing the kappa carrageenan content. The study also revealed that dye adsorption capacity of the hydrogels went down on increasing the ionic strength of the dye solution.

Zhou *et al* studied the MB adsorption on amino-functionalized attapulgite clay nanoparticle adsorbent (ATP@CCS) onto hydrothermal carbonized chitosan [101, 102]. The composites demonstrated excellent results in terms of faster attainment of equilibrium (less than 120 min) with more than 80% removal and reusability for up to five consecutive cycles.

Alginate is yet another naturally occurring anionic polymer typically obtained from brown seaweed which has potential in the field of bioremediation of water. It is biocompatible, cheap and can be converted to gel form on addition of divalent cations like Ca^{2+} [103]. Ely *et al* [104] studied the adsorption of 4-nitrophenol on Na-montmorillonite/alginate composites microcapsules. The proportion of montmorillonite in the composite materials determined the adsorption capacity of the composite and results indicated that phenolic adsorption increases with the montmorillonite concentration. The study also emphasized that encapsulation has a great influence on the adsorption capacity.

Along similar lines Lezehari *et al* [105] investigated pentachlorophenol adsorption on composite beads of aluminum-pillared clay (Al-Mont-PILC) and surfactant-modified pillared clay (CTAB-Al-Mont-PILC) onto alginate. The studies showed that the modification of clay by surfactant enhances the removal of organic pollutant in comparison to encapsulated aluminum-pillared clay.

Chitosan can be effectively compounded with montmorillonite composites for the removal of emerging pollutants like tannic acid etc. An and Dultz [106] used Chitosan/MMT composite for tannic acid removal. The composite was found to be very stable for temperature and change in pH. It is also concluded from the study that at low concentration the adsorption takes place by electrostatic mechanism. Celis *et al* [107] used montmorillonite–chitosan bionanocomposites to remove 3,6-dichloropyridine-2-carboxylic acid (clopyralid) which was found to be greater for nanocomposites with higher chitosan component.

7.6 Conclusion and prospects

In this chapter, we have discussed in part the complex problem of water pollution and its treatment. As we have explained earlier, this is an ever increasing challenge which might get worse for upcoming generations. The concern is indeed global, and it is reflected in the documentation available on the websites of various national environmental agencies of different countries. Keeping this in mind, the United Nations Environment Programme started an initiative of 'Greening the Blue' in order to employ and encourage the UN System for development of greater environmental sustainability. These methods can be employed and mimicked in terms of water resources as well. Water can be thought as a 'blue resource'. The world is facing severe water crisis and we need to come up with sustainable solutions for treating it. These solutions can be termed as 'greening the blue', various methods for 'greening the blue' are compiled in figure 7.5.

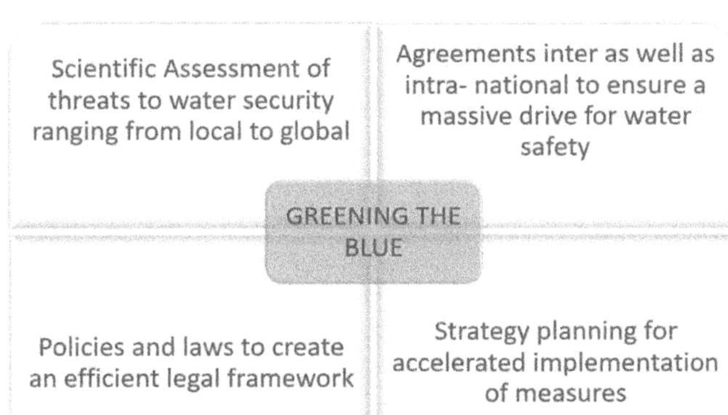

Figure 7.5. Methods of greening the blue.

Another important aspect this work brings forth is the development of greener methods for water remediation. Working with sustainable technologies is indeed the need of the hour as it is important to understand that we are creating a long-term solution to the problem. We need to emphasise the fact that the remediation process should not lead to other environmental issues. Biopolymers like cellulose and chitosan are freely and abundantly available and, most importantly, Nature compatible. It is important to bear in mind that the technologies we develop are benign, and bionanocomposites are an excellent choice of materials for this purpose. They have great potential, and the research done in this field can flourish in developing cost-effective, large-scale industrial applications in the area of water remediation.

Acknowledgment

Dr Tanvi Vats would like to thank Gautam Buddh University and Dr Pratibha Tiwari would like to thank Hansraj College, University of Delhi for providing a conducive environment for carrying out this work.

References

[1] Dar O A, Malik M A, Talukdar M and Hashmi A 2020 Bionanocomposites in water treatment *Bionanocomposites* (Amsterdam: Elsevier) pp 505–18

[2] Meybeck M 2003 Global analysis of river systems: from Earth system controls to Anthropocene syndromes *Philos. Trans. R. Soc. Lond., B, Biol. Sci.* **358** 1935–55

[3] Karl T R, Melillo J M, Peterson T C and Hassol S J 2009 *Global Climate Change Impacts in the United States* (Cambridge: Cambridge University Press)

[4] Abell R, Thieme M L, Revenga C, Bryer M, Kottelat M and Bogutskaya N *et al* 2008 Freshwater ecoregions of the world: a new map of biogeographic units for freshwater biodiversity conservation *BioScience* **58** 403–14

[5] Gleick P H 2003 Global freshwater resources: soft-path solutions for the 21st century *Science* **302** 1524–8

[6] Ruiz-Hitzky E, Darder M and Aranda P 2005 Functional biopolymer nanocomposites based on layered solids *J. Mater. Chem.* **15** 3650–62

[7] Mittal V, Kim J K and Pal K 2011 *Recent Advances in Elastomeric Nanocomposites* (Berlin: Springer)

[8] Alexandre M and Dubois P 2000 Polymer-layered silicate nanocomposites: preparation, properties and uses of a new class of materials *Mater. Sci. Eng. R. Rep.* **28** 1–63

[9] Kalsoom Khan A, Saba A U, Nawazish S, Akhtar F, Rashid R and Mir S *et al* 2017 Carrageenan based bionanocomposites as drug delivery tool with special emphasis on the influence of ferromagnetic nanoparticles *Oxid. Med. Cell. Longev.* **2017** 8158315

[10] Letaief S and Detellier C 2007 Functionalized nanohybrid materials obtained from the interlayer grafting of aminoalcohols on kaolinite *Chem. Commun.* **2007** 2613–5

[11] Li X, Chang W-C, Chao Y J, Wang R and Chang M 2004 Nanoscale structural and mechanical characterization of a natural nanocomposite material: the shell of red abalone *Nano Lett.* **4** 613–7

[12] Sun G, Liang R, Lu Z, Shi T, Geng P and Zongjin L 2017 Remarkable mechanical enhancement achieved by interfacial strengthening of organic/inorganic/fiber composites *Constr. Build. Mater.* **142** 7–10

[13] Sun Z and Mingming W 2019 Effects of sol-gel modification on the interfacial and mechanical properties of sisal fiber reinforced polypropylene composites *Ind. Crop. Prod.* **137** 89–97

[14] Rahman R and Putra S Z F S 2019 Tensile properties of natural and synthetic fiber-reinforced polymer composites *Mechanical and Physical Testing of Biocomposites, Fibre-Reinforced Composites and Hybrid Composites* (Amsterdam: Elsevier) pp 81–102

[15] Hasan K, Horváth P G and Alpár T 2020 Potential natural fiber polymeric nano-biocomposites: a review *Polymers.* **12** 1072

[16] Thakur V K and Voicu S I 2016 Recent advances in cellulose and chitosan based membranes for water purification: a concise review *Carbohydr. Polym.* **146** 148–65

[17] Thakur V K, Thakur M K and Gupta R K 2014 Raw natural fiber–based polymer composites *Int. J. Polym. Anal. Charact.* **19** 256–71

[18] Zhou H-Y, Zhou D-J, Zhang W-F, Jiang L-J, Li J-B and Chen X-G 2011 Biocompatibility and characteristics of chitosan/cellulose acetate microspheres for drug delivery *Front. Mater. Sci.* **5** 367–78

[19] Qian L and Zhang H 2010 Green synthesis of chitosan-based nanofibers and their applications *Green. Chem.* **12** 1207–14

[20] Croisier F and Jérôme C 2013 Chitosan-based biomaterials for tissue engineering *Eur. Polym. J.* **49** 780–92

[21] Rinaudo M 2006 Chitin and chitosan: properties and applications *Prog. Polym. Sci.* **31** 603–32

[22] Nthunya L N, Masheane M L, Malinga S P, Nxumalo E N, Mamba B B and Mhlanga S D 2017 Thermally and mechanically stable β-cyclodextrin/cellulose acetate nanofibers synthesized using an environmentally benign procedure *Int. J. Smart Nano Mater.* **8** 1–19

[23] Ulrikh D V, Timofeeva S S and Timofeev S S 2018 Choice of sorbents for the pretreatment in the complex technology of phytoremediation *IOP Conf. Series: Materials Science and Engineering* **451** 012235

[24] Zha F, Li S and Chang Y 2008 Preparation and adsorption property of chitosan beads bearing β-cyclodextrin cross-linked by 1, 6-hexamethylene diisocyanate *Carbohydr. Polym.* **72** 456–61

[25] Ali Z M, Laghari A J, Ansari A K and Khuhawar M Y 2013 Extraction and characterization of chitosan from Indian prawn (Fenneropenaeus indicus) and its applications on waste water treatment of local ghee industry *Extraction.* **3** 28–37

[26] Misra M, Singh N and Gupta R K 2017 Enhanced visible-light-driven photocatalytic activity of Au@ Ag core–shell bimetallic nanoparticles immobilized on electrospun TiO_2 nanofibers for degradation of organic compounds *Catal. Sci. Technol.* **7** 570–80

[27] Zhuang X, Cheng B, Kang W and Xu X 2010 Electrospun chitosan/gelatin nanofibers containing silver nanoparticles *Carbohydr. Poly.* **82** 524–7

[28] Rai M, Yadav A and Gade A 2009 Silver nanoparticles as a new generation of antimicrobials *Biotechnol. Adv.* **27** 76–83

[29] Masheane M, Nthunya L, Malinga S, Nxumalo E, Barnard T and Mhlanga S 2016 Antimicrobial properties of chitosan-alumina/f-MWCNT nanocomposites *J. Nanotechnol.* **2016** 5404529

[30] Tikhonov V E, Radigina L A and Yamskov Y A 1996 Metal-chelating chitin derivatives via reaction of chitosan with nitrilotriacetic acid *Carbohydr. Res.* **290** 33–41
[31] Dutta J, Tripathi S and Dutta P K 2012 Progress in antimicrobial activities of chitin, chitosan and its oligosaccharides: a systematic study needs for food applications *Food Sci. Technol. Int.* **18** 3–34
[32] Rajendran R, Abirami M, Prabhavathi P, Premasudha P, Kanimozhi B and Manikandan A 2015 Biological treatment of drinking water by chitosan based nanocomposites *Afr. J. Biotechnol.* **14** 930–6
[33] Abebe L S, Chen X and Sobsey M D 2016 Chitosan coagulation to improve microbial and turbidity removal by ceramic water filtration for household drinking water treatment *Int. J. Environ. Res. Public Health.* **13** 269
[34] Hadwiger L, Kendra D, Fristensky B and Wagoner W 1986 Chitosan both activates genes in plants and inhibits RNA synthesis in fungi *Chitin in Nature and Technology* (Berlin: Springer) pp 209–14
[35] Shahidi F, Arachchi J K V and Jeon Y-J 1999 Food applications of chitin and chitosans *Trends Food Sci. Technol.* **10** 37–51
[36] Papineau A M, Hoover D G, Knorr D and Farkas D F 1991 Antimicrobial effect of water-soluble chitosans with high hydrostatic pressure *Food Biotechnol.* **5** 45–57
[37] Chen C-S, Liau W-Y and Tsai G-J 1998 Antibacterial effects of N-sulfonated and N-sulfobenzoyl chitosan and application to oyster preservation *J. Food Protect.* **61** 1124–8
[38] Chen S, Wu G, Long D and Liu Y 2006 Preparation, characterization and antibacterial activity of chitosan–$Ca_3V_{10}O_{28}$ complex membrane *Carbohydr. Polym.* **64** 92–7
[39] Goy R C, Britto D and Assis O B 2009 A review of the antimicrobial activity of chitosan *Polímeros.* **19** 241–7
[40] García M, Díaz R, Puerta F, Beldarraín T, González J and González I 2010 Influence of chitosan addition on quality properties of vacuum-packaged pork sausages *Food Sci. Technol.* **30** 560–4
[41] Yadav S K, Mahapatra S S, Yadav M K and Dutta P K 2013 Mechanically robust biocomposite films of chitosan grafted carbon nanotubes via the (2+ 1) cycloaddition of nitrenes *RSC Adv.* **3** 23631–7
[42] Nthunya L N, Masheane M L, Malinga S P, Nxumalo E N and Mhlanga S D 2017 Environmentally benign chitosan-based nanofibres for potential use in water treatment *Cogent Chem.* **3** 1357865
[43] Alba M D, Cota A, Osuna F J, Pavón E, Perdigón A C and Raffin F 2019 Bionanocomposites based on chitosan intercalation in designed swelling high-charged micas *Sci. Rep.* **9** 1–9
[44] Dehaghi S M, Rahmanifar B, Moradi A M and Azar P A 2014 Removal of permethrin pesticide from water by chitosan–zinc oxide nanoparticles composite as an adsorbent *J. Saudi Chem. Society.* **18** 348–55
[45] Sureshkumar V, Daniel S K, Ruckmani K and Sivakumar M 2016 Fabrication of chitosan–magnetite nanocomposite strip for chromium removal *Appl. Nanosci.* **6** 277–85
[46] Abd-Elhakeem M A and Alkhulaqi T A 2014 Simple, rapid and efficient water purification by chitosan coated magnetite nanoparticles *J. Environ. Nanotechnol.* **3** 17–20
[47] Hritcu D, Dodi G and Popa M I 2012 Heavy metal ions adsorption on chitosan-magnetite microspheres *Int. Rev. Chem. Eng.* **4** 364–8

[48] Saifuddin N and Dinara S 2011 Pretreatment of palm oil mill effluent (POME) using magnetic chitosan *E-J. Chem.* **8** S67–78

[49] Siddeeg S M, Tahoon M A, Mnif W and Ben Rebah F 2020 Iron oxide/chitosan magnetic nanocomposite immobilized manganese peroxidase for decolorization of textile wastewater *Processes.* **8** 5

[50] Morsi R E, Alsabagh A M, Nasr S A and Zaki M M 2017 Multifunctional nanocomposites of chitosan, silver nanoparticles, copper nanoparticles and carbon nanotubes for water treatment: antimicrobial characteristics *Int. J. Biol. Macromol.* **97** 264–9

[51] Azlan K, Saime W N W and Liew L 2009 Chitosan and chemically modified chitosan beads for acid dyes sorption *J. Environ. Sci.* **21** 296–302

[52] Tjong S C 2006 Structural and mechanical properties of polymer nanocomposites *Mater. Sci. Eng.: R: Rep.* **53** 73–197

[53] Ray S S and Iroegbu A O C 2021 Nanocellulosics: benign, sustainable, and ubiquitous biomaterials for water remediation *ACS Omega.* **6** 4511–26

[54] Sain M and Oksman K 2006 *Introduction to Cellulose Nanocomposites* (Washington DC: ACS Publications)

[55] Mahmoud M E, Abdou A E, Sobhy M E and Fekry N A 2017 Solid–solid crosslinking of carboxymethyl cellulose nanolayer on titanium oxide nanoparticles as a novel biocomposite for efficient removal of toxic heavy metals from water *Int. J. Biol. Macromol.* **105** 1269–78

[56] Luo X, Lei X, Cai N, Xie X, Xue Y and Yu F 2016 Removal of heavy metal ions from water by magnetic cellulose-based beads with embedded chemically modified magnetite nanoparticles and activated carbon *ACS Sustain. Chem. Eng.* **4** 3960–9

[57] Zhu C, Liu P and Mathew A P 2017 Self-assembled TEMPO cellulose nanofibers: graphene oxide-based biohybrids for water purification *ACS Appl. Mater. Interfaces.* **9** 21048–58

[58] Oyewo O A, Elemike E E, Onwudiwe D C and Onyango M S 2020 Metal oxide-cellulose nanocomposites for the removal of toxic metals and dyes from wastewater *Int. J. Biol. Macromol.* **164** 2477–96

[59] Oyewo O A, Adeniyi A, Sithole B B and Onyango M S 2020 Sawdust-based cellulose nanocrystals incorporated with ZnO nanoparticles as efficient adsorption media in the removal of methylene blue dye *ACS Omega.* **5** 18798–807

[60] Kamel S, El-Gendy A A, Hassan M A, El-Sakhawy M and Kelnar I 2020 Carboxymethyl cellulose-hydrogel embedded with modified magnetite nanoparticles and porous carbon: effective environmental adsorbent *Carbohydr. Polym.* **242** 116402

[61] Fu F, Li L, Liu L, Cai J, Zhang Y and Zhou J *et al* 2015 Construction of cellulose based ZnO nanocomposite films with antibacterial properties through one-step coagulation *ACS Appl. Mater. Interfaces.* **7** 2597–606

[62] Guo X and Chen F 2005 Removal of arsenic by bead cellulose loaded with iron oxyhydroxide from groundwater *Environ. Sci. Technol.* **39** 6808–18

[63] Fan H, Ma X, Zhou S, Huang J, Liu Y and Liu Y 2019 Highly efficient removal of heavy metal ions by carboxymethyl cellulose-immobilized Fe_3O_4 nanoparticles prepared via high-gravity technology *Carbohydr. Polym.* **213** 39–49

[64] Dong F, Xu X, Shaghaleh H, Guo J, Guo L and Qian Y *et al* 2020 Factors influencing the morphology and adsorption performance of cellulose nanocrystal/iron oxide nanorod composites for the removal of arsenic during water treatment *Int. J. Biol. Macromol.* **156** 1418–24

[65] Li Y, Zhang J, Zhan C, Kong F, Li W and Yang C *et al* 2020 Facile synthesis of TiO_2/CNC nanocomposites for enhanced Cr (VI) photoreduction: synergistic roles of cellulose nanocrystals *Carbohydr. Polym.* **233** 115838

[66] Lin Z and Huang J 2021 A hierarchical $H_3PW_{12}O_{40}$/TiO_2 nanocomposite with cellulose as scaffold for photocatalytic degradation of organic pollutants *Sep. Purif. Technol.* **264** 118427

[67] Althomali R H, Alamry K A, Hussein M A and Khan A 2021 A green nancomposite based modified cellulose/TiO_2/cinnamon bark for the reduction of toxic organic compounds using spectrophotometric technique *J. Mater. Res. Technol.* **12** 947-66

[68] Sirajudheen P, Nikitha M R, Karthikeyan P and Meenakshi S 2020 Perceptive removal of toxic azo dyes from water using magnetic Fe_3O_4 reinforced graphene oxide–carboxymethyl cellulose recyclable composite: adsorption investigation of parametric studies and their mechanisms *Surf. Interfaces.* **21** 100648

[69] Wei X, Huang T, Yang J-H, Zhang N, Wang Y and Zhou Z-W 2017 Green synthesis of hybrid graphene oxide/microcrystalline cellulose aerogels and their use as superabsorbents *J. Hazard. Mater.* **335** 28–38

[70] Ma T, Chang P R, Zheng P, Zhao F and Ma X 2014 Fabrication of ultra-light graphene-based gels and their adsorption of methylene blue *Chem. Eng. J.* **240** 595–600

[71] Mittal H, Ballav N and Mishra S B 2014 Gum ghatti and Fe3O4 magnetic nanoparticles based nanocomposites for the effective adsorption of methylene blue from aqueous solution *J. Ind. Eng. Chem.* **20** 2184–92

[72] Sharma B, Thakur S, Mamba G, Gupta R K, Gupta V K and Thakur V K 2021 Titania modified gum tragacanth based hydrogel nanocomposite for water remediation *J. Environ. Chem. Eng.* **9** 104608

[73] Bag J, Mukherjee S, Ghosh S K, Das A, Mukherjee A and Sahoo J K *et al* 2020 Fe_3O_4 coated guargum nanoparticles as non-genotoxic materials for biological application *Int. J. Biol. Macromol.* **165** 333–45

[74] Palem R R, Shimoga G, Kang T J and Lee S-H 2020 Fabrication of multifunctional Guar gum-silver nanocomposite hydrogels for biomedical and environmental applications *Int. J. Biol. Macromol.* **159** 474–86

[75] Thakur S, Sharma B, Verma A, Chaudhary J, Tamulevicius S and Thakur V K 2018 Recent approaches in guar gum hydrogel synthesis for water purification *Int. J. Polym. Anal. Charact.* **23** 621–32

[76] Thombare N, Jha U, Mishra S and Siddiqui M 2016 Guar gum as a promising starting material for diverse applications: a review *Int. J. Biol. Macromol.* **88** 361–72

[77] Jing G, Wang L, Yu H, Amer W A and Zhang L 2013 Recent progress on study of hybrid hydrogels for water treatment *Colloids Surf. A: Physicochem. Eng. Aspects.* **416** 86–94

[78] Saya L, Malik V, Singh A, Singh S, Gambhir G and Singh W R *et al* 2021 Guar gum based nanocomposites: role in water purification through efficient removal of dyes and metal ions *Carbohydr. Polymers.* 117851

[79] Torres M G 2020 3D-composite scaffolds from radiation-induced chitosan grafted poly (3-hydroxybutyrate) polyurethane *Mater. Today Commun.* **23** 100902

[80] Mittal H, Al Alili A, Morajkar P P and Alhassan S M 2021 Graphene oxide crosslinked hydrogel nanocomposites of xanthan gum for the adsorption of crystal violet dye *J. Mol. Liquids.* **323** 115034

[81] Balachandramohan J, Sivasankar T and Sivakumar M 2020 Facile sonochemical synthesis of Ag_2O-guar gum nanocomposite as a visible light photocatalyst for the organic transformation reactions *J. Hazard. Mater.* **385** 121621

[82] Sahraei R, Pour Z S and Ghaemy M 2017 Novel magnetic bio-sorbent hydrogel beads based on modified gum tragacanth/graphene oxide: removal of heavy metals and dyes from water *J. Clean. Prod.* **142** 2973–84

[83] Kurecic M and Sfiligoj M 2012 Polymer nanocomposite hydrogels for water purification *Nanocomposites-New Trends Developments* (Books on Demand) pp 161–85

[84] Gil A, Assis F, Albeniz S and Korili S 2011 Removal of dyes from wastewaters by adsorption on pillared clays *Chem. Eng. J.* **168** 1032–40

[85] Guo F, Aryana S, Han Y and Jiao Y 2018 A review of the synthesis and applications of polymer–nanoclay composites *Appl. Sci.* **8** 1696

[86] Lee S M and Tiwari D 2012 Organo and inorgano-organo-modified clays in the remediation of aqueous solutions: an overview *Appl. Clay Sci.* **59** 84–102

[87] Uddin M K 2017 A review on the adsorption of heavy metals by clay minerals, with special focus on the past decade *Chem. Eng. J.* **308** 438–62

[88] Jawaid M, Qaiss A and Bouhfid R 2016 *Nanoclay Reinforced Polymer Composites: Nanocomposites and Bionanocomposites* (Berlin: Springer)

[89] Jlassi K, Krupa I and Chehimi M M 2017 Overview: clay preparation, properties, modification *Clay-Polymer Nanocomposites* (Amsterdam: Elsevier) pp 1–28

[90] Wesley L R (ed.) 2014 *Clays and Clay Minerals: Geological Origin, Mechanical Properties and Industrial* Applications (New York: Nova Science Publishers)

[91] Vo V-S, Mahouche-Chergui S, Babinot J, Nguyen V-H, Naili S and Carbonnier B 2016 Photo-induced SI-ATRP for the synthesis of photoclickable intercalated clay nanofillers *RSC Adv.* **6** 89322–7

[92] Guerrouache M, Mahouche-Chergui S, Chehimi M M and Carbonnier B 2012 Site-specific immobilisation of gold nanoparticles on a porous monolith surface by using a thiol–yne click photopatterning approach *Chem. Commun.* **48** 7486–8

[93] Georgiadou V, Kokotidou C, Le Droumaguet B, Carbonnier B, Choli-Papadopoulou T and Dendrinou-Samara C 2014 Oleylamine as a beneficial agent for the synthesis of $CoFe_2O_4$ nanoparticles with potential biomedical uses *Dalton Trans.* **43** 6377–88

[94] Nikolaidis A K, Achilias D S and Karayannidis G P 2011 Synthesis and characterization of PMMA/organomodified montmorillonite nanocomposites prepared by *in situ* bulk polymerization *Ind. Eng. Chem. Res.* **50** 571–9

[95] Dadashi-Silab S, Atilla Tasdelen M and Yagci Y 2014 Photoinitiated atom transfer radical polymerization: current status and future perspectives *J. Polym. Sci. Part. A: Polym. Chem.* **52** 2878–88

[96] Chakrabarty A, Zhang L, Cavicchi K A, Weiss R and Singha N K 2015 Tailor-made fluorinated copolymer/clay nanocomposite by cationic RAFT assisted pickering miniemulsion polymerization *Langmuir* **31** 12472–80

[97] Peng Y, Huang H, Zhang Y, Kang C, Chen S and Song L *et al* 2018 A versatile MOF-based trap for heavy metal ion capture and dispersion *Nat. Commun.* **9** 1–9

[98] Auta M and Hameed B 2014 Chitosan–clay composite as highly effective and low-cost dsorbent for batch and fixed-bed adsorption of methylene blue *Chem. Eng. J.* **237** 352–61

[99] Mahdavinia G R, Aghaie H, Sheykhloie H, Vardini M T and Etemadi H 2013 Synthesis of CarAlg/MMt nanocomposite hydrogels and adsorption of cationic crystal violet *Carbohydr. Polym.* **98** 358–65

[100] Mahdavinia G R, Massoudi A, Baghban A and Shokri E 2014 Study of adsorption of cationic dye on magnetic kappa-carrageenan/PVA nanocomposite hydrogels *J. Environ. Chem. Eng.* **2** 1578–87

[101] Zhou Q, Gao Q, Luo W, Yan C, Ji Z and Duan P 2015 One-step synthesis of amino-functionalized attapulgite clay nanoparticles adsorbent by hydrothermal carbonization of chitosan for removal of methylene blue from wastewater *Colloids Surf. A: Physicochem. Eng. Aspects.* **470** 248–57

[102] Zhou Y, Dai Z, Zhai D and Gao C 2015 Surface modification of polypiperazine-amide membrane by self-assembled method for dye wastewater treatment *Chin. J. Chem. Eng.* **23** 912–8

[103] Lee K Y and Mooney D J 2012 Alginate: properties and biomedical applications *Prog. Polym. Sci.* **37** 106–26

[104] Ely A, Baudu M, Basly J-P and Kankou M O S A O 2009 Copper and nitrophenol pollutants removal by Na-montmorillonite/alginate microcapsules *J. Hazard. Mater.* **171** 405–9

[105] Lezehari M, Baudu M, Bouras O and Basly J-P 2012 Fixed-bed column studies of pentachlorophenol removal by use of alginate-encapsulated pillared clay microbeads *J. Colloid Interface Sci.* **379** 101–6

[106] Özkan H, Dişli A, Yıldırır Y and Türker L 2007 A novel synthesis of bromobenzenes using molecular bromine *Molecules* **12** 2478–83

[107] Celis R, Adelino M, Hermosín M and Cornejo J 2012 Montmorillonite–chitosan bionanocomposites as adsorbents of the herbicide clopyralid in aqueous solution and soil/water suspensions *J. Hazard. Mater.* **209** 67–76

IOP Publishing

Bionanomaterials for Environmental and Agricultural Applications

Ravindra Pratap Singh and Kshitij RB Singh

Chapter 8

Utility of bionanocomposites for wastewater treatment

Uyiosa Osagie Aigbe, Kingsley Eghonghon Ukhurebor, Robert Birundu Onyancha, Onoyivwe Monday Ama, Benedict Okundaye, Ese Lawrence Esiekpe, Otolorin Adelaja Osibote, Heri Septya Kusuma and Peter Ogbemudia Osifo

Owing to the growing water scarcity crisis globally and the significant amount of contaminants released from natural and anthropogenic activities and from various industrial wastewater sources into water bodies without proper treatment, which subsequently enters the soil and disrupts aquatic and terrestrial life, there is a growing need for innovative methods utilizing nanoadsorbtion such as bionanocomposites for the sequestration of contaminants owing to the renewable nature, bio-degradability, and bio-compatibility of these materials. Hence, this chapter reviews the literature on various bionanocomposite applications in the treatment of industrial effluents containing various pollutants, factors and mechanisms affecting the sorption of these pollutants.

8.1 Introduction

Water is life and there is a greater need for its protection and preservation for the sake of future generations. Today's world is confronted with environmental degradation challenges, especially water contamination through activities of industries like mining, tanneries, textiles, electroplating, pharmaceuticals, petrochemicals, metallurgy and others. Additional causes of water contamination include agricultural processes, septic and sewage tank seepage, flooding due to worldwide warming and change in climate [1]. The threat to the pristine environment in search of industrial development is a major global concern. Owing to this development, the effects of environmental waste comprising noxious and hazardous substances is becoming ever more a key global fear. Among the various substances harmful to the health of humans are heavy metals (HMs), pharmaceuticals, pesticides, and dyes [2].

doi:10.1088/978-0-7503-3863-9ch8

A worldwide and intensifying issue that has emerged in recent times is the pollution of water and its remediation, and of specific concern is heavy metal(loid)s presence [3]. Contamination of water by noxious HMs is a severe global nuisance due to their harmful impacts on social wellbeing and the natural environment. The world today is confronted with severe threats of contamination of water, air, and land [4].

Though HMs occur naturally in the Earth's crust, their increased content resulting in environmental pollution with considerable influence on human health is due to anthropogenic activities relating to domestic, transport, and industrial expansion linked with the growing population [5, 6]. In particular, contamination of water has raised serious fears and in addition to the dearth of water resources, the creation of huge volumes of effluents has also put huge pressure on humanity [2]. As per the 2004 French Plan of National Sant Environment (PNSE), the detection of pollutants in water testing is mandatory before water is distributed to the public [7].

In organic and inorganic pollutants treatment, the widely employed techniques employed are physical, chemical and biological [8]. A wide variety of conventional techniques employed in the treatment of noxious HM-containing effluents are insufficient to deal with the wastewater treatment issues [2, 4]. They include electrochemical reduction [9], coagulation–flocculation [10], chemical-precipitation [11], ion exchange [12], photocatalysis [13], and membrane technologies [14]. But most of these treatment techniques are plagued by high operational cost and low treatment effectiveness [2].

Owing to these drawbacks, a lot of attention has been focussed on using sorption technology [15]. Sorption is a well-known efficient and financially viable technique for the treatment of HM effluent. It offers design flexibility and operation ease, extraordinary output, minimal cost, and ease of control and in several instances, the production of high-quality treated wastewater [4, 16]. The sorption is attributed to the appropriate oxygen-containing functional groups and ample reaction sites. To evaluate the key mechanism of sorption, theoretical calculations, sophisticated spectroscopic analysis, and external complexation models are utilized to explore the sorption behaviour. The general sorption mechanisms comprise explicit surface bonding, ion-exchange, electrostatic sorption, hydrogen bonding, etc [17].

Chemisorption and physisorption are two types of sorption processes. In chemisorption, the sorbate is attached to the sorbent material via electrostatic attraction or covalent bonding. Unreversible sorption with a high sorption enthalpy (H) of 200 $KJ.mol^{-1}$, elevated energy of activation, and sturdier electrostatic forces including monolayer sorption and covalent bonding define it. Chemisorption has outstanding bond energy and the bond length is short, while in physisorption, the affixation of the sorbate to the sorbent surface is via van der Waals forces. It is defined as reversible sorption, with a minimum H value of 20 $KJ.mol^{-1}$, multi-layered sorption, a minimum temperature that is below the critical temperature, and minimal energy of activation [18].

The sorption process is an outstanding method utilized for the extraction of HM ions from water-soluble solutions by accurately positioning them directly on the selected sorbent surface [19]. Over the past years, nanomaaterials (NMs) have gained huge consideration and have been used in various fields such as health care, energy, electronic devices, environmental application etc. Under the nanoscale (NS), NMs show some unique properties like quantum effect, macro quantum tunnel effects, surface effect, and small size effect. These properties promote reactivity and remarkable sorption capacity, both of which are favourable for heavy metal ions (HMIs) sequestration [9].

In solving water and wastewater issues, NMs play an important role owing to their chemical and physical properties like a high activity for sorption and photocatalysis, antimicrobial properties for disinfection, improved functional sites, and on their surface they have profuse functional groups. Also, magnetism and other distinctive electrical and optical properties make these materials exceptional candidates for the treatment of wastewater [20].

There has been recent immeasurable development in NS science and engineering with regards to the purification of water for safe drinking employing nanocomposites (NCs), nanoadsorbents, nanocatalysts etc [2]. Among the several sorbent materials, bionanocomposites (BNCs) is a new area of academic research evidenced by the few but increasing numbers of papers published in indexed journals with the key phrase 'bionanocomposite for HMs removal' over the last decade, as shown in figure 8.1, obtained from SCOPUS. Hence, this chapter highlights the application of BNCs for the sorption of HMI from a water-soluble solution owing to their availability, biodegradable nature, selectivity, sorption capacity, stability over nanoparticles (NPs) and ecological-friendly tendencies leading to green chemistry. Also discussed in detail are HM noxiousness and its impact on individual wellbeing

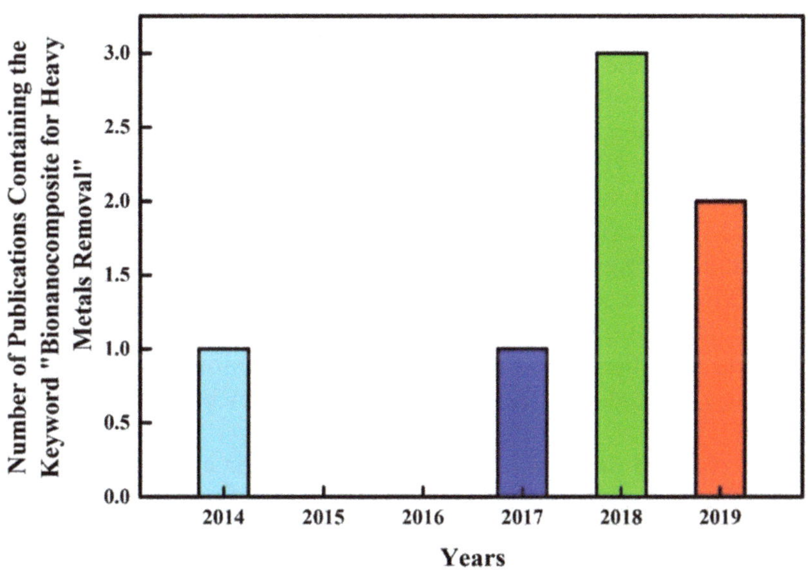

Figure 8.1. Illustration of the number of publications comprising the key phrase 'bionanocomposite for heavy metals removal' published from 2014–2019. The data was acquired from 'SCOPUS'.

and the ecosystem. This chapter also looks at the use of sorption technique in the sequestration of various HMI under varying factors such as pH, sorbent dosage, initial concentration, temperature, co-existing ions etc. Also explored in this chapter are the mechanism of sorption of HMI onto BNCs using isotherm, and kinetics models and thermodynamics.

8.2 HMs and their toxicity

A worldwide and mounting problem that has become evident is the contamination of water and its treatment. The existence of heavy metal(loid)s is of exceptional interest [3]. The major water pollutants (including groundwater, tap-water and surface water) primarily consist of organic contaminants like pesticide, detergents, pharmaceuticals and biomaterials and inorganic contaminants like HMs (metals with atomic number above 20), and arsenic (As) ions [10].

Over the last twenty years, the word ‘HMs’ has been utilized progressively in several publications and regulations relating to their chemical risks and safe use. It is regularly utilized as a group term for metallic elements and semi-metals (metalloids) linked with possible toxicity or ecotoxicity and pollution. Also, permissible guidelines regularly indicate an HMs list that they refer to. The lists of HMs may vary from one set of established guidelines to the next, and the word may be applied by not indicating which HMs are included [21].

Unlike organic pollutants, HMs are not recyclable, have a tendency to accumulate in living creatures and are persistent in Nature. HMs can effortlessly enter the food chain via several pathways and cause gradual noxious impacts with slow build-up over the life span of existing organisms. Hence, they are very challenging to eradicate spontaneously from the ecosystem, even for residue amounts [6, 12, 16]. They occur in a variety of forms like anions, cations and hydroxide complexes when they enter the water, and this depends on the dissolved oxygen, pH and the content of organic matter [22]. Some of the types of noxious materials that exist in most water resources are mercury (Mg), zinc (Zn), As, lead (Pb), nickel (Ni), chromium (Cr), and cadmium (Cd), [22, 23].

From the International Agency for Research on Cancer (IARC) and the Environmental Protection Agency of the United States (EPAUS) epidemiological and experimental findings, the probable human carcinogens are Hg, Cr, Cd, As and Pd. These HMs are also classified as prime concern metals and are crucial to community health. Thus, they can induce multiple organ impairment on exposure to regular low concentration owing to their high-grade noxiousness. Due to the degree of their toxicity, damage to various organs is caused by exposure at even minimal concentration [5].

They have received unusual attention owing to their carcinogenic nature, their selective noxiousness even at low concentration and their lack of degradability. Over-exposure to these noxious HMs can cause cardiovascular, kidney, brain, liver, nerve, and endocrine structure impairment and in grave circumstances, death. Given the detrimental impacts of HMs, they must be removed from liquefied wastes at acceptable regulation levels [2]. An overview of different HMs, their effects and the

Table 8.1. Outline of various noxious metal, their effects and recommended allowed levels in water.

HM	Effects	Recommended allowed levels (WHO and USEPA)	References
Pb	Anaemia, brain injury, anorexia, malaise, loss of appetite, damage of liver, kidney, and gastrointestinal, mental retardation in children, and nervous system syndrome.	15 $\mu g\ L^{-1}$	[23–25]
Ni	Long-lasting bronchitis, decreased lung function, cancer of the lungs, contact dermatitis, cardiovascular disease, asthma, lung fibrosis, and cancer of the respiratory tract	0.02 ppm	[24, 26]
Cr	Acute diarrhoea, epigastric pain, nausea, vomiting, lung tumours production	100 $\mu g\ L^{-1}$	[24, 25]
Hg	Nervous system damage, poisoning of the protoplasm, harsh to skin, eyes, muscles, dermatitis, damage of the kidney, and gastrointestinal (GI)	1 ppm	[23, 24]
Cu	Neurotoxicity, and severe noxiousness, dizziness, diarrhoea	1300 $\mu g\ L^{-1}$	[24, 25]
Cd	Harm to the kidneys, respiratory systems, and skeleton.	0.005 ppm	[15, 27]
As	Initiates cancers of the urinary tract, bladder and skin, bronchitis, dermatitis, bone marrow depression, haemolysis, and hepatomegaly	10 ppm (WHO)	[23]
Zn	Triggers fleeting term metal-fume fever, gastrointestinal pain	3 ppm	[24, 28]

USEPA and the World Health Organization (WHO) recommended allowed levels of HMI in water are given in table 8.1

8.3 Nanomaterials as sorbents for wastewater treatment

NMs have an extensive array of applications, in environmental and technological tasks in the fields of medicine, catalysis, conversion of energy and treatment of contaminated water [29]. A promising method to advance eco-friendly remediation procedures via ‘nanotechnology’ (NT), is described as a consortium of developing technologies functioning in the nanometer (nm) scale (1–100 nm range) to create devices, systems and materials with essentially novel properties, and works by regulating the shape and size of matter. Comprehensive NT, owing to its prospective applications in the food industry, energy, and treatment of contaminations, is presenting an advancing potential in the development and transformation of traditional remediation technologies [30].

NT growth offers a favourable alternative to improve water treatment efficiency. NMs (1–100 nm) have exclusive chemical and physical properties like specific affinity, large surface area and surface activity. They have attracted considerable interest in the identification and sequestration of inorganic HMs and organic pollutants from water [10].

NT plays a key part in the development of nanosorbents, and the large specific surface area that directly improves the sorption capacity is offered by NMs. But their colloidal suspension stability, accumulation and reusability are the key bottlenecks for their industrial use, with NMs aggregation having a negative impact on the removal performance. As a result, their surface functionalization is discovered to cause self-repulsion features and successively prevent aggregation. Also, specific functional groups are imparted by the surface functionalization which also plays a part in HMs enhanced removal and changes the sorption mechanism from physical sorption to chemisorption. The colloidal suspension's improved stability also allows for the desorption of sorbed MIs and the successive reuse of the nanosorbent. Numerous intermolecular forces like the electrostatic forces, hydrogen bonding, and van der Waals forces control the chemisorption of HMs on the modified sorbents. While the complex sorption process is controlled by surface-MIs, MI–MI and surface–surface interactions. The sorption capacity of HMs is directly impacted by the surface structures (porosity, precise surface area, and size of pore) [13].

In recent times, NT has developed its roots in the remediation of the environment [5]. To decontaminate water polluted with organic and inorganic contaminants, the various sorbents (bare or surface modified) mostly utilize and display huge capacities as nanosorbents are activated, with silica gel, graphite oxide (GO), metal oxides (MeO), biopolymers (BPs), activated carbon (AC), chelating minerals, engineered NMs and other NCs [29, 31, 32]. The subsequent sections will focus on BNCs as sorbents for HMI removal under the impacts of varying factors like time of contact, solution pH, the dosage of the sorbent applied, and temperature.

8.4 BNCs as a sorbent for wastewater treatment

During the last decades, BNCs have transformed into a regular term specified by naturally occurring polymer (biopolymer) in combination with inorganic moiety and which appears no less than one measurement on the nanometer scale (less than 1000 nm) [33]. Owing to their availability in large amounts, they have increased global attention and their ecological-friendly tendencies lead to green chemistry [34]. BNCs, (bio-hybrid materials) include natural species (mostly BPs) and mineral solids of wide-ranging morphology and structures bonded at the nanometer scale. They differ from NCs in terms of their preparation technique, applications, properties, functionalities, biodegradability, and biocompatibility. While the term BNCs is just coming into public use, they are often referred to as green-composites, bio-composites, NCs, bio-hybrids or bio-based plastics (bioplastics) and nano BNCs [32].

BPs are polymers that exist in Nature and show prospective applications in various fields like agricultural science, sanitary products, treatment of effluents and delivery of drugs [35]. The BPs can generally be divided into natural BPs like plant and animal source proteins like casein, whey-protein, soy-protein, corn-zein, wheat-gluten, gelatin, collagen, etc, and plant carbohydrate (cellulose, starch, alginate, agar, carrageenan, chitosan, etc), synthetic biodegradable BPs like poly-lactic acid (poly(L-lactide) (PLA), poly(glycolic acid) (PGA), poly(butylene succinate) (PBS), poly(vinyl alcohol) (PVA), poly(ε-caprolactone) (PCL), etc) and BPs created by microbial fermentation (poly(hydroxyalkanoates) (PHAs) comprising poly(β-hydroxybutyrate) (PHB), poly(3-hydroxybutyrate-co-3-hydroxyvalerate) (PHBV), etc) and microbial polysaccharides, like curdlan and pullulan [36].

BNCs are generally synthesized employing *in situ* polymerization, solution blending or casting and melt processing. Lately, under supercritical condition, other synthesis methods like electrospinning and processing have gained interest. Characteristically, *in situ* polymerization, melt processing or blending and solution casting or blending are implemented in the preparation of polymer NCs as the goal during processing is the proper dispersion of the NPs [35].

The all-purpose procedures generally employed for the synthesis of BNCs include: (i) *in situ* intercalation polymerization: NPs are dispersed in a liquid monomer solution for polymer formation among the intercalated sheets, and polymerization is achieved by the diffusion of an appropriate initiator or catalyst via heat or radiation. It is among the basic methods employed in the synthesis of NCs, nonetheless, it is impossible to grasp it in the case of bipolymers; (ii) melt-intercalation: to acquire a molten mass, the polymer is heated at a precise temperature and combined with NPs to dispense them appropriately in the polymer matrix. Their preparation is generally accomplished utilizing the extruder, and this method can be employed just for thermoplastic as well as proteins that meet the criterion; (iii) template synthesis: bio-molecules, parts and entire cells, and micro-organisms help as a model to create inorganics from a precursor and the templating bio-organics (nanosize particles) are captured in a mesoporous matrix. They can be prepared as nano-sized particles, with covering and mesoporous matrix in which templating bio-organic is trapped. This method is very resourceful and in various BNCs synthesis, it is being adapted. The sol–gel chemistry techniques are regularly used in the preparation of the inorganic component; and (iv) solution intercalation: BPs like starch and protein are inserted into a solvent that is entirely water-soluble in the solvent. The ensuing mixture of initial solutions of bio-material and dispersed NS particles results in the thickening, precipitation or jellification ascertained by the interactions between the elements and their association [33, 35, 37]. With the mixture of the BPs and the solution of swollen NPs (silicate platelets), after the solvent has been extracted, the solvent within the interlayer of the silicate is displaced and the polymer chains intercalate, the resulting intercalated structure forms a biopolymer [33, 38, 39].

BNCs are categorized based on their origin, forms of matrix utilized, reinforcements size and shape. While based on the shape of particles reinforcement, they are also classified as layered structures, particulates, and elongated particles [32, 40, 41]. For particular BNCs, iso-dimensional particles are mainly employed as supports and the strengthening force is moderate owing to the minimal aspect ratio. This form of support is used to decrease permeability, composites cost and improve resistance to flammability. While the elongated particle BNCs (carbon nanotubes (CNTs) and cellulose nanofibrils) are utilized as support in the elongated particle BNCs [42].

Owing to their high aspect ratio of this support, they have enhanced mechanical behaviour. The layer supported BNCs (layered polymer NCs) are grouped into three secondary groups based on particles dispersion in the matrix. The flocculated separated NCs are created when there is no barrier between sheets due to the interactions of particle–particle and as specific laminar are not separated, they are referred to as micro-composites. Hence, they act as micro-particles dissolved in the polymeric matrix. Intercalated NCs are obtained from the intercalation of polymer chains between sheets of layered NPs and exfoliated NCs are created when individual layers are separated. They have superb mechanical properties, bio-compatibility, and bio-degradability. They are increasingly being employed as an emerging sorbent for dyes and HMI removal, for applications in biomedicine such as vaccination, dressing of wounds and tissue engineering, and drug delivery systems, and BNCs films which are cost-effective, eco-friendly, and renewable materials for food packaging with enhanced antimicrobial activity [32].

For instance, novel chitosan-g-PMMA/silica biopolymer-based hybrid NCs synthesized through the emulsifier-free emulsion polymerization technique were employed for Cr(VI) removal. As the solution pH was amplified from pH1–4/5, Cr (VI) removal became more effective, with ideal elimination of Cr(VI) being detected at pH 4. At higher pH, the increased removal of Cr(VI) was due to the ionization of functional groups on the chelating sorbent surface, which involved the protonation and complex binding reactions competition. While at a low pH range (acidic medium), the reduced sorption of the sorbate to the sorbent was due to a large number of hydrogen ions contending with the MI for the active sites on the sorbent surface [34].

Also, magnetic layered double hydroxide/guargum BNCs prepared using facile and green methods were applied in the sequestration of Cd(II) from a water-soluble solution. A substantial increase in the removal of Cd(II) was noticed from pH 7–10, with optimum removal of Cd(II) onto the BNCs observed at pH 10 and was attributed to sorption. The pseudo-second-order (PSO) and Langmuir (LNR) models were most suitable to explain Cd(II) sorption onto the BNCs from the sorption studies, with an observed sorption capacity of 258 $mg.g^{-1}$ [43].

The *in situ* creation of AgNPs on the surface of *S. cerevisiae* using fulvic acids as reductants under virtual sunlight resulted in a novel BNC, *S.*cerevisiae-AgNPs (*S*/AgNPs). It was observed that the enhanced concentration of *S*/AgNPs hindered the high concentration of *E. coli* growth and the sorption of Cd(II) was noticed to be reliant on the pH with the ideal removal efficiency detected at pH 6 and was attributed to the negatively charged cell surface and the positively charged Cd(II) interaction. The sorption process of Cd(II) was best fitted using the LNR and PSO models. The observed sorption capacity was 15.01 $mg.g^{-1}$. Using the *S*/AgNPs to treat simulated pollution with various HMs, it was discovered to be effective for the sequestration of these HMs with a removal efficiency of 12.47, 6.25, 91.75, 43.58, 19.72, 22.56, and 9.65% for Mn(II), Ni(II), Cr(VI), Pb(II), Cu(II), Hg(II), Zn(II), and As (V), respectively. This was indicative of S/AgNPs BNCs being effective antibacterial agents and a bioremediation agent for the treatment of water [44].

Starch and sodium montmorillonite (starch/Na-MMT) was prepared using the intercalation method in various starch to nano-clay (SN) ratios of 5:1,10:1, and 10:3. The synthesized NCs were employed for the removal of Ni (II) and Co (II). The effective experimental conditions determined for the removal of 100 and 140 mg.L^{-1} with a removal efficiency of 97.1 and 78.07% were utilizing pH 4.5 and 6 and at an SN ratio of 10:1 and 10:3 for Ni (II) and Co (II). The Freundlich isotherm matched

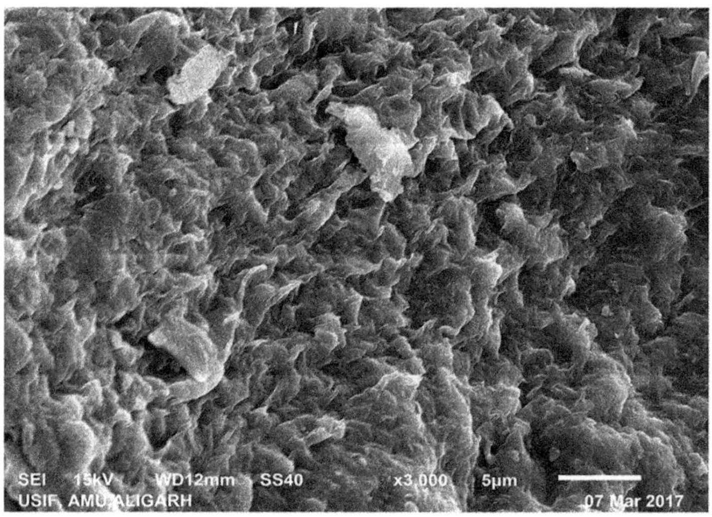

(a)

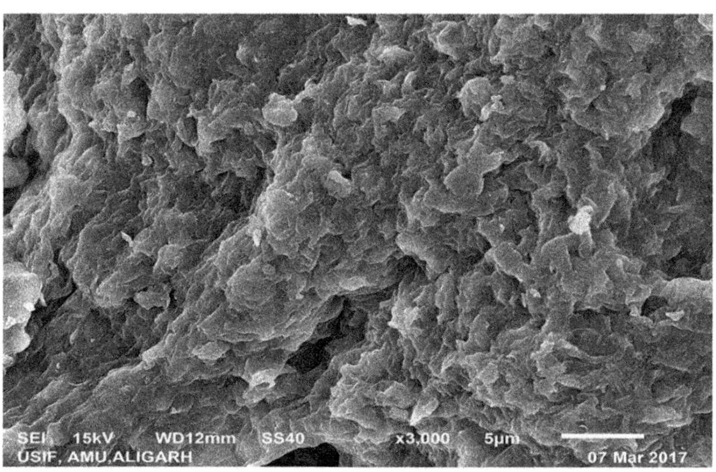

(b)

Figure 8.2. SEM images of the biosorbent (a) before and (b) Pb (II) loaded biosorbent. Reproduced with permission from Mirza and Ahmad, 2018 [47].

the experimental data well, which was indicative of a multilayer sorption process for both HMs [45].

Prepared environmentally-friendly xanthan gum/n-acetyl cysteine modified mica BNCs were employed as a scavenging sorbent for Pb (II), Cu (II) and Ni (II). Ideal sorption of Pb (II), and Cu (II) and Ni (II) was realized at pH 4.5 and 4 and at an experimental time of 60, and 120 min, respectively. The sorption process was dependably portrayed using the LNR model for all MIs at a range of temperature of 303–323 K and Pb (II), Cu (II) and Ni (II) had determined sorption capacities of 530.54, 177.2 and 51.48 $mg.g^{-1}$, respectively. Assessment of the thermodynamic results shows that the sorption of MIs to the BNCs was spontaneous, endothermic and with enhanced arbitrariness at the solid/liquid boundary [46].

Mirza and Ahmed [47] employed bio-degradable xanthan gum/montmorillonite (XG/MMT) BNCs for the sequestration of Pb (II) from manufacturing and synthetic effluents. The SEM images of the biosorbent displayed in figure 8.2(a) and (b), shows a tightly enclosed lattice structure having a porous and intermittent surface and a new continuous surface with pores being covered with MIs after sorption. The sorption process was ideally described using the Freundlich and PSO models. While Pb (II) sorption was determined from the thermodynamics study to be endothermic and spontaneous. The sorbent was effectively rejuvenated using 0.05 M HCl for up to five cycles, showing that this BNC was a promising and potential sorbent for practical applications (figure 8.3).

Chitin BNCs obtained from carb shells were employed in the removal of HMIs. An increase in pH was found to increase the percentage of HMIs extracted. The removal of the HMIs was ideal at pH 10 and was ascribed to the negatively charged

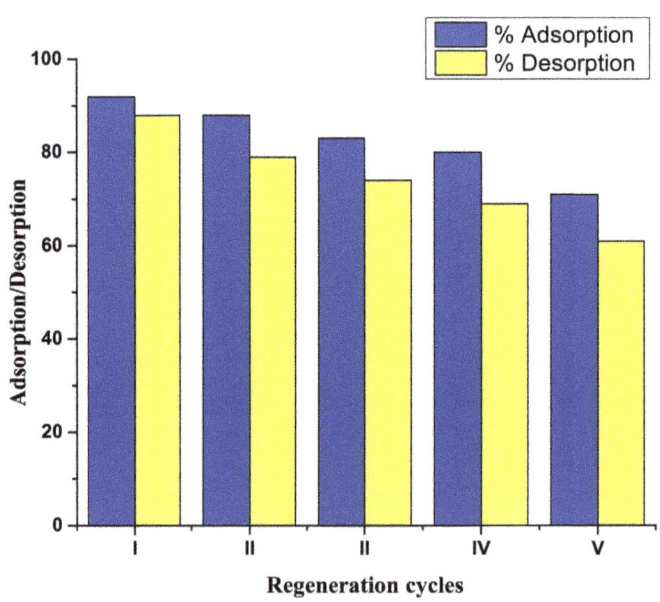

Figure 8.3. Desorption cycle. Reproduced with permission from Mirza and Ahmad, 2018 [47].

sites on the sorbent surface and the positively charged MIs interaction. While the reduced removal at low pH was also owing to the significant concentrations and high mobility of H^+ ions on the sorbent surface leading to a reduced interaction between the HMIs and the sites on the sorbent surface. The sorption process for these HMIs were best fitted using the PSO and Freundlich and Temkin models, with the determined sorption capacities of 142 (Pb (II)), 166.6 (Cu (II)), 138.8 (Ni (II)), 100 (Cd (II)) and 142.85 mg.g^{-1} (Zn (II)). The thermodynamic result of the sorption of HMIs was spontaneous, endothermic, and improved arbitrariness at the solid–liquid interface [48]. Based on the experimental results from previous studies reviewed, BNCs are promising sorbents for the removal of HMIs. A list of BNMs utilized in the sorption of different HMs with their optimum pH, isotherm and kinetics models, thermodynamics and sorption capacity values is summarized in table 8.2.

8.5 Factors affecting sorption of HMs using BNCs

The sorption of HMs onto various sorbents happens through coordination, micro-precipitation, chelation, ion-exchange, mechanism combination, complexation, and electrostatic interaction. The mechanism of sorption is considered by sorption conditions (temperature, pH, the dosage of sorbent used, sorption system types, and volume treated), the sorbent physicochemical characteristics (chemical and texture functionality) and the properties of the sorbate [18]. The three main factors that influence the performance of NMs for HMs sorption which are the pH, sorbent dosage, and contact time are discussed in the subsections that follow.

8.5.1 Effect of pH

In the process of sorption, the most important variable and a controlling factor is the solution pH. During the sorption process, the sorbents surface charge, ionization extent and sorbate speciation are affected by it [46]. It impacts not only the prepared materials surface charge but also alters the HMI distribution, and as a consequence lead to the interaction between contaminants and new NMs electrostatically [17].

The value of pH indicates the H^+ level of concentration in water and electrostatic interaction and ion exchange are considerably affected by the H^+ concentration [22]. The interactions between the sorbent surface and HMIs electrostatically are enhanced by the alteration of the point of zero charge (pH_{pzc}) of the modified sorbents by the surface modification [13]. Remarkably, when the $pH < pH_{pzc}$ and $pH > pH_{pzc}$, the sorbent displays a positive and negative charge [17].

For individual MIs, there is a specific pH value at which intense sorption of the MIs take place and this pH is distributed at high pH for cationic metals and positioned at low pH for anionic MIs. The MI sorption efficiency is dominated by the solution initial pH, as it impacts the catalyst surface chemistry and the ionization and dissociation of the molecules of the sorbent [58].

In this study, the impact of solution pH (2–10) on the sequestration of Cu (II), Cd (II), Zn (II) and Ni(II) was explored applying multi-walled CNTs/chitosan NCs. It was observed that the removal of these MIs was substantially low between the pH

Table 8.2. A summary of different experimental conditions such as optimum pH, isotherm and kinetic models, thermodynamics, and sorption capacity of numerous BNMs for HMs removal from water.

BNCs	HMI	Optimum pH	Maximum sorption capacity (mg. g^{-1})	Isotherm/kinetic models	Thermodynamics ($\Delta H°$and$\Delta G°$-KJ. mol^{-1})	References
Chitosan and gum arabic NPs	Cu (II)	10	303/344.8	LNR/PSO		[49]
Alumina/chitosan (AlCs) composite	Cr (VI)	3	8.62	D-R/PSO and intraparticle diffusion	Spontaneous, endothermic and randomness	[50]
Chitosan/Ag BNCs	Cu (II), Pb (II), and Cd (VI),	5.5–6.5	97, 88, and 89%			[51]
BNCs of eggshell/Ag–Fe	As (V), Pb (II) and Hg (II)		17.7, 27.8 and 15.9	Freundlich, Sip, and Redlich–Peterson/ Elovich, PSO and Elovich		[52]
Biopolymer (chitosan)/nanoclay composites	Cu (II) and Ni (II)		176 and 144			[53]
l-methionine montmorillonite encapsulated guar gum-g-polyacrylonitrile copolymer hybrid NCs	Pb (II) and Cu (II)	5	125 and 90.91	LNR/PSO	Spontaneous, endothermic and randomness	[54]
Chitosan/Mg–Al-layered double hydroxide NCs	Pb (II) and Cd (II)	3	333.3 and 140.8	LNR/PSO		[55]
Functionalised mesoporous KCC–1 and chitosan–oleic acid	Pb (II)	9	68	LNR/PSO	Spontaneous, endothermic and randomness	[56]
Starch-graft-acrylic acid/montmorillonite superabsorbent NC hydrogels	Pb (II) and Cu (II)	4		Freundlich/PSO		[57]

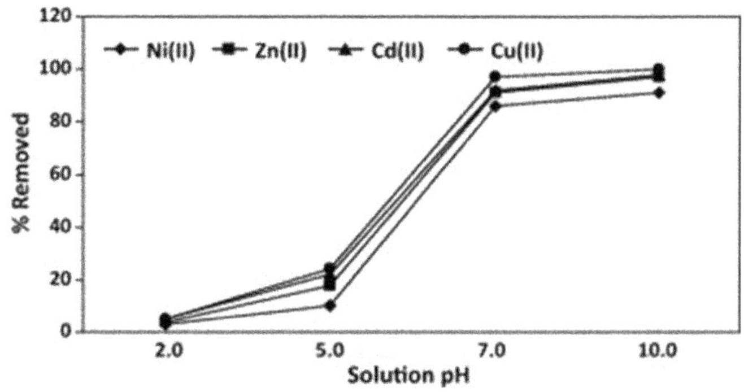

Figure 8.4. pH effect on the biosorption of Ni (II), Zn (II), Cd (II) and Cu (II). Reproduced with permission from Salam *et al*, 2011 [59].

range of 2–5 owing to the contest for the sorbent active sites between protons (H^+) and the target MIs. Enhanced removal of individual MIs was also noticed at elevated pH values, which was attributed to MI hydrolysis (figure 8.4) [59].

In this study, a decrease in the percentage removal of MI was noticed using cross-linked chitosan beads not grafted at lower pH values owing to the amine group of the chitosan protonation that generates an electrostatic repulsion between the MIs and the amine group on the chitosan. With the enhancement in the pH, the beads' surface becomes deprotonated and they are negatively charged. Hence, there is increased removal of these HMIs until an optimum value is attained (pH 5-Pb (II) and Cu (II), pH 6-Zn (II) and Cd (II) and pH 7-Ni (II)). Above the ideal value of removal, the insoluble metal hydroxide starts to precipitate from the solution ensuing in a reduction in the MIs removal [60].

In the study by Ahmed and Mirza [61], BNCs were used to evaluate the impact of pH on the sorption of anionic dye and HMIs. The BNCs' pH_{pzc} was established to be 8.2. when the $pH > pH_{pzc}$, sorption was favourable for positively charged sorbate, and $pH < pH_{pzc}$ when sorption was favourable for negatively charged sorbate. The determined sorption capacity observed at pH 5, 4 and 2.1 for Pb (II), Ni (II) and congo red (CR) were 47.98, 42.91 and 40 $mg.g^{-1}$. An increase in pH led to an improved sorption capacity due to reduced competition between H^+ and the positively charged MIs. The attraction between the positive CR solution in the acidic pH and the negatively charged sites electrostatically also led to improved sorption capacity. Overall, from the research papers reviewed, the sorption of most HMIs using BNCs was observed to be in the pH range of 2–10 as summarized in table 8.2. Although the attraction between the positively charged CR solution and the negatively charged sites electrostatically improves sorption potential in acidic pH.

8.5.2 Effect of sorbent dosage

An important factor that impacts the removal capacities of HMIs is the dosage of NMs. To achieve anticipated removal efficiency, in most publications the ideal dosage of the sorbent is selected and is valuable for the cost-effective application of

NMs [62]. The amount of sorbent utilized in the sequestration of MIs is very crucial owing to it governing the equilibrium between the sorbent and sorbate and the sorption cost for an individual unit of water-soluble solution [58].

A key factor that impacts the removal capacities is the sorbent dosage and the optimum sorbent dosage is used to improve the removal efficiency of HMIs [17]. The amount of MIs sorbed tend to intensify with enhanced sorbent dosage until equilibrium is achieved and it becomes constant. This increase in the quantity of MIs sorbed with enhancing sorbent dosage is ascribed to the increased amount of energetic sites and the enhanced sorbent surface area with rising sorbent dosage [58].

In the study by Salam *et al* [59] and Boulaiche *et al* [48], the percentage of MIs removal using BNCs was swiftly enhanced with an increase in the NCs dosage. The intensification in the NCs dosage ensued in a rise in the amount of active binding sites and enhanced surface area for MI removal. Further increase in the dosage of the NCs resulted in an equilibrium between the NCs/MIs and the non-sorbed MIs in the solution.

Equally, Nithya and Sudha reported an increased removal of MIs with an upsurge in the dosage of BNCs, owing to the accessible NCs binding sites for MI complexation. Equilibrium was attained after 5 g of the sorbent was used, which resulted in reduction in the accessible sites in the sorbent [63]. Wang *et al* [64] indicated in their research paper an increased uptake in the sequestration of Pb (II) (78.3%) and Cu (II) (63%) with an increase in the dosage of the BNCs. The improvement in the removal of these MIs was because of more sorption sites accessibility and surface area increase with increasing dosage of BNCs.

In this study, there was a strong increase in the percentage of MIs removed using sorbent with enhanced sorbent dosage from 2–6 $g.L^{-1}$. This was ascribed to appropriate binding sites for metals complexation. With sorbent dosage above 6 $g.L^{-1}$, there was no further improvement in the removal percentage which ensued in the creation of uniformity between the MIs hemmed into the sorbent and the residual unadsorbed in the mixture [60].

8.5.3 Effect of contact time

The rate of removal during the sorption and redox process is substantially influenced by the contact time between the NPs and the HMIs. To study the impact of contact time, various kinetics models are applied and the rate of sorption of HMIs to NMs rapidly reaches the highest point in the starting stage and gradually slows down with the time until equilibrium is attained [62]. In this study, owing to the profusion of effective sites on the sorbent surface, the sorption of Pb (II) and Cu (II) to BNCs was rapid during the first 150 min, After 150 min, the sorption of these HMIs became slow, and no further change was observed due to the reduced active sites on the sorbent surface. The determined sorption capacities for Pb (II) and Cu (II) at 150 min were 32.65 and 25.74 $mg.g^{-1}$ [54].

In the study of Ahmed and Mirza, [61], the sorption of Ni (II), and Pb (II) and CR were swift within the first stage, lasting 120 and 240 min, respectively. It was also noticed that the sorption of the sorbate after the first stage became slow after equilibrium was attained. The swift sorption of the HMI and CR was because of the huge amount of unoccupied active sites on the sorbent surface that led to a rapid mass transfer, while the reduced rate of sorption was because of the residual active sites on the sorbent surface becoming less available.

Rapid sorption of MIs to chitin was also noticed in the initial phase for all studied MIs owing to the well supported accessible sites. Subsequently, the sorption process of these MIs slowed down until equilibrium was achieved within 30 min owing to the saturation of the binding sites on the sorbent surface [48].

In Khalek *et al* [65] and Igberase *et al* [60] studies, swift sorption of various HMIs onto BNCs was noticed within the initial 30, 45 and 55 min of the sorption activity. A further change was not observed after the initial phase of the sorption process and equilibrium was attained afterwards. These phenomena were due to the sufficient and well-supported accessible sites for HMI binding and binding sites were coated with MIs resulting in the repulsion of the MIs from the surface of the sorbent with time.

8.6 Mechanism of sorption of HMs using BNCs

The sorption isotherms is crucial for the sorption mechanism routes optimization, the surface properties, sorbent capacities expression and the constructive design of the sorption describes the sorbent materials' interrelation with the contaminants. Knowledge of the sorption mechanism, surface features and the degree of sorbent affinity is provided by the physicochemical factors complemented by the basic thermodynamic theories [66].

8.6.1 Isotherms models

The sorption isotherm and kinetics are utilized regularly to describe the process of sorption, mechanism, and performance of sorption processes. The isotherm provides information about the types of sorption system, sorption capacities, properties of sorbents surface, and the association between sorbents and sorbates. The kinetics provides knowledge on the sorption rate information, stages of rate-limiting and the diffusion processes of the sorbate [25].

The sorption isotherm is usually employed to define the association between the in-solution sorbate concentration (liquid-phase) and the sorbent (solid-phase) at a steady temperature under specific conditions like ionic strength, sorbent particle size, static mass, and solution pH [67].

The well-known isotherm models employed for the sorption process are Koble–Corrigan, Khan, Hill, Flory–Huggins, Dubinin–Radushkevich (DB), Temkin, Frenkel–Hasley–Hill, Henderson, Giles-Smith, LNR, Sips, Freundlich, Redlich–Peterson, Temkin, Toth, Brunauer–Emmett–Teller (BET), and Radke–Prausnitz isotherm. The extensively employed isotherms to explain the sorption equilibrium between the sorbent surface and the sorbate are LNR and Freundlich models [68].

The theoretical LNR equation was first utilized in the sorption of gases on a solid surface and created based on the resulting theories which are that a static amount of existing sites are accessible on the surface of the sorbent and all active sites have equal energy, sorption is reversible, no further sorption can happen on a site once it is occupied by a sorbate, and between sorbate species there is no interaction [67, 69]. The LNR isotherm adopts monolayer sorption on a corresponding surface [25].

One of the original experimental calculations utilized to explain equilibrium statistics and the sorption attributes of an inhomogeneous surface is the Freundlich equation. The Freundlich isotherm cannot define the (calculations) one-dimensionality scale at an extremely minimal concentration or the saturation impact at extremely elevated concentrations. But also, it cannot define the saturation behaviour of a sorbent [70]. The Freundlich model assumes multilayer sorption on a different surface [25].

The DB isotherm is an experimental sorption model that is mostly utilized to articulate the mechanism of sorption with Gaussian energy distribution to diverse surfaces. This model is just appropriate for the transitional array of sorbate concentrations owing to the idealistic asymptotic behaviour shown and at low pressure does not predict Henry's law. Also, the model is a semi-empirical calculation in which sorption is controlled by a mechanism of the pore-filling process and assumes a multilayer character comprising van der Waals forces, and is appropriate for physical sorption procedures. Gases and vapours sorption on microporous sorbents are qualitatively described using this fundamental equation and it is generally employed to differentiate between the physical and chemical sorption of MIs [71].

The Temkin model is built on the assumption that because of the reduced sorbent–sorbate interactions during gas-phase sorption, the heat of sorption due to sorbate interactions decreases linearly over time. It is an extension of the Gibbs relation to sorbents with actively homogeneous surfaces [72].

The Freundlich and LNR isotherms are combined in the Sips model. It predicts the heterogeneous sorption systems and can be employed to sidestep the drawback of the increasing sorbate concentration linked with the Freundlich model. At low sorbate concentrations, the model is successfully reduced to the Freundlich model, while at elevated sorbate concentrations, the model forecasts a monolayer sorption capacity feature of the LNR model. The linearized and non-linearized forms of the various isotherm equations and their notations are illustrated in table 8.3 [9].

The error structure, error insertion into the independent variable, and the weight imposed on each data point can all be influenced by modifying the linearized data, resulting in variations in the values of the fitted parameters between linear and non-linear isotherm models [70]. The sorption isotherm can be examined dependably by the regression coefficient value (R^2) and the sorption capacity are ascertained from the isotherm plots [73]. The energy of the reaction system, the rate of removal, and the quantity of sorption and the sorption isotherm are curves that illustrate the effect of temperature on the behaviour of sorption. Remarkably, the sorption of most HMIs onto BNCs is largely due to the LNR and Freundlich models, demonstrating that the removal behaviour of HMIs on BNCs are evenly covered by a specific molecular layer or multilayers (table 8.2). The maximum sorption capacity obtained using BNCs for HMIs is also shown in table 8.2.

Table 8.3. General forms and notations of isotherm and kinetic models and thermodynamics.

Models	Linearized	Non-linearized	Notations	References
Isotherm				
LNR	$\frac{C_e}{q_e} = \left(\frac{1}{Q_{\max}}\right)C_e + \frac{1}{Q_{\max}K_L}$ (Type 1) $\frac{1}{q_e} = \left(\frac{1}{Q_{\max}K_L}\right)\frac{1}{C_e} + \frac{1}{Q_{\max}}$ Lineweaver–Burk linearization of Langmuir (Type 2). $q_e = \left(\frac{-1}{K_L}\right)\frac{q_e}{C_e} + Q_{\max}$ (Eadie–Hoffsie linearization of Langmuir (Type 3). $\frac{q_e}{C_e} = -K_L q_e + Q_{\max}K_L$ Scatchard linearization of Langmuir (Type 4).	$q_e = \frac{Q_{\max}K_L C_e}{1 + K_L C_e}$	C_e (mg L^{-1}), $Q_{\max}$ (mg g^{-1}) and K_L (L mg^{-1}) indicate the sorbate concentration in the solution at equilibrium, the maximum saturated monolayer adsorption capacity of the sorbent under the given conditions, and the LNR constant related to the affinity between an adsorbent and adsorbate.	[67]
Freundlich	$\log q_e = \frac{1}{n}\log C_e + \log K_F$	$q_e = K_F C_e^{1/n}$	K_F (mg g^{-1})(L mg^{-1})$^{1/n}$ and 1/n (dimensionless; $0 < n < 10$) represents the Freundlich constant and Freundlich intensity parameter, indicating the magnitude of the sorption driving force or surface heterogeneity.	[70]
D–R	$\ln q_e = \ln q_m - K\varepsilon^2$ where $\varepsilon = RT\ln\left(1 + \frac{1}{C_e}\right)$	$q_e = q_m.\ \exp(-K.\ \varepsilon^2)$	ε and q_e represent the Polanyi potential and extent of sorbate adsorbed per unit mass of adsorbent. q_m, C_e, K, R and T indicate the theoretical sorption capacity, the equilibrium concentration, the constants related to sorption energy, the universal gas constant and the temperature in Kelvin (K).	[78]
Sips	$ln\left(\frac{q_e}{q_m - q_e}\right) = \frac{1}{n}ln(C_e) \quad + \ln(a_s)^{\frac{1}{n}}$	$q_e = \frac{q_m a_S C_e^{\frac{1}{n}}}{\left(1 + a_S C_e^{\frac{1}{n}}\right)}$	q_e, C_e and a_s signify the equilibrium sorption capacity of the sorbent (mg.g^{-1}), the equilibrium concentration of the sorbate (mg.L^{-1}), and the Sips isotherm model constant relating to the energy of sorption (L.g^{-1})	[79]

(Continued)

Table 8.3. (*Continued*)

Models	Linearized	Non-linearized	Notations	References
Kinetic				
PFO	$\log(q_e - q_t) = \log q_e \frac{K_1}{2.3030} t$	$\frac{dq_t}{dt} = K(q_e - q_t)$	q_e, and q_t signifies the equilibrium sorption capacity of the sorbent (mg.g^{-1}), and the sorption capacity (mg.g^{-1}). The contact time is denoted by t. K and K_1 and K_2 are the rate constants for PFO and PSO models.	[9]
PSO	$\frac{t}{q_t} = \frac{1}{K_2 q_e^2} + \frac{1}{q_e} t$	$\frac{dq_t}{dt} = K_2(q_e - q_t)^2$		[80]
Elovich	$q_t = \frac{1}{\beta}(\alpha\beta) + \frac{1}{\beta} In(t)$	$\frac{dq_t}{dt} = \alpha e^{-\beta t}$	α represents the initial rate (mg g^{-1} min^{-1}) and the parameter β (g mg^{-1}) relates to the degree of surface coverage and activation energy for chemisorption.	[81]
Thermodynamics				
Thermodynamics	$\Delta G^\circ = -RT\ln K_c,$ The change in the Gibbs energy ($\boldsymbol{\Delta G^\circ}$) is directly evaluated utilizing $\Delta G^\circ = -RT\ln K_c,$ The link between ΔG° and ΔH° and ΔS° is described by $\Delta G^\circ = \Delta H^\circ - T\Delta S^\circ$ Substituting $\Delta G^\circ = -RT\ln K_c,$ for $\Delta G^\circ = \Delta H^\circ - T\Delta S^\circ$, the van 't Hoff equation is achieved. The change in enthalpy ($\boldsymbol{\Delta H^\circ}$) and entropy ($\boldsymbol{\Delta S^\circ}$) are ascertained from the gradient and intercept of $\ln K_C = \frac{-\Delta H^\circ}{R} \times \frac{1}{T} + \frac{\Delta S^\circ}{R}$		R represents the universal gas constant (8.3144 J (mol^{-1} × K^{-1})) and T is the absolute temperature in Kelvin (K).	[70]

8.6.2 Kinetic models

To evaluate the rate of reaction and the functional application prospective employing novel NMs for HMI sorption, the PFO, PSO, Zeldowitsch, Elovich and Lagergren kinetic models are employed to investigate the impact of contact time on MIs sorption [17]. The sorption kinetics are described by examining the change of the sorptive capacity or the sorbent sorptive rate [22].

The PFO model describes the association between the rate at which the sorbent sorption sites are filled and the total of uninhabited sites. It is well-specified utilizing the Lagergren calculation. The PSO kinetic model defines the sorbent sorption capacity dependence on time [74]. The Elovich model defines the chemical nature of sorption [75] and it was initially introduced in 1939 and is suitable for systems with heterogeneous sorbing surfaces in the chemical sorption processes. The common kinetic models used in the sorption process in their linearized and non-linearized forms are depicted in table 8.3 [76].

In general, the rate of HMIs sorption onto BNCs varies with time, from fast to slow and with the sorption process ultimately attaining equilibrium. The point for most sorption processes to reach equilibrium is generally ambiguous for various sorbent materials and the PSO kinetic model indicates that chemical sorption is the prevailing mechanism in most studies reviewed using BNCs for MI sorption as observed in table 8.2 [17].

8.6.3 Thermodynamics

Thermodynamics studies play a crucial part in forecasting sorption mechanisms and the major difference between chemical and physical sorption [70]. The physical sorption (physisorption) results from the comparative feeble interactions like the van der Waals force. While the chemical sorption (chemisorption) comprises sturdier chemical interactions (chemical bonding) with the related exchange of electrons between the sorbate and sorbents [67].

The thermodynamic factors are assessed according to the laws of thermodynamics using the following equations depicted in table 8.3. On the whole, a temperature rise often corresponds with a reduction in the sorption capacity, which indicates that the sorption process was exothermic ($\Delta H^\circ < 0$). In the exothermic sorption process ($-\Delta H^\circ$), energy is released in the form of heat to the surroundings and comprises either chemisorption or physisorption or a combination of both processes (comprehensive sorption). In divergence, in the endothermic process ($+\Delta H^\circ$), energy is sorbed from the surroundings in the form of heat and ascribed to chemisorption. A key factor in establishing the extent of the sorption process spontaneity is the ΔG° and a higher negative value of ΔG° signifies additional actively favourable sorption and it is spontaneous. In contrast, the sorption process is non-viable and non-spontaneous when ΔG° is positive and the extent of ΔG° also confirms the sorption forms (physical or chemical). While the structure of the sorbate at the solid/solution boundary during the sorption process becoming less random ($\Delta S^\circ < 0$) or more random ($\Delta S^\circ > 0$) is shown by the magnitude and sign of ΔS°. The sorption phenomenon comprises associative and dissociative mechanisms when ΔS° is negative and positive [67]. The sorption is mostly physical when the ΔG° value is between −20 and 0 $KJ.mol^{-1}$, chemisorption when ΔG° value is between −80 and −400 $KJ.mol^{-1}$ and both physical sorption and chemisorption if the ΔG° value is between −20 and −80 $KJ.mol^{-1}$ [77]. As observed in the computed variables of thermodynamics in table 8.2, the sorption of most HMIs to BNCs was noticed to be spontaneous, endothermic and there was improved randomness at the solid–liquid interface.

8.7 Conclusion and prospects

A predominant global issue is the lack of access to drinking water and sanitation. Effective effluents treatment is imperative to bring it to an environmentally friendly form for impending needs. In recent times, the purification of water using bionanocomposites has been the focus and interest of various researchers and agencies of government, with the sorption technique offering thought-provoking benefits due to simplicity and the most versatile method available to extract the organic and mineral pollutants from industrial wastewater. Bionanocomposites sorbents owing to their environmental and recyclable nature are promising water remediation materials for the sequestration of microorganisms, and inorganic and organic pollutants (metal ions) from water and industrial effluent. Therefore, the applications of various bionanocomposites in the sequestration of metals ions are discussed in this chapter and also important factors affecting the sorption process and mechanisms of sorption using bionanocomposite are reviewed. To design more eco-responsive management of waste techniques to tackle the global water pollution problem, it is required for scientists and manufacturers to collaborate on creating cost-efficient techniques for producing and assessing the effectiveness of BNCs to a sizeable scale for effective field use.

Acknowledgement

The authors express their sincere appreciation to their respective affiliated institutions, the editors, anonymous reviewers, the IOP publishing team, and authors whose research materials and publications have been used during writing this book chapter.

References

[1] Song C, Wu S, Cheng M, Tao P, Shao M and Gao G 2014 Adsorption studies of coconut shell carbons prepared by KOH activation for removal of lead (II) from aqueous solutions *Sustainability* **6** 86–98

[2] Lee S, Eom Y, Park J, Lee J and Kim S 2017 Micro-hydrogel particles consisting of hyperbranched polyamidoamine for the removal of heavy metal ions from water *Sci. Rep.* **7** 1–9

[3] Hassan M, Naidu R, Du J, Liu Y and Qi F 2020 Critical review of magnetic biosorbents: their preparation, application, and regeneration for wastewater treatment *Sci. Total Environ.* **702** 134893

[4] Olivera S, Muralidhara H, Venkatesh K, Guna V, Gopalakrishna K and Kumar Y 2016 Potential applications of cellulose and chitosan nanoparticles/composites in wastewater treatment: a review *Carbohydrate Polym.* **153** 600–18
[5] Bhateria R and Singh R 2019 A review on nanotechnological application of magnetic iron oxides for heavy metal removal *J. Water Process Eng.* **31** 100845
[6] Ramrakhiani L, Halder A, Majumder A, Mandal A, Majumdar S and Ghosh S 2017 Industrial waste derived biosorbent for toxic metal remediation: mechanism studies and spent biosorbent management *Chem. Eng. J.* **308** 1048–64
[7] Dar O, Malik M, Talukdar M and Hashmi A 2020 Bionanocomposites in water treatment *Bionanocomposites: Micro and Nano Technologies* ed K Zia, F Jabeen, M Anjum and S Ikram (Amsterdam: Elsevier) 505–18
[8] Carolin C, Kumar P, Saravanan A, Joshiba G and Naushad M 2017 Efficient techniques for the removal of toxic heavy metals from aquatic environment: a review *J. Environ. Chem. Eng.* **5** 2782–99
[9] Yang J, Hou B, Wang J, Tian B, Bi J, Wang N, Li X and Huang X 2019 Nanomaterials for the removal of heavy metals from wastewaterNanomaterials **9** 424
[10] Lu F and Astruc D 2018 Nanomaterials for removal of toxic elements from water *Coord. Chem. Rev.* **356** 147–64
[11] Aigbe U, Das R, Ho W, Srinivasu V and Maity A 2018 A novel method for removal of Cr (VI) using polypyrrole magnetic nanocomposite in the presence of unsteady magnetic fields *Sep. Purif. Technol.* **194** 377–87
[12] Ge F, Li M, Ye H and Zhao B 2012 Effective removal of heavy metal ions Cd^{2+}, Zn^{2+}, Pb^{2+}, Cu^{2+} from aqueous solution by polymer-modified magnetic nanoparticles *J. Hazard. Mater.* **211** 366–72
[13] Pandey L 2020 Surface engineering of nano-sorbents for the removal of heavy metals: Interfacial aspects *J. Environ. Chem. Eng.* **9** 104586
[14] Aigbe U, Ho W, Maity A, Khenfouch M and Srinivasu V 2018 Removal of hexavalent chromium from wastewater using PPy/Fe_3O_4 magnetic nanocomposite influenced by rotating magnetic field from two pole three-phase induction motor *J. Phys.: Conf. Ser.* **984** 012008
[15] Asuquo E and Martin A 2016 Sorption of cadmium (II) ion from aqueous solution onto sweet potato (Ipomoea batatas L.) peel adsorbent: characterisation, kinetic and isotherm studies *J. Environ. Chem. Eng.* **4** 4207–28
[16] Fu F and Wang Q 2011 Removal of heavy metal ions from wastewaters: a review *J. Environ. Manage.* **92** 407–18
[17] Wu Y, Pang H, Liu Y, Wang X, Yu S, Fu D, Chen J and Wang X 2019 Environmental remediation of heavy metal ions by novel-nanomaterials: a review *Environ. Pollut.* **246** 608–20
[18] Aigbe U and Osibote O 2020 A review of hexavalent chromium removal from aqueous solutions by sorption technique using nanomaterials *J. Environ. Chem. Eng.* **8** 104503
[19] Aigbe U, Onyancha R, Ukhurebor K and Obodo K 2020 Removal of fluoride ions using a polypyrrole magnetic nanocomposite influenced by a rotating magnetic field *RSC Adv.* **10** 595–609
[20] Vunain E, Mishra A and Mamba B 2016 Dendrimers, mesoporous silicas and chitosan-based nanosorbents for the removal of heavy-metal ions: a review *Int. J. Biol. Macromol.* **86** 570–86

[21] Duffus J 2002 Heavy metals a meaningless term *Pure Appl. Chem.* **74** 793–807
[22] Duan C, Ma T, Wang J and Zhou Y 2020 Removal of heavy metals from aqueous solution using carbon-based adsorbents: a review *J. Water Process Eng.* **37** 101339
[23] Fiyadh S, AlSaadi M, Jaafar W, AlOmar M, Fayaed S, Mohd N, Hin L and El-Shafie A 2019 Review on heavy metal adsorption processes by carbon nanotubes *J. Clean. Prod.* **230** 783–93
[24] Kejriwal R, Mandke M and Ingle P 2018 Bio-sorption of heavy metals: a review *Int. J. Adv. Res. Chem. Sci.* **5** 32–42
[25] Yu G *et al* 2018 Carbon nanotubes, graphene, and their derivatives for heavy metal removal *Adv. Compos. Hybrid Mater.* **1** 56–78
[26] Elumalai V, Brindha K and Lakshmanan E 2017 Human exposure risk assessment due to heavy metals in groundwater by pollution index and multivariate statistical methods: a case study from South Africa *Water* **9** 234
[27] Baby R, Saifullah B and Hussein M 2019 Carbon nanomaterials for the treatment of heavy metal-contaminated water and environmental remediation *Nanoscale Res. Lett.* **14** 1–17
[28] Chubaka C, Whiley H, Edwards J and Ross K 2018 Lead, zinc, copper, and cadmium content of water from south Australian rainwater tanks *Int. J. Environ. Res. Public Health* **15** 2018
[29] Sheet I, Kabbani A and Holail H 2014 Removal of heavy metals using nanostructured graphite oxide, silica nanoparticles and silica/graphite oxide composite *Energy Procedia* **50** 130–38
[30] Ibrahim R, Hayyan M, AlSaadi M, Hayyan A and Ibrahim S 2016 Environmental application of nanotechnology: air, soil, and water *Environ. Sci. Pollut. Res.* **23** 13754–88
[31] Jawed A, Saxena V and Pandey L 2020 Engineered nanomaterials and their surface functionalization for the removal of heavy metals: a review *J. Water Process Eng.* **33** 101009
[32] Zafar R, Zia K, Tabasum S, Jabeen F, Noreen A and Zuber M 2016 Polysaccharide based bionanocomposites, properties and applications: a review *Int. J. Biol. Macromol.* **92** 1012–24
[33] Pande V 2017 Bionanocomposite: a review *Austin J. Nanomed. Nanotechnol.* **5** 1–3
[34] Sethy T and Sahoo P 2019 Highly toxic Cr (VI) adsorption by (chitosan-g-PMMA)/silica bionanocomposite prepared via emulsifier-free emulsion polymerisation *Int. J. Biol. Macromol.* **122** 1184–90
[35] Saini R, Bajpai A and Jain E 2017 Fundamentals of bionanocomposites *Biodegradable and Biocompatible Polymer Composites: Processing, Properties and ApplicationsSeries in Composites* (Cambridge: Woodhead Publishing)
[36] Rhim J, Park H and Ha C 2013 Bio-nanocomposites for food packaging applications *Prog. Polym. Sci.* **38** 1629–52
[37] Shchipunov Y 2012 Bionanocomposites: Green sustainable materials for the near future *Pure Appl. Chem.* **84** 2579–607
[38] Sharma B, Malik P and Jain P 2018 Biopolymer reinforced nanocomposites: a comprehensive review *Mater. Today Commun.* **16** 353–63
[39] Mousa M, Dong Y and Davies I 2016 Recent advances in bionanocomposites: preparation, properties, and applications *Int. J. Polym. Mater. Polym. Biomater.* **65** 225–54
[40] Oboh J, Okafor J, Kovo A and Abdulrahman A 2017 Investigation of eco-friendly cellulosic nanoparticles potential as reinforcement agent in the production of natural rubber composites *Niger. J. Technol.* **36** 1078–87

[41] Motaung T and Linganiso L 2018 Critical review on agrowaste cellulose applications for biopolymers *Int. J. Plast. Technol.* **22** 185–216
[42] Bhattacharya M 2016 Polymer nanocomposites—a comparison between carbon nanotubes, graphene, and clay as nanofillers *Materials* **9** 262
[43] Dinari M and Tabatabaeian R 2018 Ultra-fast and highly efficient removal of cadmium ions by magnetic layered double hydroxide/guargum bionanocomposites *Carbohydr. Polym.* **192** 317–26
[44] Chen Z, Li Z, Chen G, Zhu J, Liu Q and Feng T 2017 In situ formation of AgNPs on S. cerevisiae surface as bionanocomposites for bacteria killing and heavy metal removal *Int. J. Environ. Sci. Technol.* **14** 1635–42
[45] García-Padilla Á, Moreno-Sader K, Realpe Á, Acevedo-Morantes M and Soares J 2020 Evaluation of adsorption capacities of nanocomposites prepared from bean starch and montmorillonite *Sustain. Chem. Pharm.* **17** 100292
[46] Ahmad R and Mirza A 2018 Application of Xanthan gum/n-acetyl cysteine modified mica bionanocomposite as an adsorbent for the removal of toxic heavy metals *Groundw. Sustain. Dev.* **7** 101–8
[47] Mirza A and Ahmad R 2018 Novel recyclable (Xanthan gum/montmorillonite) bionanocomposite for the removal of Pb (II) from synthetic and industrial wastewater *Environ. Technol. Innov.* **11** 241–52
[48] Boulaiche W, Hamdi B and Trari M 2019 Removal of heavy metals by chitin: equilibrium, kinetic and thermodynamic studies *Appl. Water Sci.* **9** 1–10
[49] Abreu F, Silva N, Sipauba M, Pires T, Bomfim T, Monteiro Junior O and Forte M 2018 Chitosan and gum arabic nanoparticles for heavy metal adsorption *Polímeros* **28** 231–38
[50] Gandhi M, Viswanathan N and Meenakshi S 2010 Preparation and application of alumina/chitosan biocomposite *Int. J. Biol. Macromol.* **47** 146–54
[51] Al-Sherbini A, Ghannam H, El-Ghanam G, El-Ella A and Youssef A 2019 Utilization of chitosan/Ag bionanocomposites as eco-friendly photocatalytic reactor for bactericidal effect and heavy metals removal *Heliyon* **5** E01980
[52] Alamillo-López V, Sánchez-Mendieta V, Olea-Mejía O, González-Pedroza M and Morales-Luckie R 2020 Efficient removal of heavy metals from aqueous solutions using a bionanocomposite of eggshell/Ag-Fe *Catalysts* **10** 727
[53] Malayoglu U 2018 Removal of heavy metals by biopolymer (chitosan)/nanoclay composites *Sep. Sci. Technol.* **53** 2741–49
[54] Ahmad R and Hasan I 2017 L-methionine montmorillonite encapsulated guar gum-g-polyacrylonitrile copolymer hybrid nanocomposite for removal of heavy metals *Groundw. Sustain. Develop.* **5** 75–84
[55] Lyu F, Yu H, Hou T, Yan L, Zhang X and Du B 2019 Efficient and fast removal of Pb^{2+} and Cd^{2+} from an aqueous solution using a chitosan/Mg-Al-layered double hydroxide nanocomposite *J. Colloid Interface Sci.* **539** 184–93
[56] Zarei F, Marjani A and Soltani R 2019 Novel and green nanocomposite-based adsorbents from functionalised mesoporous KCC-1 and chitosan-oleic acid for adsorption of Pb (II) *Eur. Polym. J.* **119** 400–9
[57] Güçlü G, Al E, Emik S, İyim T, Özgümüş S and Özyürek M 2010 Removal of Cu^{2+} and Pb^{2+} ions from aqueous solutions by starch-graft-acrylic acid/montmorillonite superabsorbent nanocomposite hydrogels *Polym. Bull.* **65** 333–46

[58] Soliman N and Moustafa A 2020 Industrial solid waste for heavy metals adsorption features and challenges: a review *J. Mater. Res. Technol.* **9** 10235–53

[59] Salam M, Makki M and Abdelaal M 2011 Preparation and characterization of multi-walled carbon nanotubes/chitosan nanocomposite and its application for the removal of heavy metals from aqueous solution *J. Alloys Compd.* **509** 2582–87

[60] Igberase E, Osifo P and Ofomaja A 2017 The adsorption of Pb, Zn, Cu, Ni, and Cd by modified ligand in a single component aqueous solution: equilibrium, kinetic, thermodynamic, and desorption studies *Int. J. Anal. Chem.* **2017** 6150209

[61] Ahmad R and Mirza A 2018 Adsorptive removal of heavy metals and anionic dye from aqueous solution using novel Xanthan gum-Glutathione/Zeolite bionanocomposite *Groundw. Sustain. Dev.* **7** 305–12

[62] Yu G, Wang X, Liu J, Jiang P, You S, Ding N, Guo Q and Lin F 2021 Applications of nanomaterials for heavy metal removal from water and soil: a review *Sustainability* **13** 713

[63] Nithya R and Sudha P 2017 Removal of heavy metals from tannery effluent using chitosan-g-poly (butyl acrylate)/bentonite nanocomposite as an adsorbent *Textiles Cloth. Sustain.* **2** 1–8

[64] Wang S, Zhang C and Chang Q 2017 Synthesis of magnetic crosslinked starch-graft-poly (acrylamide)-co-sodium xanthate and its application in removing heavy metal ions *J. Exp. Nanosci.* **12** 270–84

[65] Khalek M, Mahmoud G and El-Kelesh N 2012 Synthesis and characterization of poly-methacrylic acid grafted chitosan-bentonite composite and its application for heavy metals recovery *Chem. Mater. Res* **2** 1–12

[66] Kyzas G and Matis K 2015 Nanoadsorbents for pollutants removal: a review *J. Mol. Liq.* **203** 159–68

[67] Tran H, You S and Chao H 2016 Thermodynamic parameters of cadmium adsorption onto orange peel calculated from various methods: a comparison study *J. Environ. Chem. Eng.* **4** 2671–82

[68] Lim J, Mubarak N, Abdullah E, Nizamuddin S and Khalid M 2018 Recent trends in the synthesis of graphene and graphene oxide based nanomaterials for removal of heavy metals —a review *J. Ind. Eng. Chem.* **66** 29–44

[69] Desta M 2013 Batch sorption experiments: Langmuir and Freundlich isotherm studies for the adsorption of textile metal ions onto teff straw (Eragrostis tef) agricultural waste *J. Thermody.* **2013** 375830

[70] Tran H, You S, Hosseini-Bandegharaei A and Chao H 2017 Mistakes and inconsistencies regarding adsorption of contaminants from aqueous solutions: a critical review *Water Res.* **120** 88–116

[71] Ayawei N, Ebelegi A and Wankasi D 2017 Modelling and interpretation of adsorption isotherms *J. Chem.* **2017** 3039817

[72] Labied R, Benturki O, Eddine Hamitouche A and Donnot A 2018 Adsorption of hexavalent chromium by activated carbon obtained from a waste lignocellulosic material (Ziziphus jujuba cores): kinetic, equilibrium, and thermodynamic study *Adsorpt. Sci. Technol.* **36** 1066–99

[73] Alamin N, Khan A, Nasrullah A, Iqbal J, Ullah Z, Din I, Muhammad N and Khan S 2021 Activated carbon-alginate beads impregnated with surfactant as sustainable adsorbent for efficient removal of methylene blue *Int. J. Biol. Macromol.* **176** 233–43

[74] Edet U and Ifelebuegu A 2020 Kinetics, isotherms, and thermodynamic modeling of the adsorption of phosphates from model wastewater using recycled brick waste *Processes* **8** 1251

[75] Tutu H, Bakatula E, Dlamini S, Rosenberg E, Kailasam V and Cukrowska E 2013 Kinetic, equilibrium and thermodynamic modelling of the sorption of metals from aqueous solution by a silica polyamine composite *Water SA* **39** 437–44

[76] Wu F, Tseng R and Juang R 2009 Characteristics of Elovich equation used for the analysis of adsorption kinetics in dye-chitosan systems *Chem. Eng. J.* **150** 366–73

[77] Chen J, Shu C, Wang N, Feng J, Ma H and Yan W 2017 Adsorbent synthesis of polypyrrole/TiO2 for effective fluoride removal from aqueous solution for drinking water purification: adsorbent characterization and adsorption mechanism *J. Colloid Interface Sci.* **495** 44–52

[78] Nagy B, Mânzatu C, Măicăneanu A, Indolean C, Barbu-Tudoran L and Majdik C 2017 Linear and nonlinear regression analysis for heavy metals removal using *Agaricus bisporus* macrofungus *Arabian J. Chem.* **10** S3569–79

[79] Popoola L 2019 Characterization and adsorptive behaviour of snail shell-rice husk (SS-RH) calcined particles (CPs) towards cationic dye *Heliyon* **5** e01153

[80] Elwakeel K, Elgarahy A and Khan Z 2020 Perspectives regarding metals/minerals-incorporated materials for water purification: with special focus on Cr (VI) removal *Mater Adv.* **1** 1546–74

[81] Singh K, Lataye D and Wasewar K 2016 'Removal of fluoride from aqueous solution by using low-cost sugarcane bagasse: Kinetic study and equilibrium isotherm analyses *J. Hazard. Toxic, Radioactive Waste* **20** 309

IOP Publishing

Bionanomaterials for Environmental and Agricultural Applications

Ravindra Pratap Singh and Kshitij RB Singh

Chapter 9

Potentialities of bionanocomposite hydrogels for wastewater treatment

Madhavi Vemula and Vijay Bhaskar Reddy Ambavaram

Hydrogels have been utilized as potential adsorbents for water pollutants owing to their innovative characteristics including swellability, resilience and hydrophilicity. Nanomaterials with ultrahigh specific surface area combine with natural (bio) hydrogels and represent the most feasible method to enhance the hydrogel properties for diverse applications. The coalescence of natural hydrogels and nanomaterials elicits bionanocomposite hydrogels (BNCHs) that show exceptional adsorptive performance for water contaminants such as heavy metals, toxic dyes, organic pollutants etc. This chapter reviews the recent progress in devising BNCHs that opens new avenues for the resilient materials significantly on wastewater treatment.

9.1 Introduction

Access to safe water has been disrupted by the increase in industrial operations, agricultural practices, extreme weather conditions etc. Contaminated water with chemical pollutants that are of concern includes organic impurities, pesticides, heavy metals, pathogens etc. The release of pollutants by various modes into water can cause environmental pollution as some of them are not readily biodegradable and considered to be lethal and carcinogenic. Therefore, the presence of these pollutants in the environment is controlled and emphasizes its necessity before their release into the aquatic environment. The continuous research has developed many techniques to remove water pollutants such as chemical precipitation, ion exchange, adsorption, coagulation, reverse osmosis, electrochemical treatment etc. Among all, the adsorption process is significantly considered as one of the most appropriate and relevant approaches to remove water pollutants due to its economics and ease in operation [1].

doi:10.1088/978-0-7503-3863-9ch9

Sorbent materials showed their capability in sorbing organic and inorganic pollutants. Gels, a nonfluid colloidal or polymer network are used as adsorbents for water pollutants as they can expand their whole volume by a fluid. Hydrogels, with porous three-dimensional network structures, containing hydrophilic and hydrophobic parts in a definite proportion, are water-swollen materials that allow diffusion of solute particles through their structural configuration. They swell intensively by increasing their initial volume several times without dissolving or changing their shape due to their hydrophilic functional groups such as –OH, -COOH etc. The water diffuses into hydrogels by the difference in the osmotic pressure and capillary action [2].

The flexibility of hydrogels due to their swelling nature, makes them ideal for different applications. The presence of hydrophilic functional groups, crosslinking polymerization methods etc control the swelling property of hydrogels. Some other factors like ionic strength, pH, temperature etc can also enhance the swelling ability of hydrogels. However, the characteristic properties of hydrogels including superabsorptivity, flexibility, elasticity, biocompatibility and permeability are significantly attributed to their potential use in various applications in industries, medicine, agriculture, environmental remediation etc. The three-dimensional configuration of hydrogel is normally originated by physico-chemical crosslinking (figure 9.1). However, the cost, tunability and biodegradability of hydrogels are the limitations. In this context, biobased hydrogels are currently a highlighted topic and represent their adaptivity in various applications. Added to this, nanomaterial-supported biobased hydrogels have become a cornerstone in scientific research due to their advanced inherent properties. This chapter will primarily

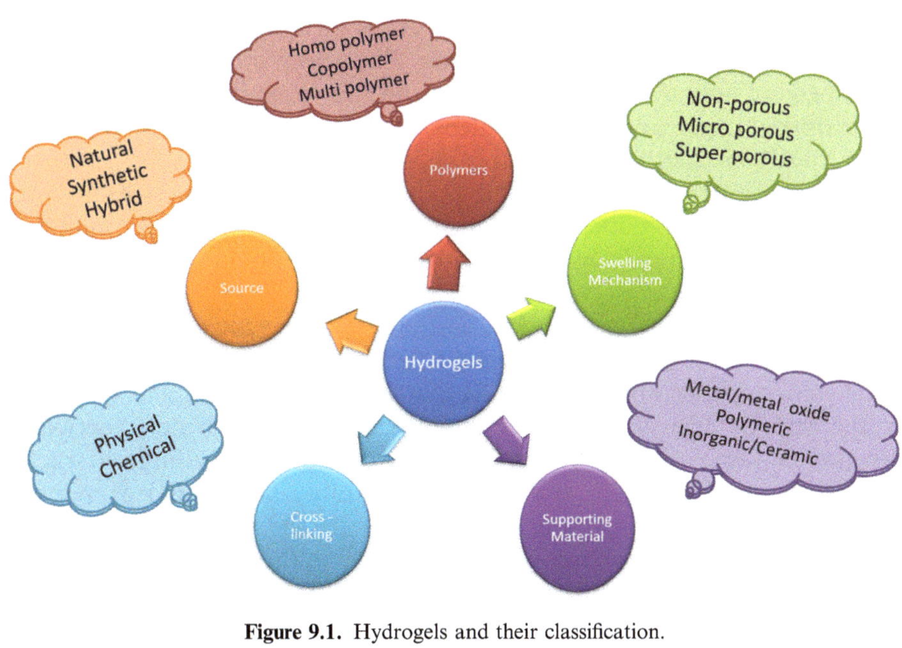

Figure 9.1. Hydrogels and their classification.

emphasize the recent research carried out in biobased nanocomposite hydrogels for water treatment.

9.2 Bionanocomposite hydrogels

The wide spectrum of applications of hydrogels depends on the porosity, swelling capability and network of hydrogel that in turn depend on the kind and density of crosslinker, polymerization methods and polar functional groups used in the formation of a hydrogel. The diffusion property of hydrogels depends on the thermodynamic behaviour of their components used and their structure [3]. The porosity of the hydrogel structure allows the adsorption of pollutants from aqueous solution [4, 5]. Nanotechnology is emerging in its implementation in water treatment technology based on adsorption. The high surface area, adsorption efficiency attributable to more adsorption sites, high chemical stability, high permeability, longevity, lower toxicity and cost are accountable for the high throughput efficiency of nanoparticles. The scientific research in hydrogel technology with new improved and tailored properties including swelling, thermomechanical and adsorption properties motivates researchers to fabricate nanocomposite hydrogels. Biomaterials are easily available and have excellent properties like biodegradability, non-toxicity, relative affordability, and they are renewable and recyclable. Hence the recent research emphasized nanotechnology and biomaterials in the preparation of hydrogels with synergistic effects; and thus prepared bionanocomposite hydrogels (BNCHs) have gained interest in water treatment.

BNCHs are classified according to their source, composition, crosslinking and network electrical charge. Crosslinking is considered as one of the significant factors that affect the configurational properties of hydrogels, considerably their mechanical strength and degree of swelling. Physically crosslinked hydrogels are often formed by noncovalent interactions and are reversible [6], whereas chemically crosslinked hydrogels utilize covalent interactions and are stable [7]. Natural polymers based on the nanomaterial composite, physical/chemical crosslinking, presence of nature of charge on the polymer backbone and side groups are imperative for their practical use and versatility in extensive applications. Natural polymers, particularly cellulose, starch, gelatin, chitosan, lignin etc have been gaining more interest in research due to their nontoxicity, availability, biodegradability and biocompatibility. Different nanoparticles supporting biopolymeric hydrogels have been investigated for the development of BNCHs. The hydrogel matrix either absorbs the nanoparticles or disperses and entraps them inside the matrix. The synergistic effects of crosslinked polymeric network and diverse nanomaterials improve the properties, including swelling, characteristic interactions to a great extent and thus, they exhibit exceptional properties and applications have been prevailing comparable to the conventional hydrogels. Various nanomaterials such as carbon-based, metal/metal oxide based, inorganic/ceramic nanoparticles have been exploited for their superior properties and drawn a surge of attention and are believed to be a revolutionary type of materials for the development of BNCHs.

9.3 Adsorptive removal of aqueous contaminants by bionanocomposite hydrogels

Adsorption of contaminants by biobased hydrogels with unique functionalities (polar functional groups) strongly interact with aqueous molecules to promote water solubility. These biobased hydrogels tailored with suitable crosslinking agents can increase the gel hydrophilicity, network structure and functional properties. Biohydrogels with specific functional groups in monomers form electrostatic interactions with pollutant and thus assist removing pollutants [8]. Ion exchange is also reported as the other removal mechanism where hydrogels contain acidic or basic functionalities [9]. Further, to ameliorate the mechanical properties, these biobased hydrogels are reinforced with nanoparticles that could be incorporated into three-dimensional polymer networks. The inclusion of nanoparticles to the hydrogel structure contributes to the physico-chemical interactions with novel and value-added properties. It is noteworthy that swelling capacity can be modulated towards the suitable swelling features by various methodological approaches in the synthesis of bionanocomposite hydrogels. The outstanding characteristics of BNCHs made them pioneering technology to be employed in wastewater treatment.

9.4 Biopolymers in nanocomposite hydrogels

The nanocomposites with bio/natural polymers have been considered as promising materials compared to synthetic polymers due to their ideal properties like biodegradability, nontoxicity and compatibility (figure 9.2). Hence, they are significantly used for different applications as they not only fulfil explicitly desirable properties, enhanced recovery, recyclability but also are available in substantial quantities and are cost effective [10]. Various natural polymer based nanocomposite hydrogels like cellulose, chitosan, pectin, carrageenan, etc and many more are good examples with interesting properties such as hydrophilic properties and non-toxic have become most desirable materials in recent research.

9.4.1 Cellulose based BNCHs

Cellulose, a natural polymer obtained substantially in plant cell walls, bacteria, fungi, algae etc that consists of polysaccharides with long chains of β-D-glucopyranose units. Cellulose has hydroxyl groups on the surface providing chemical functionalization reactions, such as oxidation, esterification, grafting etc. Carboxy methyl cellulose (CMC), the most common derivative of cellulose is known for its water solubility. Yang *et al* [11] successfully prepared CMC based hydrogel beads by inverse suspension crosslinking using epichlorohydrin (ECH) crosslinker and evaluated on the removal of heavy metal ions from aqueous solution. The adsorption of Pb^{2+}, Ni^{2+} and Cu^{2+} was high and the results confirmed that formation of coordinate bonds with the oxygen atoms in their hydrophilic carboxyl groups are responsible for the elimination of metal ions. Peng *et al* [12] fabricated cellulose-clay hydrogel nanocomposites by chemical crosslinking of cellulose and carboxy methyl cellulose to remove methylene blue (MB) dye from water. The

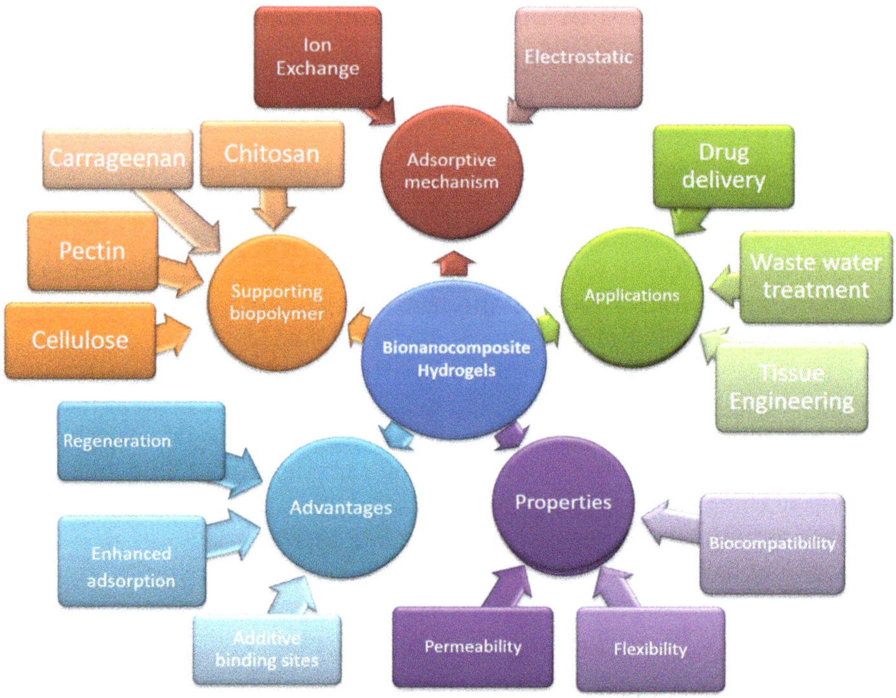

Figure 9.2. Bionanocpmposite hydrogels used for water treatment.

bionanohydrogel demonstrated superadsorbent properties for its ability to remove MB with removal efficiency of 98% through a physico-sorption process. In another report, bionanocomposite hydrogels were produced from hydroxypropyl methylcellulose grafted with acrylamide and natural sodium montmorillonite [13]. The prepared bionanohydrogels removed CV, a cationic dye from wastewater with adsorption capacity of 67.2 mg g^{-1}. CMC based magnetite nanocomposite functionalised with a crosslinker, 3-aminopropyltrimethoxysilane was fabricated by Uva *et al* [14] and investigated to remove MB and $CdCl_2$. Thus prepared BNCHs exhibited the adsorption capacity of 620 and 100 mg g^{-1} of MB and Cd, respectively.

Graphene oxide based cellulose hydrogel was employed for adsorptive removal of metal ions such as Cu^{2+}, Zn^{2+}, Fe^{3+} and Pb^{2+}. It was reported that the adsorption efficiency was significantly increased by the presence of –OH and -COOH groups onto the surface of cellulose by the addition of GO for the retention of nanoporous property and gel structure. The calculated adsorption capacity of GO/cellulose hydrogel was up to 94.34 mg g^{-1} [15]. Melilli *et al* [16] fabricated methacrylated CMC hydrogel, MCMC/nGO and M-CMC/M-nGO nanocomposite hydrogels exhibited good adsorption properties for MB and Cu(II) and is recyclable for almost three cycles. The outstanding adsorption capacity of these BNCHs was up to 350 mg g^{-1} and 145 mg g^{-1} for MB dye and Cu(II) adsorption, respectively. In another study, Wang *et al* [17] prepared composite hydrogel formulated on hydroxypropyl cellulose and graphene oxide and employed for the removal of phenol. The single network composite hydrogel (SNCH) and double network composite hydrogel (DNCH) were prepared and reported that DNCH demonstrated enhanced adsorption ability over SNCH due to more functional groups and larger surface area.

9.4.2 Chitosan based BNCHs

Chitosan, a high molecular weight polysaccharide is the second most widely available biopolymer that can be present in exoskeleton of crustaceans. The conformational flexibility of linear chains in chitosan assists the effective adsorption enhancing property for pollutant removal. Molecular weight, chain length, charge distribution, charge density are some of the factors affecting physicochemical properties of chitosan. The presence of –OH and $–NH_2$ functional groups on the chitosan make it attractive in various applications. Its high adsorption capacity, chelation activity, stability and versatility made it a potential green reactant in various scientific and engineering applications. Zhu *et al* [18] fabricated a macroporous magnetic chitosan-g-poly (acrylic acid) hydrogel stabilized by Fe_3O_4 nanoparticles. With the existence of abundant functional groups and tunable pore structure, the magnetic microporous hydrogel demonstrated a good potential and an excellent adsorbent for the fast removal of Cd(II) and Pb(II). The maximum adsorption efficiency was determined to be 308.84 and 695.22 mg g^{-1} for Cd(II) and Pb (II), respectively. Due to its magnetic property, it can be easily regenerated suggesting its reusability.

Chen *et al* [19] synthesized graphene oxide-chitosan (GO–CS) nanocomposite hydrogels by self-assembly method that were utilized as adsorbents for the adsorptive removal of heavy metal ions and organic dyes from waste water. It was reported that strong hydrogen bonds as well as electrostatic interaction between GO and CS make these hydrogels mechanically strong. The maximum adsorption capacities for MB and anionic Eosin Y were more than 300 mg g^{-1} for both dyes. Furthermore, the reported maximum adsorption capacity for Cu(II) and Pb(II) ions were to be 70 and 90 mg g^{-1}, respectively. Puspitasari *et al* [20] prepared nanocomposite hydrogel by chitosan-activated carbon nanoparticles (chitosan/ACNPs) and tested for water purification. Chitosan/ACNP hydrogels were able to adsorb Fe and Zn efficiently when compared to Pb and Cu. Akbarzadeh *et al* [21] synthesized magnetic nanocomposite hydrogels using an *in situ* synthesis of carboxymethyl chitosan/poly (acrylamide)(CMC-PAM) hydrogels in the presence of Fe_3O_4-montmorillonite nanoparticles. The magnetic CMC-PAM hydrogels were employed to remove crystal violet (CV) dye. The results exhibited that the maximum adsorption efficiency of these hydrogels was 72.4 mg g^{-1} of CV. A hybrid nanocomposite made of chitosan-silica was prepared by sol–gel method by hydrolysis of tetraethoxysilane in the chitosan solution by Budnyak *et al* [22] The adsorption capacity of the synthesized composite towards V, Mo and Cr was found to be 121, 37.4 and 35.2 mg g^{-1}, respectively. Zhou *et al* [23] prepared nano TiO_2/chitosan/poly(N-isopropylacrylamide) hydrogel composite by two-step polymerisation method and investigated for its adsorption of acid fuchsin dye from water. The chitosan based nanocomposite hydrogel showed high efficiency of photocatalytic degradation for AF dyes and the removal rate under UV radiation was 90.5% within 160 min

Liu *et al* [24] fabricated graphene oxide functionalised with three-dimensional porous β-cyclodextrin/chitosan hydrogel in the presence of sodium ascorbate by chemical reduction method. The nanocomposite hydrogel showed ultrahigh adsorption for MB of 1134 mg g^{-1} capacity. Melo *et al* [25] proposed cellulose nanowhiskers to improve the biosorption efficiency of chitosan-g-poly (acrylic acid) hydrogel. The nanocomposite hydrogel showed an enhanced biosorption capacity towards MB and presented the maximum removal of 1968 mg g^{-1}. The mechanism proposed for the monolayer adsorption is electrostatic attraction.

9.4.3 Pectin based nanocomposite hydrogels

Pectin is a naturally occurring biopolymer composed of D-galacturonic acid. Owing to the quintessential properties of pectin such as chemical structure, crosslinking mechanism, low-cost, renewable nature and abundance, it has been utilized in diverse applications. Kodoth *et al* [26] incorporated silver nanoparticles into pectin based copolymer grafted with poly(2-acrylamido-2-methyl-1-propanesulfonic acid)-polyacrylamide) hydrogel and utilized as an adsorbent for the removal of divalent metal ions and dyes. The Pec-g-poly(AMPS-co-AAm)/Ag nanocomposite hydrogel was able to remove 1950 mg g^{-1} of CV dye,111 mg g^{-1} $C^{-1}u^{-1}$(II) and 130 mg g^{-1} Pb(II) from wastewater. The desorption studies revealed the possible regeneration of the hydrogel at the stripping pH 1–2. Pectin (acrylic acid-co-acrylamide)(PAA) with metal nanoparticle (Cu and CO) hydrogels were prepared by Faten *et al* [27]. These hydrogels were investigated for the reduction of 2-nitrophenols and it was reported that PAA-Cu exhibited greater reduction capacity than PAA-Co.

Babaladimath *et al* [28] made pectin grafted with 2-acrylamido-2-methylpropanesulfonic acid (Pec-g-PAMPS) hydrogel and stabilized the silver nanoparticles. They reported that the presence of Ag nanoparticles in hydrogel enhanced the swelling ability and binding efficiency significantly. The Ag nanocomposite Pec-g-PAMPS hydrogel exhibited maximum adsorption of 89 mg g^{-1} for MB dye and is attributed to the inclusion of Ag Nps made more binding sites available for adsorption. Pectin-crosslinked guargum/superparamagnetic iron oxide (Pec-GG/SPION) nanocomposite hydrogel has been prepared by Sharma *et al* [29] and utilised for the adsorptive removal of m-cresol and o-chlorophenol from aqueous system. Experimental results showed that Pec-GG/SPION nanocomposite hydrogel exhibited maximum adsorption capacity of 176.1 and 75.6 mg g^{-1} for m-cresoland o-chlorophenol, respectively.

9.4.4 Carrageenan based nanocomposite hydrogels

Carrageenan is linear water soluble sulfated polysaccharide obtained by alkaline extraction from red seaweed. These are highly used as gelling agent in biomedical, food and non-food applications. Carrageenan exists in three different forms, namely, lambda (non-gelling), iota (weak gelling) and kappa (strong gelling). The presence of hydrophilic sulfate groups with high ionization tendency property of gelatinization gives carrageenan its wide use in commercial applications. Carrageenan is a negatively charged polysaccharide that can react with positively charged species. Mahdavinia *et al* [30] synthesized a nanocomposite hydrogel by grafting acrylamide onto k-carrageenan biopolymer with sepiolite clay. The maximum adsorption efficiency of the prepared nanocomposite hydrogel is 33.1 mg g^{-1} for CV dye.

In another study, kappa-carrageenan/polyvinyl alcohol/montorillonite nanocomposite hydrogels were developed by Hosseinzedah *et al* [31] and employed to remove CV from water. The nanocomposite hydrogel demonstrated maximum adsorption capacity of 151 mg g^{-1}. The magnetic nanocomposite hydrogel consisting of Fe_3O_4 nanoparticles embedded in κ-carrageenan grafted with N-hydroxyethylacrylamide (κC-g-PHEAA) hydrogel adsorbent was prepared by Kulal *et al* [32] and is utilised to remove MB and R6G, cationic dyes and Cu(II) and Hg(II) from water. The maximum adsorption capacity of the prepared nanocomposite hydrogel was found to be 196.1, 307.8 mg g^{-1}, 243.4 and 229.9 mg g^{-1} for MB, R6G, Cu(II) and Hg(II), respectively. It was observed that electrostatic interaction and formation of coordination complex are the possible mechanisms in adsorption of cationic dyes and metal ions respectively. Hosseinzadeh [33] prepared nanocomposite hydrogel by grafting acrylic acid onto κ-carrageenan with multi-walled carbon nanotube (MWCNT) and was evaluated for the removal of CV. The adsorption capacity of κ-carrageenan based bio-hydrogel nanocomposite was found to be 118 mg g^{-1}.

Mahdavinia *et al* [34] prepared nanocomposite hydrogel by polymerization of acrylamide with carrageenan and laponite RD nanoclay and was investigated for the removal a cationic CV dye from water. It was found that degree of swelling was high in presence of carrageenan. The results showed that the maximum adsorption capacity for carrageenan based hydrogel was 79.8 mg g^{-1}. Mahdavinia *et al* [35] investigated magnetic iron ions and chitosan-crosslinked κ-carrageenan bionanocomposite and used them as adsorbent to remove MB from water. The experimental adsorption capacities of magnetic and non-magnetic adsorbents for MB were obtained as 123.1 and 130.4 mg g^{-1}, respectively.

9.5 Removal of heavy metals in aqueous solution

Heavy metals are distinct among aqueous contaminants that can be extremely hazardous and cause several ecological, health and environmental problems. Regardless of the use of heavy metals in industrial processes and consumer products, the improper disposal into the environment without any prior treatment has become challenging and has gained great importance in recent years. A summary of adsorption capacities of bionanocomposite hydrogels for various metal ions is presented in table 9.1.

Zulfiqar *et al* [36] removed Pb(II) from wastewater using oil palm bio-waste/MWCNTs onto PVA hydrogel (OPB/PVA/MWCNTs) composites. It was observed that the reinforced nanocomposites showed enhanced mechanical properties and reported that the mechanism involved was chemical sorption mechanism. The

Table 9.1. The adsorption capacities of bionanocomposite hydrogels for various metal ions removal from water.

Biobased nanocomposite hydrogel (adsorbent)	Metal ion adsorbed	Adsorption capacity (mg/g)	Isotherm	Reference
CMC-magnetite	Cd(II)	100	Langmuir	[14]
Magnetic chitosan-g-poly (acrylic acid)	Cd(II)	308.84	Langmuir	[18]
	Pb(II)	695.22		
GO–chitosan	Cu(II)	70	Freundlich	[19]
	Pb(II)	90		
κC-g-PHEAA	Cu(II)	234.4	Freundlich	[32]
	Hg(II)	229.9		
OPB/PVA/MWCNTs	Pb(II)	30.031	Langmuir	[36]
wheat bran-g-poly (methacrylic acid) and nanosized clinoptilolite	Pb(II)	166.7	Langmuir	[37]
	Cu(II)	243.9		
	Cd(II)	175.4		
	Ni(II)	166.6		
alginate/reduced graphene oxide double network	Cu(II)	169.5	Langmuir	[42]
	Cr(VI)	72.5		
NPGESNC-AcA	Cu(II)	147.10	Langmuir	[44]

electrostatic attractions between Pb(II) and OPB/PVA/MWCNTs could successfully enhance the adsorption capacity. The experimental adsorption results for the as-prepared nanocomposite hydrogel showed the highest adsorption efficiency up to 30.031 mg g^{-1}.

Barati *et al* [37] examined the synthesis and utilization of low cost wheat bran-g-poly (methacrylic acid) and nanosized clinoptilolite hydrogel for the adsorptive removal of Pb(II), Cu(II), Cd(II) and Ni(II) cations from wastewater. The kinetic report revealed that the rate of adsorption on nanocomposite hydrogel was more than 80% within 15 min. The maximum monolayer adsorption capacity of the nanocomposite hydrogel was found to be 166.7, 243.9, 175.4 and 166.6 mg g^{-1} for Pb(II), Cu(II), Cd(II) and Ni(II) cations, respectively. In another study, Saberi *et al* [38] aimed to remove the Cu(II) and Pb(II), heavy metal ions and methylene violet (MV) and congo red (CR), toxic dyes using the magnetic Fe_3O_4-based starch-poly (acrylic acid) hydrogel nanocomposite (Fe_3O_4@St-AcANCH). The results indicated that the maximum adsorption capacity was 93.83, 99.32, 96.7 and 97.5% for Pb(II), Cu(II),CR and MV, respectively. This is attributed to the strong dipole interactions of strongly polarized surface on magnetic nanocomposite hydrogel. Kaşgöz *et al* [39] synthesized acrylamide (AAm)–2-acrylamido2-methylpropane sulfonic acid (AMPS) sodium salt with clay. The synthesized nanocomposite hydrogel was effectively used to remove heavy metal ions Cu(II), Cd(II) and Pb(II) from water. The maximum adsorption capacity of nanocomposite hydrogel was 1.28, 1.07 and 1.03 mmol g^{-1} for Cd(II), Cu(II) and Pb(II), respectively.

A bionanocomposite hydrogel was synthesized by grafting acrylamide with starch by using nanohematite by Alimardan *et al* [40] and used to remove Ni(II), Pb(II) and Cu(II) metal ions from industrial wastewater. It was found that the concentrations of metal ions adsorbed onto the hydrogel were 42.10, 35.93 and 25% for Pb(II), Ni(II) and Cu(II), respectively. In another study, Sekhavat Pour *et al* [41] magnetic hydrogel beads based on poly(vinyl alcohol)/carboxymethylstarch-g-poly(vinyl imidazole) and used as an eco-friendly adsorbent to remove Pb(II),Cu(II) and Cd (II) from water with maximum adsorption efficiency of 65, 83.60 and 53.20 mg g^{-1}, respectively. Zhuang *et al* [42] synthesized alginate/reduced GO double network (GAD) hydrogel through a feasible method and investigated on the mechanical and adsorption properties. The resulting GAD exhibited lower swelling ratio that enhanced gel stability in highly concentrated alkali solutions. The GAD beads showed a remarkable adsorption capacity of 169.5 and 72.5 mg g^{-1} of Cu (II) and Cr (VI), respectively. The authors demonstrated that GAD can still retain its adsorption capability even after 10 regeneration cycles.

In another study, Fosso-Kankeu *et al* [43] demonstrated the utilization of gelatine hybrid nanocomposite hydrogel by ameliorating clinoptilolite within the hydrogel matrix. The thermal stability of the hydrogel was increased with the inclusion of clinoptilolite. The synthesized hybrid(bio)nanocomposite hydrogel showed promising results for the adsorption of Cd (II) from aqueous solution with maximum adsorption capacity of 78.13 mg g^{-1}. Dil *et al* [44] prepared (NPGESNC-AcA) nanocomposite hydrogel as biosorbent and utilised to remove Cu (II) in wastewater. It was demonstrated that the incorporation of Ag nanoparticles in the polymer matrix enhances thermal stability of the hydrogel. It was also reported that the maximum adsorption capacity of 147.10 mg g^{-1} at pH 5.5 within 40 min The results exhibited that the biosorbent loaded by Cu (II) ions could be easily recycled wherein Cu (II) ions were recovered above 84.3% with a single wash by 0.1MHCl.

9.6 Removal of organic dyes in aqueous solution

Organic dyes, similar to heavy metals are hazardous and increase the burden of unwanted pollutants in aqueous systems. Dyes extend unwanted colour into water systems and decrease the penetration of sunlight into water impeding photosynthetic activities. Hence, these dyes are extremely toxic to aquatic flora and fauna even at low concentrations due to their complex composition, high solubility in water and their persistent bioaccumulation. Therefore, the removal of these organic dyes has become of utmost concern for water researchers. A heightened interest for removal of organic dyes by bio-based nanocomposite hydrogels is envisaged owing to the enhanced properties and improved adsorption capacities (table 9.2). Chen *et al* [45] synthesized eco-friendly graphene oxide-chitosan hydrogel adsorbent to remove MB and Eosin Y dyes. It was found that the BNCHs present electrostatic interaction mechanism in the removal of dyes. The maximum adsorption capacity towards these ionic dyes was higher than 300 mg g^{-1} for both.

Peng *et al* [46] fabricated cellulose-clay nanocomposite hydrogel with super-adsorbent properties and high dye removal efficiency. Thus prepared BNCH

Table 9.2. The adsorption capacities of bionanocomposite hydrogels for various organic dyes removal from water.

Biobased nanocomposite hydrogel (adsorbent)	Dye adsorbed	Adsorption capacity (mg/g)	Isotherm	Reference
CMC-PAM-Fe_3O_4	CV	72.4	Langmuir	[21]
β-cyclodextrin/chitosan functionalized graphene oxide	MB	1134	Freundlich	[24]
cellulose nanowhiskers-chitosan-g-poly (acrylic acid)	MB	1968	Langmuir	[25]
Pec-g-poly(AMPS-co-AAm)/Ag nanocomposite	CV	111	Langmuir	[26]
Pec-g-PAMPS	MB	89	Freundlich	[28]
kappa-carrageenan-g-poly(acrylamide)/ sepiolite	CV	33.1	Langmuir	[30]
polyacrylamide grafted gumkaraya -NiS nanoparticle	R6G	1244.71	Langmuir	[47]
CMC-AM/nHA-CS	Acid Blue	29.30	—	[48]
CMC-AM/nHA-MS		29.52		

exhibited spontaneous physico-sorption process and the results exhibited that the cellulose-clay nanocomposite hydrogels with maximum adsorption capacity of 98% at 100 mg L^{-1} of MB dye. The findings reported that hydrophobic interaction was the main reason to remove MB, besides electrostatic attraction. Kumar *et al* [47] reported the fabrication and use of polyacrylamide grafted gumkaraya and NiS nanoparticle based hydrogel to successfully remove rhodamine 6G dye from water. The enhanced adsorption efficiency of the hydrogel bionanocomposites is attributed to the improved porosity, surface area, hydrodynamic volume and chelating binding sites on the surface due to the presence of nanoparticles. The adsorption of R6G followed monolayer adsorption with maximum adsorption capacity of 1244.71 g L^{-1}. The experiments showed that this adsorbent can be regenerated for three cycles. Carboxymethyl cellulose/acrylamide/nano-hydroxyapatite Clam (CMC-AM/nHA-CS) and Magellan (CMC-AM/nHA-MS) shell composite hydrogels were synthesized by Varaprasad *et al* [48] and evaluated as an adsorbent for Acid blue 113(AB) from wastewater. The adsorption of AB on CMC-AM/nHA-MS was high compared to CMC-AM/nHA-CS and this is attributed to the percentage of Ca(II) was higher in nHA-MS that enhanced swelling capacity of the nHA hydrogels. The highest adsorption capacities of CMC-AM/nHA-CS and CMC-AM/nHA-MS for AB dye were found to be 29.30 and 29.52 mg g^{-1}, respectively.

Soleimani *et al* [49] successfully synthesized nanocomposite hydrogel by grafting cellulose nanowhiskers onto graphene oxide. The biocomposite graphene oxide hydrogel demonstrated high performance to remove organic dyes such as MB and RhB from aqueous solution. It was reported that the BNCH showed fast rate and adsorption capacity for removal of dye from textile wastewater due to the presence of plenty of functional groups on hydrogel. The removal efficiency was found to be 100 and 90% for MB and RhB, respectively. The removal of MB cationic dye from wastewater using organic/inorganic hydrogel nanocomposite of TiO_2 incorporated sodium alginate crosslinked polyacrylic acid (SA-cl-poly(AA)-TiO_2). A very high adsorption capacity of 2257.36 mg g^{-1} was reported for the adsorptive removal of MB [50]. The high adsorption capacity of SA-cl-poly(AA)-TiO_2 nanocomposite hydrogel was ascribed to the distinct structure and functional groups on the surface that provide large surface area and good affinity for the feasible and fast adsorption of dyes.

The graphene oxide-poly acrylic acid hydrogel grafted onto natural salep biopolymer was prepared by Rezanejade *et al* [51]. The prepared BNCH was applied for the adsorptive removal of MV, a cationic dye in different media. It was employed on real samples such as milk, wastewater, laboratory water, Caspean sea water and green tea under optimized conditions. The results showed that the MV removals are 97.65, 90.48, 96.25, 97.12 and 94.31%, respectively. The maximum adsorption was attributed to the better formation of hydrophilic polymer networks with dangling PAA chains. A simple and cost-effective approach for the preparation of silver nanocomposite hydrogel (SNC) based on sodium alginate was reported by Karthika Devi *et al* [52]. The prepared bio-SNC demonstrated extraordinary adsorbent properties and was effectively utilised for the removal of MB dye from water. The maximum monolayer adsorption capacity of the prepared SNC was calculated as 213.7 mg g^{-1}.

9.7 Conclusions

Focussing on wastewater treatment, this chapter provides recent progress in the evaluation of bionanocomposite hydrogels (BNCHs) for their adsorptive removal of pollutants. Nanocomposite hydrogels' qualities allow them to demonstrate extraordinary characteristics that enable them to be utilized as essential materials in water treatment. From time to time, substantial modifications are made to improve the field of hydrogels for their comprehensive applications. The inclusion of biomaterials into the network of hydrogel is an innovative method to acquire hydrogels with good degradation, biocompatibility, cost effectiveness and durable properties for advanced applications. These properties make bionanocomposite hydrogels potential materials for water treatment. Some insights on the utilization of bio-based nanocomposite hydrogels for numerous cycles of adsorption and desorption have been provided. The literature emphasized in this chapter is expected to give innovative ideas in the field of bionanocomposite hydrogels for several other applications.

9.8 Future prospects

This chapter provides the detailed introduction towards the emergence of bionanocomposite hydrogels as conventional adsorbents for removal of water contaminants. As far as the future research is concerned, more comprehensive understanding is required to address the interaction mechanisms between pollutant and adsorbent to direct the synthesis of relevant bionanocomposite hydrogel adsorbents with definite properties. While extensive research is ongoing in order to provide new kinds of hydrogel adsorbents for future necessities, there is a need for proper exploration of economic viability to determine the suitability of bionanocomposite hydrogels to be implemented in large scale. Hence, there is a long way to assimilate biomaterials into nanocomposite hydrogel that lead to tailored properties and transforming into innovative applications.

Acknowledgement

The authors acknowledge research support and facilities of BVRIT Hyderabad College of Engineering for Women, Hyderabad, India.

References

[1] Vemula M G P 2021 Modelling of experimental Cr(VI) adsorption on acid and alkali pretreated rice husk-graphene oxide *Mater. Today Proc.* **43** 3394–6
[2] Wang J, Gao S, Tian J, Cui F and Shi W 2020 Recent developments and future challenges of hydrogels as draw solutes in forward osmosis process *Water* **12** 692
[3] Khan S, Ullah A, Ullah K and Rehman N-U 2016 Insight into hydrogels *Designed Monom. Polym.* **19** 456–78
[4] Thakur S and Arotiba O 2017 Synthesis, characterization and adsorption studies of an acrylic acid-grafted sodium alginate-based TiO_2 hydrogel nanocomposite *Adsorp. Sci. Technol.* **36** 458–77
[5] Singhal R and Gupta K 2015 A review: Tailor-made hydrogel structures (classifications and synthesis parameters) *Polym.-Plast. Technol. Eng.* **55** 54–70
[6] Slaughter B V, Khurshid S S, Fisher O Z, Khademhosseini A and Peppas N A 2009 Hydrogels in regenerative medicine *Adv. Mater.* **21** 3307–29
[7] Chung H J and Park T G 2009 Self-assembled and nanostructured hydrogels for drug delivery and tissue engineering *Nano Today.* **4** 429–37
[8] Sahiner N, Ozay O and Aktas N 2012 The Removal of cyanide ions from aquatic environments by quaternizable p(4-VP) hydrogels of different dimensions *Water Air Soil Pollut.* **224** 1
[9] Sahiner N, Ozay O, Aktas N, Blake D A and John V T 2011 Arsenic (V) removal with modifiable bulk and nano p(4-vinylpyridine)-based hydrogels: the effect of hydrogel sizes and quarternization agents *Desalination* **279** 344–52
[10] Ojijo V, Sinha and Ray S 2013 Processing strategies in bionanocomposites *Prog. Polym. Sci.* **38** 1543–89
[11] Yang S, Fu S, Liu H, Zhou Y and Li X 2010 Hydrogel beads based on carboxymethyl cellulose for removal heavy metal ions *J. Appl. Polym. Sci.* **119** 1204–10
[12] Peng N, Hu D, Zeng J, Li Y, Liang L and Chang C 2016 Superabsorbent cellulose–clay nanocomposite hydrogels for highly efficient removal of dye in water *ACS Sustain. Chem. Eng.* **4** 7217–24
[13] Mahdavinia G R, Hasanpour J, Rahmani Z, Karami S and Etemadi H 2013 Nanocomposite hydrogel from grafting of acrylamide onto HPMC using sodium montmorillonite nanoclay and removal of crystal violet dye *Cellulose.* **20** 2591–604

[14] Uva M, Tambasco M, Grassi G, Corsi I, Protano G and Atrei A 2017 Carboxymethylcellulose hydrogels cross-linked with magnetite nanoparticles for the removal of organic and inorganic pollutants from water *J. Environ. Chem. Eng.* **5** 3632–9
[15] Chen X, Zhou S, Zhang L, You T and Xu F 2016 Adsorption of heavy metals by graphene oxide/cellulose hydrogel prepared from NaOH/urea aqueous solution *Materials* **9** 582
[16] Melilli G, Yao J, Chiappone A, Sangermano M and Hakkarainen M 2021 Photocurable 'all-lignocellulose' derived hydrogel nanocomposites for adsorption of cationic contaminants *Sustain. Mater. Technol.* **27** e00243
[17] Wang J, Zhang N, Jiang C and Zhang C 2018 Adsorptive removal of phenol by single and double network composite hydrogels based on hydroxypropyl cellulose and graphene oxide *J. Mater. Res.* **33** 3898–905
[18] Zhu Y, Zheng Y, Wang F and Wang A 2016 Fabrication of magnetic macroporous chitosan- g -poly (acrylic acid) hydrogel for removal of Cd^{2+} and Pb^{2+} *Int. J. Biol. Macromol.* **93** 483–92
[19] Chen Y, Chen L, Bai H and Li L 2013 Graphene oxide–chitosan composite hydrogels as broad-spectrum adsorbents for water purification *J. Mater. Chem.* A **1** 1992–2001
[20] Puspitasari F H, Salamah U, Sari N R, Maddu A and Solikhin A 2020 Potential of chitosan hydrogel based activated carbon nanoparticles and non-activated carbon nanoparticles for water purification *Fibers Polym.* **21** 701–8
[21] Meysam A, Mohammad T V, Gholam and Reza M 2018 Preparation of a novel magnetic nanocomposite hydrogel based on carboxymethyl chitosan for the adsorption of crystal violet as cationic dye *J. Chem. Health Risks* **8** 289–304
[22] Budnyak T M, Pylypchuk I V, Tertykh V A, Yanovska E S and Kolodynska D 2015 Synthesis and adsorption properties of chitosan-silica nanocomposite prepared by sol–gel method *Nanoscale Res. Lett.* **10** 1–10
[23] Zhou J, Hao B, Wang L, Ma J and Cheng W 2017 Preparation and characterization of nano-TiO_2/chitosan/poly(N-isopropylacrylamide) composite hydrogel and its application for removal of ionic dyes *Sep. Purif. Technol.* **176** 193–9
[24] Liu Y, Huang S, Zhao X and Zhang Y 2018 Fabrication of three-dimensional porous β-cyclodextrin/chitosan functionalized graphene oxide hydrogel for methylene blue removal from aqueous solution *Colloids Surf.*, A **539** 1–10
[25] Melo B C, Paulino F A A, Cardoso V A, Pereira A G B, Fajardo A R and Rodrigues F H A 2018 Cellulose nanowhiskers improve the methylene blue adsorption capacity of chitosan-g-poly(acrylic acid) hydrogel *Carbohydr. Polym.* **181** 358–67
[26] Kodoth A K and Badalamoole V 2019 Silver nanoparticle-embedded pectin-based hydrogel for adsorptive removal of dyes and metal ions *Polym. Bull.* **77** 541–64
[27] El Fadl F I, Mahmoud G A, Badawy N A, Kamal F H and Mohamed A A 2017 Pectin based hydrogels and its ferrite nanocomposites for removal of nitro compounds *Desalin. Water Treat.* **90** 283–93
[28] Babaladimath G and Badalamoole V 2018 Silver nanoparticles embedded pectin-based hydrogel: a novel adsorbent material for separation of cationic dyes *Polym. Bull.* **76** 4215–36
[29] Sharma G, Kumar A, Chauhan C, Okram A, Sharma S and Pathania D *et al* 2017 Pectin-c rosslinked -guar gum/SPION nanocomposite hydrogel for adsorption of m-cresol and o-chlorophenol *Sustain. Chem. Pharm.* **6** 96–106
[30] Mahdavinia G R and Asgari A 2013 Synthesis of kappa-carrageenan-g-poly(acrylamide)/sepiolite nanocomposite hydrogels and adsorption of cationic dye *Polym. Bull.* **70** 2451–70

[31] Hosseinzadeh H, Zoroufi S and Mahdavinia G R 2015 Study on adsorption of cationic dye on novel kappa-carrageenan/poly(vinyl alcohol)/montmorillonite nanocomposite hydrogels *Polym. Bull.* **72** 1339–63

[32] Kulal P and Badalamoole V 2020 Hybrid nanocomposite of kappa-carrageenan and magnetite as adsorbent material for water purification *Int. J. Biol. Macromol.* **165** 542–53

[33] Hosseinzadeh H 2015 Synthesis of carrageenan/multi-walled carbon nanotube hybrid hydrogel nanocomposite for adsorption of crystal violet from aqueous solution *Pol. J. Chem. Technol.* **17** 70–6

[34] Mahdavinia G R, Massoudi A, Baghban A and Massoumi B 2012 Novel carrageenan-based hydrogel nanocomposites containing laponite RD and their application to remove cationic dye *Iran. Polym. J.* **21** 609–19

[35] Mahdavinia G R and Mosallanezhad A 2016 Facile and green rout to prepare magnetic and chitosan-crosslinked κ-carrageenan bionanocomposites for removal of methylene blue *J. Water Process Eng.* **10** 143–55

[36] Zulfiqar M, Lee S Y, Mafize A A, Kahar N A, Johari K and Rabat N E 2020 Efficient removal of Pb(II) from aqueous solutions by using oil palm bio-waste/MWCNTs reinforced PVA Hydrogel composites: kinetic, isotherm and thermodynamic modeling *Polymers.* **12** 430

[37] Barati A, Moghadam E A, Miri T and Asgari M 2014 Rapid removal of heavy metal cations by novel nanocomposite hydrogels based on wheat bran and clinoptilolite: kinetics, thermodynamics, and isotherms *Water Air Soil Pollut.* **225** 2096

[38] Saberi A, Alipour E and Sadeghi M 2019 Superabsorbent magnetic Fe_3O_4-based starch-poly (acrylic acid) nanocomposite hydrogel for efficient removal of dyes and heavy metal ions from water *J. Polym. Res.* **26**

[39] Kaşgöz H, Durmuş A and Kaşgöz A 2008 Enhanced swelling and adsorption properties of AAm-AMPSNa/clay hydrogel nanocomposites for heavy metal ion removal *Polym. Adv. Technol.* **19** 213–20

[40] Alimardan Z and Darabi M 2015 Removal of Pb(II), Ni(II) and Cu(II) metal ions from industrial wastewater by nanocomposite hydrogel base starch *J. Bio. Env. Sci.* **6** 487–91

[41] Sekhavat Pour Z and Ghaemy M 2015 Removal of dyes and heavy metal ions from water by magnetic hydrogel beads based on poly(vinyl alcohol)/carboxymethyl starch-g-poly(vinyl imidazole) *RSC Adv.* **5** 64106–18

[42] Zhuang Y, Yu F, Chen H, Zheng J, Ma J and Chen J 2016 Alginate/graphene double-network nanocomposite hydrogel beads with low-swelling, enhanced mechanical properties, and enhanced adsorption capacity *J. Mater. Chem.* A **4** 10885–92

[43] Fosso-Kankeu E, Mittal H, Waanders F and Ray S S 2017 Thermodynamic properties and adsorption behaviour of hydrogel nanocomposites for cadmium removal from mine effluents *J. Ind. Eng. Chem.* **48** 151–61

[44] Dil N N and Sadeghi M 2018 Free radical synthesis of nanosilver/gelatin-poly (acrylic acid) nanocomposite hydrogels employed for antibacterial activity and removal of Cu(II) metal ions *J. Hazard. Mater.* **351** 38–53

[45] Chen Y, Chen L, Bai H and Li L 2013 Graphene oxide–chitosan composite hydrogels as broad-spectrum adsorbents for water purification *J. Mater. Chem.* A **1** 1992–2001

[46] Peng N, Hu D, Zeng J, Li Y, Liang L and Chang C 2016 Superabsorbent cellulose–clay nanocomposite hydrogels for highly efficient removal of dye in water *ACS Sustain. Chem. Eng.* **4** 7217–24

[47] Kumar N, Mittal H, Parashar V, Ray S S and Ngila J C 2016 Efficient removal of rhodamine 6G dye from aqueous solution using nickel sulphide incorporated polyacrylamide grafted gum karaya bionanocomposite hydrogel *RSC Adv.* **6** 21929–39

[48] Varaprasad K, Nunez D, Yallapu M M, Jayaramudu T, Elgueta E and Oyarzun P 2018 Nano-hydroxyapatite polymeric hydrogels for dye removal *RSC Adv.* **8** 18118–27

[49] Soleimani K, Tehrani A D and Adeli M 2018 Bioconjugated graphene oxide hydrogel as an effective adsorbent for cationic dyes removal *Ecotoxicol. Environ. Saf.* **147** 34–42

[50] Thakur S, Pandey S and Arotiba O A 2016 Development of a sodium alginate-based organic/inorganic superabsorbent composite hydrogel for adsorption of methylene blue *Carbohydr. Polym.* **153** 34–46

[51] Rezanejade Bardajee G, Sadat Hosseini S and Vancaeyzeele C 2019 Graphene oxide nanocomposite hydrogel based on poly(acrylic acid) grafted onto salep: an adsorbent for the removal of noxious dyes from water *New J. Chem.* **43** 3572–82

[52] Karthiga Devi G, Senthil Kumar P and Sathish Kumar K 2016 Green synthesis of novel silver nanocomposite hydrogel based on sodium alginate as an efficient biosorbent for the dye wastewater treatment: prediction of isotherm and kinetic parameters *Desalin. Water Treat.* **57** 27686–99

IOP Publishing

Bionanomaterials for Environmental and Agricultural Applications

Ravindra Pratap Singh and Kshitij RB Singh

Chapter 10

Bionanomaterials utility in food industry and its challenges

Shweta Rathee, Eneyew Tadesse Melaku and Ankur Ojha

Nanotechnology is a rapidly growing toolbox that has affected the many facets of the food industry in the last decade. It has opened roads to develop various nanoproduct formulations, nanoencapsulation, fabrication of a wide range of nanobiosensors, and composite packaging films. It has impacted the product shelf life, bioavailability, environmental sustainability, processing, and detection methods. Potential risks, safety, toxicity assessments, and exposure routes of bionanomaterials in food products are summarized in this book chapter, considering the effect at the cellular level and animal models. Moreover, the worldwide regulatory frameworks, significant challenges, and prospects of this fast-growing field are also highlighted.

10.1 Introduction

Globally, the food industry is as much an integral part of economic growth as other industries, i.e., pharmaceutical, material science, electronics, aerospace, and the defense industry. It is growing continuously to meet the increasing and changing demands of consumers worldwide. According to the research estimates, it contributes to around 10% of the world's GDP. With the increase in worldwide population, there is a heavy burden on the food industry, which indirectly puts pressure on economic system. Even the Food and Agricultural Organization (FAO) says that the world population will reach 8 billion by 2025 and 9.6 billion by 2050 [1]. To provide sustainable nutrition, secure and safe food, there is a need for new technologies that could help us to meet these challenges without compromising on the ecosystem. Nanotechnology is an emerging field that promises to revolutionize the food industry to achieve sustainable development goals target 12.3 (to halve global per capita food waste). The problem of climate change, food safety, and security could

doi:10.1088/978-0-7503-3863-9ch10

be addressed. However, the nanofood market is in its infancy but is gathering pace slowly and steadily. The market share of nanofood is increasing from 7 billion U.S. dollars in 2015 to 20.4 billion U.S. dollars in 2020 [2]. In the production of nanofoods, the USA is at the top, followed by Japan and China.

When talking about nanotechnology, it is about manipulating the materials at an atomic level [3, 4]. It is an interdisciplinary field enabled by relatively new techniques. Bionanomaterials are made from the reduction of natural compounds like carbohydrates, protein, enzymes, lipids, and phenolic compounds [5]. They exhibit superior physical, chemical, biological, nutritional, thermo-mechanical properties along with enhanced biocompatibility, bioavailability, and bioreactivity [6]. They can be of different types: polymer-based, metal-based, metal-oxides based and carbon-based bionanomaterials. There are a few examples of naturally occurring bionanomaterials like magneto bacteria that produce magnetic nanoparticles. Similarly, diatoms have silica materials. They exist in different forms like one- (nanorods, nanofibrils, and nanotubes), two- (sheets and laminates), and three-dimensional (solid core and nanodroplets with a liquid core) substances [7]. The major applications in the food industry are for new product formulation with improved sensory properties, nanoencapsulation for the development of functional foods with enhanced bioavailability, packaging materials with better thermo-mechanical properties, and biosensors for improving the safety of food [7–10].

Bionanomaterials are present in our daily diet from the very beginning but are not recognized. Some well-known examples include emulsions, micelle, dietary fiber, foams, etc [11]. Gradually, nanotechnology has become a booster for the food industry with its magical applications such as novel sensorial properties, improved shelf life, excellent bioavailability, targeted delivery of nutrients, improved digestibility, antimicrobials, improved packaging materials, nanosensors, and bionanosensors [12, 13]. This chapter takes a unique approach to integrate bionanomaterial's role in the food industry and its effect on the natural environment. Additionally, its toxicological impact on human health with its regulatory framework is also discussed. Its prospects and challenges are also highlighted in detail.

10.2 Bionanomaterials applications in food industry

Bionanomaterials can be divided into various groups depending on the requirements being considered. There are many applications of bionanomaterials in the food industry, including new product formulations, nanoencapsulation of functional ingredients (vitamins, minerals, bioactive compounds, and nutraceuticals), product safety detection, and functional packaging. All are discussed in the section below with a schematic representation in figure 10.1.

10.2.1 New product formulations

With the increasing change in the demands of consumers, there is a need for the development of new food products. Bionanomaterials have helped in bridging this gap between the demand and supply of nanofood products. Nanofoods are much

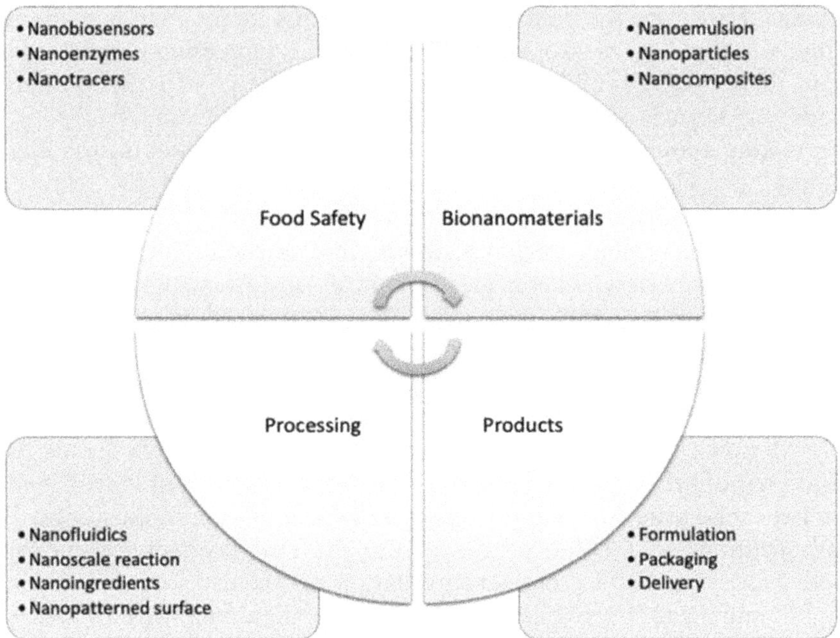

Figure 10.1. Prospects of bionanomaterial in the food industry.

better than genetically modified crops. Further, they help in acceptability by consumers due to their enhanced organoleptic properties. High-fat foods such as spreads, mayonnaise, creams, yogurts, and ice creams could be tailored to low-fat content without compromising their organoleptic properties [14]. Nanoedible coatings, along with active ingredients like antioxidants and antimicrobials could be applied to different food categories that help taste enhancement, anti-browning, and increasing the shelf life of products. Bionanomaterials also help in reducing the amount of food additives, salt, sugar, and color due to their small size to provide more healthy food [15]. 3D printing and nanotechnology combination helps in the customization of nutrition and creating new gastronomic experiences. Global food manufacturers such as Kraft, Nestle, Altria, Heinz, and Unilever have increased their research and development due to the increase in demand of nanofoods.

Nanoencapsulation of beta-carotene in starch added to juices showed better antioxidation, bioavailability, and improved shelf life [16]. In China, The Qinhuangdai Taiji Ring Nanoproducts Co. Ltd. produces six different types of nanotea products [17]. Nanofiltration is another important application of nanotechnology. It also helped to create juices with much improved antioxidant activity due to better retention of macro- and micronutrients [18, 19]. Chitosan nanoparticles used for the encapsulation of anthocyanin, helped to improve the sensory properties in beverages. Even degradation of anthocyanin is slowed down [20]. Nanopatterning is another recent technology that is gaining popularity. Nanopatterning of different biomolecules helps to form other new food products [21].

10.2.2 Nanoencapsulation of functional components

Nanoencapsulation can be defined as the entrapment of different functional components within a carrier on a nanoscale level. Functional components play an essential role in providing such dietary benefits to consumers along with nutrition. There are different functional components such as vitamins, minerals, flavors, bioactive compounds, proteins, nucleic acids, probiotics, prebiotics, and nutraceuticals [22]. They provide different health benefits, including being anti-inflammatory, antimicrobial, antioxidant, anti-cancer, anti-obese, with cardiovascular protective activity, anti-diabetic, neuroprotective activity, and immunity booster [23–28]. However, they are heat sensitive, less bioavailable, generate off-flavors, unstable during storage, rapidly digestible, and hydrophobic. With the help of different bionanomaterials, their encapsulation could be made possible. It can reduce the adverse sensory effects of these critical functional components with improved bioavailability, targeted delivery, along with an easily absorbable form. It also enhances the delivery, bioaccessibility, bioavailability of functional components for improved nutrition and health outcomes [29].

Due to lifestyle changes, there is an increase in chronic diseases like diabetes, hypertension, heart diseases, and obesity. WHO in 2018 stated that chronic diseases account for more than 70% of deaths globally. All this makes people more conscious about their dietary habits, increasing their uptake of functional foods. Development of different functional foods with improved properties could be made possible with the help of nanoencapsulation. Nanoencapsulation bionanomaterials are classified as lipid-based nanomaterials, Nature-inspired nanomaterials, special equipment-based nanomaterials, and biopolymer based-nanomaterials. They are formulated using different top-down and bottom-up approaches such as microfluidizers, ultrasonicators, solvent evaporation, emulsification, precipitation, coacervation, and spray drying [30]. Lipid-based nanoencapsulation includes the use of nanoemulsions, nanoliposomes, and solid lipid nanoparticles. There are various commercial products available in the market in different countries across the globe. NovaSol®, a nanocarrier contains two active substances, CoQ10 (for fat reduction) and alpha-lipoic acid, to enable weight management. Increasing bioavailability and potency of the target compounds is being developed by the German company Aquanova. Recently, three lipid carriers consisting of nanoemulsion, solid lipid nanoparticle, and nanostructured lipid carrier were developed from zein peptides to encapsulate lutein with much-improved bioavailability [31]. Similarly, there are many applications of nature-inspired nanomaterials. β-cyclodextrin used for the entrapment of chlorogenic acid using nanocomplexation has improved the aroma profiles of the food products like bread, cookies, and mushrooms [32]. Amylose-based nanoencapsulation of catechin and p-coumaric acid has proved to be a promising technique and preserved the bioactive compound in foods [33, 34].

Special equipment-based nanoencapsulation is gaining popularity due to its various advantages. In one study, researchers used microfluidization for the nanoencapsulation of *Thymus capitatus* essential, showing improved antibacterial activity in milk [35]. A recent approach used different biopolymers like carbohydrates, polymers, gums, and proteins for the nanoencapsulation of DHA with improved shelf life, oxidative stability, and bioavailability [36]. Another study based was on nanoencapsulation of D-limonene in the nanocomplexes made up of pectin-whey protein [37]. The most important advantage of using bionanomaterials is that they improve solubility, stability, increased bioavailability, protection from external conditions, sustained and targeted delivery without affecting textural properties [38]. A variety of nanofoods are already available in the form of fortified foods, nutraceuticals, and enriched food products containing bionanomaterials in encapsulated minerals, vitamins, antioxidants, health-promoting supplements, and antimicrobials. The examples of various vitamins, minerals, bioactive compounds, and nutraceuticals added into food products are essential fatty acids. [39, 40], carotenoids [41–44], vitamins [45–47], antioxidants [48, 49], phytosterols [50] and, bioactive peptides [51]. The recently investigated bionanomaterials for encapsulation of functional components have been tabulated in table 10.1.

10.2.3 Biosensing for food safety

A biosensor is an analytical device that gives an electronic signal by the interaction of bioreceptor with target analyte. It consists of a bioreceptor, transducer, signal processor, and display unit. It gives us results in the form of an electronic signal. They can be divided into optical, electrochemical, mass-based, and calorimetric [52, 53]. They give rapid, accurate, selective, and cost-effective detection. By 2050, the global population will reach 9 billion and require 70% more food, which will create an alarming situation for meeting the demand for safe and secure food [54]. According to a report published by WHO, in 2015, every year, around 420 000 die from eating contaminated food [55]. Safe food is a matter of global concern that has been in the limelight in recent years due to the increasing number of foodborne outbreaks, creating the need for screening of foods at every step of the food supply chain. There are various risk factors such as pathogens, toxins, heavy metals, pesticides, contaminants, allergens, and illegal drugs [56] detected by biosensors, as explained in figure 10.2. Traditional methods used for evaluating foodborne pathogens are culture-based [57], immunoassays [58], and nucleic acid-based techniques [59]. Similarly, high-performance liquid chromatography [60], mass spectrometry [61], gas chromatography [56, 57], enzyme-linked immunosorbent analysis [62], and DNA microarray technology [63] are used for the detection of other analytes. They have the disadvantages of being time-consuming, expensive, and labor-intensive. Therefore, to overcome the above limitations, there is a preferred need to use advanced biosensing methods.

Many pathogenic bacteria are detected using different types of biosensors in food, such as simultaneous detection of *Escherichia coli*, *Listeria monocytogenes*, and *Salmonella Typhimurium* [64]. Other examples are impedance biosensors for rapid and sensitive *Salmonella typhimurium* to identify target bacteria from a large volume of bacterial sample [65]. Dielectrophoretically enhanced detection of *Escherichia coli* Dh5α cells from their suspensions is a low-cost and rapid alternative to other methods [66]. Recently DNAzymes based biosensors have been used for the

Table 10.1. Bionanomaterials used as carriers of various functional ingredients.

Nanomaterials	Raw materials	Functional components	Class	Size of carrier (nm)	Purpose of encapsulation	References
Nanoemulsions	Whey protein concentrate, maltodextrin and gum arabic	Krill oil	Fatty acids	153.9–162 nm	Encapsulation of valuable oils	[106]
	Modified starch	Quercetin, curcumin	Polyphenols	175.44 nm	Suitable medium for solubilizing phytochemicals	[107]
	Sesame oil, linseed oil, walnut oil, Lactoferrin	Lycopene	Carotenoid	212.93 nm	Delivery of lipophilic bioactive compounds	[107]
	Medium chain triglyceride (MCT), lecithin, and polysorbate 80	Carvacrol	Polyphenols	105.32 nm	Targeted delivery	[108]
Coacervation	Zein, chitosan	Resveratrol	Polyphenols	556–631 nm	Improve the stability and bioavailability	[109]
	Gelatin, Sodium carboxymethyl cellulose	Zeaxanthin	Polyphenols	210.7 nm	Heat-sensitive bioactive compounds, flavor and food ingredients	[110]
	Gelatin, arabic gum	Oregano essential oil	Fatty acids	87.5–119.8 nm	Enhanced encapsulation	[111]
Liposomes/ niosomes	Lecithin, Cholesterol, Tween-80	Vitamin C, β-Carotene	Vitamins,	252.67 nm	Multicomponent nutrient delivery system	[112]
	Liposome, Chitosan	Linseed oil, Quercetin	Fatty acids, Polyphenols	262.2 nm	Improved stability	[123]
	Tween 80	α-tocopherol, γ-oryzanol	Polyphenols	200 nm	Improved controlled release and stability	[114]
Inclusion complexation	Chitosan hydrochloride, carboxymethyl chitosan and whey protein isolate	Anthocyanins	Polyphenols	332.20 nm	Improved thermal stability and sustained release	[115]
	Chitosan Zein	Quercetagetin	Polyphenols	309 nm	Promising nano-delivery systems	[116]
	β-lactoglobulin	Astaxanthin	Carotenoid	<300nm	Effective delivery process for E-Asta	[117]

Source: Upgraded and modified from [118].

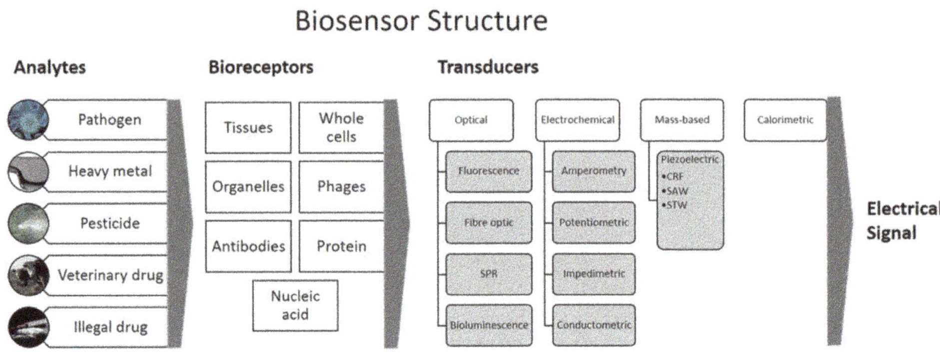

Figure 10.2. Bio-enabled nanomaterials as sensors for food safety.

Table 10.2. Summary of nanobiosensors used for detection of various analytes.

Nanomaterials	Bioreceptor	Target analyte	Limit of detection	Detection method	References
Au nanourchins/ G.O.	Aptamer	Staphylococcal Enterotoxin B	0.21 fM	Differential pulse voltametry	[119]
AuNPs	Tyramine oxidase	Tyramine	2.9 μM	Colorimetric	[120]
Zinc-up conversion nanoparticles		Octopamine	0.081 mg.L	Fluorescence	[121]
MNPs	Nanozyme, aptamer	Aflatoxin B1	5 pg mL^{-1}	Electrochemical	[122]
Au/Pt nanocluster		*S. aureus*	80 CFU mL^{-1}	Colorimetric	[123]
SnO2-SiC and AuNPs	Antibody	Acrylamide	46 ng kg^{-1}	Cyclic voltametry	[124]
rGO	Enzyme	Sterigmatocystin/ Aflatoxin B1	2.3 3 1029 mol L^{-1}	Amperometry	[125]
AuNPs and MNPs	Aptamer	Chloroamphenicol	0.13 pg mL^{-1}	Colorimetric	[126]
Au nanoflowers	Antibody	Amantadine	0.005 ng ml^{-1}	Surface-enhanced Raman scattering	[127]

detection of various pathogenic bacteria [67]. Electrochemical biosensors are also used for applications like rapid detection of bacteria found on a facile synthesis of silver wire across electrodes. It can detect bacteria in lysozyme-treated human serum and milk samples [68]. Similarly, there are many biosensors available for the detection of allergens present in food like lysozyme [69], ovomucoid [70], gliadin [71], caseins [72], hazelnut protein [69], and aflatoxin [73]. Different pesticides like carbendazim [74], carbamate [72], and organophosphorus [75] are detected using other biosensors. A summary of the various biosensors used for analysis is given in table 10.2.

10.2.4 Bionanomaterials for food packaging

Globalization and industrialization have increased the use of synthetic plastics for packaging, but their use has given a significant toll on the environment. With the help of bionanomaterials, unique food packaging materials can be designed. They are biocompatible, biodegradable, non-toxic, economical, and sustainable [76], and are schematically shown in figure 10.3. The most commonly used are nanocellulose, alginate, chitosan, polylactic acid, cellulose acetate, and chitin-based bionanomaterial in food packaging [77–81]. Imperm® and Duretham® are examples of designed and engineered packaging materials with enhanced environmental and health safety [82]. There are mainly three types of packaging materials: nano-reinforcement, active packaging, and intelligent packaging. The purpose of nano-reinforcement is to increase the tensile and burst strength of the packaging material, thereby enhancing its mechanical properties [33, 34]. Barrier properties (like oxygen permeability and water vapor transmission rate) and thermal properties are also improved [84]. Active packaging is about enhancing the shelf life of the products by incorporating active components that are antimicrobial, antioxidant, offer quality/temperature control, have scavenging properties, and release/emitting properties [85–87]. Intelligent packaging is another emerging packaging technology that can simultaneously monitor product freshness and microbial activity or other properties [88]. Its main categories are a chemical or biological indicator, freshness indicator, radio frequency identification systems, sensors (time-temperature indicator), and integrity indicator [89–91]. The design of packaging material depends on two components: receptors (biomacromolecules) and reporter. Nanomaterial (and enzymes) act as a reporter that translates molecular signals into information that can be coded in the form of color, fluorescence, or electronic signals [92]. In general, they are attached as labels, incorporated into, or printed onto food packaging materials for quality monitoring [93]. Swaroop and Shukla in 2018 developed a packaging material using polylactic acid and magnesium oxide that is effective against bacterial biofilms [94]. Sarwar *et al* [95, 96] created nanocomposite films using polyvinyl alcohol, nanocellulose, and alginate. These films are antimicrobial and effective against both *Staphylococcus aureus* and *Escherichia coli.* In another

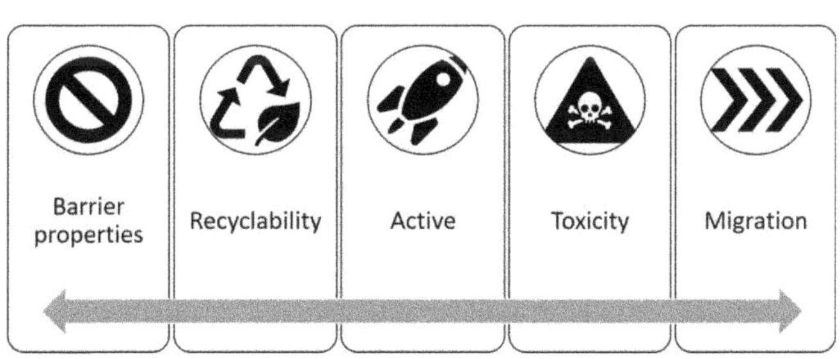

Figure 10.3. Bionanomaterials for sustainable food packaging applications.

study, long chain branching structures are introduced into the polylactic acid foam to produce eco-friendly, lightweight, and heat-resistant packaging material [97].

The use of biopolymers in food packaging has gained momentum in recent years, especially alginate and chitosan. An example is using alginate-based composite films by incorporating thymol as a good packaging material when applied on fresh-cut fruits. It also helped in retaining weight loss, nutrition, and surface color [98]. Another study based on the blend of gum kondagogu and sodium alginate exhibited good tensile strength, permeability, transparency, and biodegradability [99]. Recently, different nanoparticles were added for reinforcement in the fabrication of starch/sodium alginate bio-composite films. They are effective against many food-bacterial pathogens [100]. Alginate and chitosan-based composite aerogels based on noncovalent crosslinking enhanced the physical and antimicrobial properties [101]. Structural and antimicrobial evaluations show that the papers embedded with the sodium alginate, oxidized nanocellulose, and silver nanoparticles improved the mechanical, barrier, thermal and antibacterial properties [102].

10.3 Challenges and regulatory issues

Applications of bionanomaterials in the food industry are immense and increasing continuously. Nevertheless, there is a concern due to the challenges involved in their use. One big challenge is due to the less toxicological research in this area [82, 83]. So research is needed in this direction for assessing their risks. Cell lines, microbial, animal studies, and human trials should be done in this direction for determining responses generated by the body. Another critical challenge is designing industry-relevant materials as there is little knowledge about their structural properties and kinetics [103]. Additionally, there is a need for the training of the workforce. The change in their properties over time gives them an unpredictable effect on the environment. Even the uptake, bioaccumulation, biotransformation, and other risks of bionanomaterials in foods and packaging have not been wholly understood [104] and need lots of study in this area.

Commercially there are many products available in the market without approval. Globally there is a need for proper regulations. In the European Union, different legal acts contain several specific guidelines for products, including food. Additionally, a recent revision of the REACH regulation on chemical substances applied introduced nano-specific provisions [105]. Different organizations like European Chemical Agency (ECHA), Scientific Committee on Emerging and Newly Identified Health Risks (SCENIHR), European Food Safety Agency (EFSA), Environmental Protection Agency (EPA), Organisation for Economic Cooperation and Development (OECD), International Standard Organization (ISO), Food and Drug Administration (FDA), World Health Organization (WHO), and Scientific Committees are working in this direction in Europe and USA [12, 88]. For proper risk assessment, general regulatory procedures like the identification of nanomaterials through their size, surface characteristics, chemical composition, and stability should be used for their interaction and persistence inside the body. The above discussed parameters could be identified as critical components

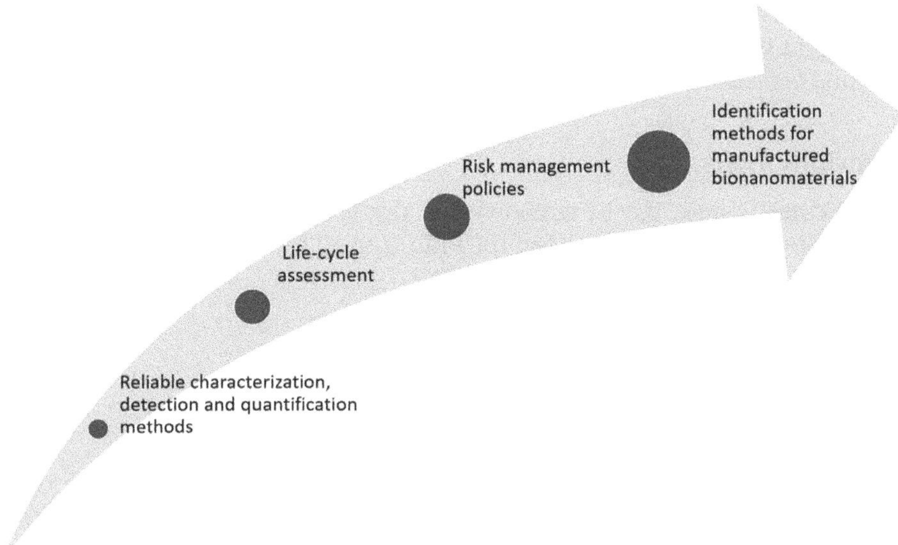

Figure 10.4. Parameters for implementation of nanostructures in the market.

for proper implementation of regulatory guidelines [86], as shown diagrammatically in figure 10.4.

10.4 Conclusion and prospects

Studies have shown that the public perception is not antagonistic towards the bionanomaterials if the benefits could be adequately explained to consumers. Researches have shown that consumers are more satisfied with bionanomaterials than other nanomaterials as they are derived from biological sources, sustainable, and less toxic. They have helped in the novel food structures that have enhanced the organoleptic properties of food, giving more choice to consumers. Nanoencapsulation helped in improving the bioavailability of functional components that have opened doors for personalized nutrition. Different packaging materials have shown more welcoming results as compared to the nanomaterials that contact the food. They help reduce the problem of food loss, the primary concern of the food industry in the form of active and smart packaging. The availability of different types of portable and economical sensors solved the detection of different types of contaminants present in the food. With the help of proper scientific research and development, product oversight and post commercialization, marketability and uses will help take the benefit of bionanomaterials to the next level.

The future of nanotechnology is very bright as it will open many avenues for the food industry. It has tremendous potential for shaking the food sector in the coming years. In our opinion, there is need of more collaborations from both public and private institutions. So that the proper commercialization of nanofoods could be done. More research on the toxicity of bionanomaterials using short- and long-duration *in vivo* studies. This information will help to regulate their amount to be used in the development of nanofoods. Proper communication about their safe use with the help of appropriate packaging and labeling will be done to create more enthusiastic and conscientious end users so that consumers will accept them with open hearts and without any doubts.

Acknowledgment

Shweta Rathee would like to acknowledge the UGC, Government of India, for providing the fellowship for pursuing the PhD, and all the other authors are thankful to their respective institutions.

References

[1] FAO - News Article: 2050: A third more mouths to feed (http://www.fao.org/news/story/en/item/35571/icode/)

[2] Nanotechnology in the food industry: 'plenty of room' to innovate - New Food Magazine (https://www.newfoodmagazine.com/article/42338/nanotechnology-food-industry-plenty-room-innovate/)

[3] Sahani S and Sharma Y C 2021 Advancements in applications of nanotechnology in global food industry *Food Chem.* **342** 128318

[4] Singh T, Shukla S, Kumar P, Wahla V and Bajpai V K 2017 Application of nanotechnology in food science: perception and overview *Front. Microbiol.* **8** 1501

[5] Tarhan Ö 2020 Safety and regulatory issues of nanomaterials in foods *Handbook of Food Nanotechnology* (Amsterdam: Elsevier) pp 655–703

[6] Anandharamakrishnan C 2021 Trends and impact of nanotechnology in agro-food sector *Innovative Food Processing Technologies* (Amsterdam: Elsevier) pp 523–31

[7] Pathakoti K and Manubolu M 2017 Nanostructures: current uses and future applications in food science *J. Food Drug Anal.* **25** 245–53

[8] Ghorani B and Tucker N 2015 Fundamentals of electrospinning as a novel delivery vehicle for bioactive compounds in food nanotechnology *Food Hydrocoll.* **51** 227–40

[9] Weidemaier K, Carruthers E, Curry A, Kuroda M, Fallows E and Thomas J *et al* 2015 Real-time pathogen monitoring during enrichment: a novel nanotechnology-based approach to food safety testing *Int. J. Food Microbiol.* **198** 19–27

[10] Mei L and Wang Q 2020 Advances in using nanotechnology structuring approaches for improving food packaging *Annu. Rev. Food Sci. Technol.* **11** 339–64

[11] Rogers M A 2016 Naturally occurring nanoparticles in food *Curr. Opin. Food Sci.* **7** 14–19

[12] He X, Deng H and Hwang H M 2019 The current application of nanotechnology in food and agriculture *J. Food Drug Anal.* **27** 1–21

[13] Joye I J, Corradini M G, Duizer L M, Bohrer B M, LaPointe G and Farber J M *et al* 2019 A comprehensive perspective of food nanomaterials *Adv. Food Nutr. Res.* **2019** 1–45

[14] Anandharamakrishnan C 2021 Trends and impact of nanotechnology in agro-food sector *Innovative Food Processing Technologies: A Comprehensive Review* (Amsterdam: Elsevier) pp 523–31

[15] Chaudhry Q and Groves K 2010 Nanotechnology applications for food ingredients, additives and supplements *Nanotechnologies in Food* (Cambridge: Royal Society of Chemistry) pp 69–85

[16] Eljeeva Emerald F M, Pushpadass H A, Joseph D and Jaya S V 2021 Impact of nanotechnology in beverage processing *Innovative Food Processing Technologies: A Comprehensive Review* (Amsterdam: Elsevier) pp 688–700
[17] Chelladurai V and Jayas D S 2018 *Nanoscience and Nanotechnology in Foods and Beverages* (Boca Raton, FL: CRC Press)
[18] Cassano A, Conidi C and Castro-Muñoz R 2018 Current and future applications of nanofiltration in food processing *Separation of Functional Molecules in Food by Membrane Technology* (Amsterdam: Elsevier) pp 305–48
[19] Nath K, Dave H K and Patel T M 2018 Revisiting the recent applications of nanofiltration in food processing industries: progress and prognosis *Trends Food Sci. Technol.* **73** 12–24
[20] He B, Ge J, Yue P, Yue X Y, Fu R and Liang J *et al* 2017 Loading of anthocyanins on chitosan nanoparticles influences anthocyanin degradation in gastrointestinal fluids and stability in a beverage *Food Chem.* **221** 1671–7
[21] Deotale S M, Dutta S, Moses J A and Anandharamakrishnan C 2021 Nanopatterning of biomolecules *Innovative Food Processing Technologies: A Comprehensive Review* (Amsterdam: Elsevier) pp 651–65
[22] Sahani S and Sharma Y C 2021 Advancements in applications of nanotechnology in global food industry *Food Chem.* **342** 128318
[23] Misra S, Pandey P and Mishra H N 2021 Novel approaches for co-encapsulation of probiotic bacteria with bioactive compounds, their health benefits and functional food product development: a review *Trends Food Sci. Technol.* **109** 340–51
[24] Fan M, Li T, Li Y, Qian H, Zhang H and Rao Z *et al* 2021 Vaccinium bracteatum Thunb. as a promising resource of bioactive compounds with health benefits: an updated review *Food Chem.* **356** 129738
[25] Pires E O, Caleja C, Garcia C C, Ferreira I C F R and Barros L 2021 Current status of genus impatiens: bioactive compounds and natural pigments with health benefits *Trends Food Sci. Technol.* (in press)
[26] Ruiz Rodríguez L G, Zamora Gasga V M, Pescuma M, Van Nieuwenhove C, Mozzi F and Sánchez Burgos J A 2021 Fruits and fruit by-products as sources of bioactive compounds. Benefits and trends of lactic acid fermentation in the development of novel fruit-based functional beverages *Food Res. Int.* **140** 109854
[27] Jimenez-Lopez C, Pereira A G, Lourenço-Lopes C, Garcia-Oliveira P, Cassani L and Fraga-Corral M *et al* 2021 Main bioactive phenolic compounds in marine algae and their mechanisms of action supporting potential health benefits *Food Chem.* **341** 128262
[28] Verma D K and Srivastav P P 2020 Bioactive compounds of rice (*Oryza sativa* L.): review on paradigm and its potential benefit in human health *Trends Food Sci. Technol.* **97** 355–65
[29] Sun-Waterhouse D and Waterhouse G I N 2016 Recent advances in the application of nanomaterials and nanotechnology in food research *Novel Approaches of Nanotechnology in Food* (Amsterdam: Elsevier) pp 21–66
[30] Sampathkumar K, Tan K X and Loo S C J 2020 Developing nano-delivery systems for agriculture and food applications with nature-derived polymers *iScience* **23** 101055
[31] Liu M, Wang F, Pu C, Tang W and Sun Q 2021 Nanoencapsulation of lutein within lipid-based delivery systems: characterization and comparison of zein peptide stabilized nano-emulsion, solid lipid nanoparticle, and nano-structured lipid carrier *Food Chem.* **358** 129840

[32] Budryn G, Zaczyńska D and Oracz J 2016 Effect of addition of green coffee extract and nanoencapsulated chlorogenic acids on aroma of different food products *LWT - Food Sci Technol.* **73** 197–204

[33] Wang Y, Zhang Y, Guan L, Wang S, Zhang J and Tan L *et al* 2021 Lipophilization and amylose inclusion complexation enhance the stability and release of catechin *Carbohydr. Polym.* **269** 118251

[34] Wang S, Kong L, Zhao Y, Tan L, Zhang J and Du Z *et al* 2019 Lipophilization and molecular encapsulation of p-coumaric acid by amylose inclusion complex *Food Hydrocoll.* **93** 270–5

[35] Ben Jemaa M, Falleh H, Neves M A, Isoda H, Nakajima M and Ksouri R 2017 Quality preservation of deliberately contaminated milk using thyme free and nanoemulsified essential oils *Food Chem.* **217** 726–34

[36] Singh H, Kumar C, Singh N, Paul S and Jain S K 2018 Nanoencapsulation of docosahexanoic acid (DHA) using combination of food grade polymeric wall materials and its application for improvement in bioavailability and oxidative stability *Food Funct.* **9** 2213–27

[37] Ghasemi S, Jafari S M, Assadpour E and Khomeiri M 2018 Nanoencapsulation of D-limonene within nanocarriers produced by pectin-whey protein complexes *Food Hydrocoll.* **77** 152–62

[38] Yin H Y and Tsai W C 2015 Advances of nanomaterials for food processing *Handbook of Food Chemistry* (Berlin: Springer) pp 1137–59

[39] da Silva Stefani F, de Campo C, Paese K, Stanisçuaski Guterres S, Haas Costa T M and Hickmann Flôres S 2019 Nanoencapsulation of linseed oil with chia mucilage as structuring material: characterization, stability and enrichment of orange juice *Food Res. Int.* **120** 872–9

[40] Ghayour N, Hosseini S M H, Eskandari M H, Esteghlal S, Nekoei A R and Hashemi Gahruie H *et al* 2019 Nanoencapsulation of quercetin and curcumin in casein-based delivery systems *Food Hydrocoll.* **87** 394–403

[41] İnanç Horuz T and Belibağlı K B 2018 Nanoencapsulation by electrospinning to improve stability and water solubility of carotenoids extracted from tomato peels *Food Chem.* **268** 86–93

[42] Rostamabadi H, Falsafi S R and Jafari S M 2019 Nanoencapsulation of carotenoids within lipid-based nanocarriers *J. Control. Rel.* **298** 38–67

[43] de Oliveira Cavalcanti Medeiros A K, de Carvalho Gomes C, de Araújo Amaral M L Q, de Medeiros L D G, Medeiros I, Porto D L, Aragão C F S, Maciel B L L, de Araújo Morais A H and Passos T S *et al* 2019 Nanoencapsulation improved water solubility and color stability of carotenoids extracted from Cantaloupe melon (*Cucumis melo* L.) *Food Chem.* **270** 562–72

[44] de Oliveira G L R, Medeiros I, da Cruz Nascimento S S, Viana R L S, Porto D L and Rocha H A O *et al* 2021 Antioxidant stability enhancement of carotenoid rich-extract from Cantaloupe melon (*Cucumis melo* L.) nanoencapsulated in gelatin under different storage conditions *Food Chem.* **348** 129055

[45] Resende D, Costa Lima S A and Reis S 2020 Nanoencapsulation approaches for oral delivery of vitamin A *Colloid. Surf. B Biointerfaces.* **193** 111121

[46] de Melo A P Z, da Rosa C G, Noronha C M, Machado M H, Sganzerla W G and da Cunha Bellinati N V *et al* 2021 Nanoencapsulation of vitamin D3 and fortification in an experimental jelly model of Acca sellowiana: Bioaccessibility in a simulated gastrointestinal system *LWT* **145** 111287

[47] Mohammadi B, Shekaari H and Zafarani-Moattar M T 2020 Synthesis of nanoencapsulated vitamin E in phase change material (PCM) shell as thermo-sensitive drug delivery purpose *J. Mol. Liq.* **320** 114429
[48] Davies S, Contri R V, Guterres S S, Pohlmann A R and Guerreiro I C K 2020 Simultaneous nanoencapsulation of lipoic acid and resveratrol with improved antioxidant properties for the skin *Colloids Surf. B Biointerfaces.* **192** 111023
[49] Ruengdech A and Siripatrawan U 2021 Application of catechin nanoencapsulation with enhanced antioxidant activity in high pressure processed catechin-fortified coconut milk *LWT* **140** 110594
[50] Feng S, Yan J, Wang D, Jiang L, Sun P and Xiang N *et al* 2021 Preparation and characterization of soybean protein isolate/pectin-based phytosterol nanodispersions and their stability in simulated digestion *Food Res. Int.* **143** 110237
[51] Hosseini S F, Ramezanzade L and McClements D J 2021 Recent advances in nanoencapsulation of hydrophobic marine bioactives: Bioavailability, safety, and sensory attributes of nano-fortified functional foods *Trends Food Sci. Technol.* **109** 322–39
[52] Zhang C, Huang L, Pu H and Sun D-W 2021 Magnetic surface-enhanced Raman scattering (MagSERS) biosensors for microbial food safety: Fundamentals and applications *Trends Food Sci. Technol.* **113** 366–81
[53] Yang J, Shen M, Luo Y, Wu T, Chen X and Wang Y *et al* 2021 Advanced applications of chitosan-based hydrogels: From biosensors to intelligent food packaging system *Trends Food Sci. Technol.* **110** 822–32
[54] Neethirajan S, Ragavan V, Weng X and Chand R 2018 Biosensors for sustainable food engineering: Challenges and perspectives *Biosensors.* **8** 8010023
[55] Turasan H, Kokini J and Turasan H 2021 Novel nondestructive biosensors for the food industry *Annu. Rev. Food Sci. Technol.* **12** 539–66
[56] Cheng W, Tang X, Zhang Y, Wu D and Yang W 2021 Applications of metal-organic framework (MOF)-based sensors for food safety: Enhancing mechanisms and recent advances *Trends Food Sci. Technol.* **112** 268–82
[57] Ozogul Y, Kuley Boğa E, Akyol I, Durmus M, Ucar Y and Regenstein J M *et al* 2020 Antimicrobial activity of thyme essential oil nanoemulsions on spoilage bacteria of fish and food-borne pathogens *Food Biosci.* **36** 100635
[58] Xiao X, Hu S, Lai X, Peng J and Lai W 2021 Developmental trend of immunoassays for monitoring hazards in food samples: a review *Trends Food Sci. Technol.* **111** 68–88
[59] Kim J H and Oh S W 2021 Pretreatment methods for nucleic acid-based rapid detection of pathogens in food: a review *Food Control.* **121** 107575
[60] Ran J, Zhang L, Yao J, Wang S, Liang P and Dong N 2021 Cucurbit[7]uril as a matrix solid-phase dispersion for the extraction of quaternary ammonium pesticides from vegetables and their determination using HPLC–UV *Food Chem.* **350** 129236
[61] Constantinou M, Louca-Christodoulou D and Agapiou A 2021 Method validation for the determination of 314 pesticide residues using tandem MS systems (GC–MS/MS and LC-MS/MS) in raisins: Focus on risk exposure assessment and respective processing factors in real samples (a pilot survey) *Food Chem.* **360** 129964
[62] Xiong Y, Leng Y, Li X, Huang X and Xiong Y 2020 Emerging strategies to enhance the sensitivity of competitive ELISA for detection of chemical contaminants in food samples *TrAC - Trends Anal. Chem.* **126** 115861

[63] Cimaglia F, Tristezza M, Saccomanno A, Rampino P, Perrotta C and Capozzi V *et al* 2018 An innovative oligonucleotide microarray to detect spoilage microorganisms in wine *Food Control.* **87** 169–79

[64] Duncan T V 2011 Applications of nanotechnology in food packaging and food safety: Barrier materials, antimicrobials and sensors *J. Colloid Interface Sci.* **363** 1–24

[65] Xue L, Guo R, Huang F, Qi W, Liu Y and Cai G *et al* 2021 An impedance biosensor based on magnetic nanobead net and MnO2 nanoflowers for rapid and sensitive detection of foodborne bacteria *Biosens. Bioelectron.* **173** 112800

[66] Petrovszki D, Valkai S, Gora E, Tanner M, Bányai A and Fürjes P *et al* 2021 An integrated electro-optical biosensor system for rapid, low-cost detection of bacteria *Microelectron. Eng.* **239–40** 111523

[67] Ma X, Ding W, Wang C, Wu H, Tian X and Lyu M *et al* 2021 DNAzyme biosensors for the detection of pathogenic bacteria *Sens. Actuators, B: Chem.* **331** 129422

[68] Feng Y, Zhou D, Gao L and He F 2020 Electrochemical biosensor for rapid detection of bacteria based on facile synthesis of silver wire across electrodes *Biosens. Bioelectron.* **168** 112527

[69] Sun Y, Lin Y, Han R, Wang X and Luo C 2019 A chemiluminescence biosensor for lysozyme detection based on aptamers and hemin/G-quadruplex DNAzyme modified sandwich-rod carbon fiber composite *Talanta* **200** 57–66

[70] Benedé S, Ruiz-Valdepeñas Montiel V, Povedano E, Villalba M, Mata L and Galán-Malo P *et al* 2018 Fast amperometric immunoplatform for ovomucoid traces determination in fresh and baked foods *Sens. Actuators B Chem.* **265** 421–8

[71] Rouf T B, Díaz-Amaya S, Stanciu L and Kokini J 2020 Application of corn zein as an anchoring molecule in a carbon nanotube enhanced electrochemical sensor for the detection of gliadin *Food Control.* **117** 107350

[72] Indyk H E, Gill B D, Wood J E, Chetikam S and Kobayashi T 2021 The determination of intact β-casein in milk products by biosensor immunoassay *J. Food Compos. Anal.* **101** 103946

[73] Yadav N, Yadav S S, Chhilar A K and Rana J S 2021 An overview of nanomaterial based biosensors for detection of Aflatoxin B1 toxicity in foods *Food Chem. Toxicol.* **152** 112201

[74] Ouyang Q, Wang L, Ahmad W, Rong Y, Li H and Hu Y *et al* 2021 A highly sensitive detection of carbendazim pesticide in food based on the upconversion-MnO_2 luminescent resonance energy transfer biosensor *Food Chem.* **349** 129157

[75] Zhao G, Zhou B, Wang X, Shen J and Zhao B 2021 Detection of organophosphorus pesticides by nanogold/mercaptomethamidophos multi-residue electrochemical biosensor *Food Chem.* **354** 129511

[76] Rawtani D, Rao P K and Hussain C M 2020 Recent advances in analytical, bioanalytical and miscellaneous applications of green nanomaterial *TrAC Trends Anal. Chem.* **133** 116109

[77] Shahabi-Ghahfarrokhi I and Babaei-Ghazvini A 2019 Using photo-modification to compatibilize nano-ZnO in development of starch-kefiran-ZnO green nanocomposite as food packaging material *Int. J. Biol. Macromol.* **124** 922–30

[78] Marrez D A, Abdelhamid A E and Darwesh O M 2019 Eco-friendly cellulose acetate green synthesized silver nano-composite as antibacterial packaging system for food safety *Food Packag Shelf Life.* **20** 100302

[79] Dong X, Liang X, Zhou Y, Bao K, Sameen D E and Ahmed S *et al* 2021 Preparation of polylactic acid/TiO_2/GO nano-fibrous films and their preservation effect on green peppers *Int. J. Biol. Macromol.* **177** 135–48

[80] Pandey V K, Upadhyay S N, Niranjan K and Mishra P K 2020 Antimicrobial biodegradable chitosan-based composite nano-layers for food packaging *Int. J. Biol. Macromol.* **157** 212–9

[81] Bose I, Singh R, Negi P and Singh Y 2021 Chitin as bio-based nanomaterial in packaging: a review *Mater. Today Proc.* **46** 11254–7

[82] Hammed A M 2018 Nanomaterial for food packaging *Preparation and Processing of Religious and Cultural Foods* (Amsterdam: Elsevier) pp 163–71

[83] Garcia C V, Shin G H and Kim J T 2018 Metal oxide-based nanocomposites in food packaging: applications, migration, and regulations *Trends Food Sci. Technol.* **82** 21–31

[84] Mei L and Wang Q 2020 Advances in using nanotechnology structuring approaches for improving food packaging *Annu. Rev. Food Sci. Technol.* **11** 339–64

[85] Mohammadi M, Mirabzadeh S, Shahvalizadeh R and Hamishehkar H 2020 Development of novel active packaging films based on whey protein isolate incorporated with chitosan nanofiber and nano-formulated cinnamon oil *Int. J. Biol. Macromol.* **149** 11–20

[86] Bi F, Qin Y, Chen D, Kan J and Liu J 2021 Development of active packaging films based on chitosan and nano-encapsulated luteolin *Int. J. Biol. Macromol.* **182** 545–53

[87] Kamkar A, Molaee-aghaee E, Khanjari A, Akhondzadeh-basti A, Noudoost B and Shariatifar N *et al* 2021 Nanocomposite active packaging based on chitosan biopolymer loaded with nano-liposomal essential oil: Its characterizations and effects on microbial, and chemical properties of refrigerated chicken breast fillet *Int. J. Food Microbiol.* **342** 109071

[88] Lu P, Yang Y, Liu R, Liu X, Ma J and Wu M *et al* 2020 Preparation of sugarcane bagasse nanocellulose hydrogel as a colourimetric freshness indicator for intelligent food packaging *Carbohydr Polym.* **249** 116831

[89] Park S, Jeon Y, Han T, Kim S, Gwon Y and Kim J 2020 Nanoscale manufacturing as an enabling strategy for the design of smart food packaging systems *Food Packag. Shelf Life.* **26** 100570

[90] Sani M A, Tavassoli M, Hamishehkar H and McClements D J 2021 Carbohydrate-based films containing pH-sensitive red barberry anthocyanins: Application as biodegradable smart food packaging materials *Carbohydr. Polym.* **255** 117488

[91] Qin Y, Yun D, Xu F, Chen D, Kan J and Liu J 2021 Smart packaging films based on starch/polyvinyl alcohol and Lycium ruthenicum anthocyanins-loaded nano-complexes: Functionality, stability and application *Food Hydrocoll.* **119** 106850

[92] Yong H and Liu J 2020 Recent advances in the preparation, physical and functional properties, and applications of anthocyanins-based active and intelligent packaging films *Food Packag. Shelf Life.* **26** 100550

[93] Kalpana S, Priyadarshini S R, Maria Leena M, Moses J A and Anandharamakrishnan C 2019 Intelligent packaging: trends and applications in food systems *Trends Food Sci. Technol.* **93** 145–57

[94] Swaroop C and Shukla M 2018 Nano-magnesium oxide reinforced polylactic acid biofilms for food packaging applications *Int. J. Biol. Macromol.* **113** 729–36

[95] Sarwar M S, Niazi M B K, Jahan Z, Ahmad T and Hussain A 2018 Preparation and characterization of PVA/nanocellulose/Ag nanocomposite films for antimicrobial food packaging *Carbohydr. Polym.* **184** 453–64

[96] Bajpai V K, Kamle M, Shukla S, Mahato D K, Chandra P and Hwang S K *et al* 2018 Prospects of using nanotechnology for food preservation, safety, and security *J. Food Drug Anal.* **26** 1201–14

[97] Li P, Zhu X, Kong M, Lv Y, Huang Y and Yang Q *et al* 2021 Fully biodegradable polylactide foams with ultrahigh expansion ratio and heat resistance for green packaging *Int. J. Biol. Macromol.* **183** 222–34

[98] Chen J, Wu A, Yang M, Ge Y, Pristijono P and Li J *et al* 2021 Characterization of sodium alginate-based films incorporated with thymol for fresh-cut apple packaging *Food Control.* **126** 108063

[99] Ramakrishnan R K, Wacławek S, Černík M and Padil V V T 2021 Biomacromolecule assembly based on gum kondagogu-sodium alginate composites and their expediency in flexible packaging films *Int. J. Biol. Macromol.* **177** 526–34

[100] Mousavi S N, Daneshvar H, Seyed Dorraji M S, Ghasempour Z, Panahi-Azar V and Ehsani A 2021 Starch/alginate/Cu-g-C_3N_4 nanocomposite film for food packaging *Mater. Chem. Phys.* **267** 124583

[101] Pan J, Li Y, Chen K, Zhang Y and Zhang H 2021 Enhanced physical and antimicrobial properties of alginate/chitosan composite aerogels based on electrostatic interactions and noncovalent crosslinking *Carbohydr. Polym.* **266** 118102

[102] Adel A M, Al-Shemy M T, Diab M A, El-Sakhawy M, Toro R G and Montanari R *et al* 2021 Fabrication of packaging paper sheets decorated with alginate/oxidized nanocellulose-silver nanoparticles bio-nanocomposite *Int. J. Biol. Macromol.* **181** 612–20

[103] Patwardhan S V, Manning J R H and Chiacchia M 2018 Bioinspired synthesis as a potential green method for the preparation of nanomaterials: opportunities and challenges *Curr. Opin. Green Sustain. Chem.* **12** 110–6

[104] McClements D J and Xiao H 2017 Is nano safe in foods? Establishing the factors impacting the gastrointestinal fate and toxicity of organic and inorganic food-grade nanoparticles *NPJ Sci. Food.* **1** 6

[105] Miernicki M, Hofmann T, Eisenberger I, von der Kammer F and Praetorius A 2019 Legal and practical challenges in classifying nanomaterials according to regulatory definitions *Nat. Nanotechnol.* **14** 208–16

[106] El-Messery T M, Altuntas U, Altin G and Özçelik B 2020 The effect of spray-drying and freeze-drying on encapsulation efficiency, *in vitro* bioaccessibility and oxidative stability of krill oil nanoemulsion system *Food Hydrocoll.* **106** 105890

[107] Iqbal R, Mehmood Z, Baig A and Khalid N 2020 Formulation and characterization of food grade O/W nanoemulsions encapsulating quercetin and curcumin: Insights on enhancing solubility characteristics *Food Bioprod. Process* **123** 304–11

[108] Khan I, Bhardwaj M, Shukla S, Lee H, Oh M W and Bajpai V K *et al* 2019 Carvacrol encapsulated nanocarrier/ nanoemulsion abrogates angiogenesis by downregulating COX-2, VEGF and CD31 *in vitro* and *in vivo* in a lung adenocarcinoma model *Colloids Surf. B Biointerfaces* **181** 612–22

[109] Ren X, Hou T, Liang Q, Zhang X, Hu D and Xu B *et al* 2019 Effects of frequency ultrasound on the properties of zein-chitosan complex coacervation for resveratrol encapsulation *Food Chem.* **279** 223–30

[110] Zhang J, Jia G, Wanbin Z, Minghao J, Wei Y and Hao J *et al* 2021 Nanoencapsulation of zeaxanthin extracted from Lycium barbarum L. by complex coacervation with gelatin and CMC *Food Hydrocoll.* **112** 106280

[111] Hernández-Nava R, López-Malo A, Palou E, Ramírez-Corona N and Jiménez-Munguía M T 2020 Encapsulation of oregano essential oil (Origanum vulgare) by complex coacervation between gelatin and chia mucilage and its properties after spray drying *Food Hydrocoll.* **109** 106077

[112] Liu X, Wang P, Zou Y X, Luo Z G and Tamer T M 2020 Co-encapsulation of Vitamin C and β-Carotene in liposomes: storage stability, antioxidant activity, and *in vitro* gastro-intestinal digestion *Food Res. Int.* **136** 109587

[113] Huang J, Wang Q, Chu L and Xia Q 2020 Liposome-chitosan hydrogel bead delivery system for the encapsulation of linseed oil and quercetin: Preparation and *in vitro* characterization studies *LWT* **117** 108615

[114] Huynh Mai C, Thanh Diep T, Le T T T and Nguyen V 2020 Advances in colloidal dispersions: a review *J. Dispers Sci. Technol.* **41** 479–94

[115] Wang S, Ye X, Sun Y, Liang J, Yue P and Gao X 2021 Nanocomplexes derived from chitosan and whey protein isolate enhance the thermal stability and slow the release of anthocyanins in simulated digestion and prepared instant coffee *Food Chem.* **336** 127707

[116] Chen S, Ma X, Han Y, Wei Y, Guo Q and Yang S *et al* 2020 Effect of chitosan molecular weight on zein-chitosan nanocomplexes: Formation, characterization, and the delivery of quercetagetin *Int. J. Biol. Macromol.* **164** 2215–23

[117] Qiao X, Yang L, Gu J, Cao Y, Li Z and Xu J *et al* 2021 Kinetic interactions of nanocomplexes between astaxanthin esters with different molecular structures and β-lactoglobulin *Food Chem.* **335** 127633

[118] Ezhilarasi P N, Karthik P, Chhanwal N and Anandharamakrishnan C 2013 Nanoencapsulation techniques for food bioactive components: a review *Food Bioproc. Technol.* **6** 628–47

[119] Mousavi Nodoushan S, Nasirizadeh N, Amani J, Halabian R and Imani Fooladi A A 2019 An electrochemical aptasensor for staphylococcal enterotoxin B detection based on reduced graphene oxide and gold nano-urchins *Biosens. Bioelectron.* **127** 221–8

[120] Navarro J, de Marcos S and Galbán J 2020 Colorimetric-enzymatic determination of tyramine by generation of gold nanoparticles *Microchim. Acta* **187** 174

[121] Cao Y, Hu X, Zhao T, Mao Y, Fang G and Wang S 2021 A core–shell molecularly imprinted optical sensor based on the upconversion nanoparticles decorated with zinc-based metal-organic framework for selective and rapid detection of octopamine *Sens. Actuators B Chem.* **326** 128838

[122] Wu L, Zhou M, Wang Y and Liu J 2020 Nanozyme and aptamer- based immunosorbent assay for aflatoxin B1 *J. Hazard. Mater.* **399** 123154

[123] Bagheri pebdeni A and Hosseini M 2020 Fast and selective whole cell detection of *Staphylococcus aureus* bacteria in food samples by paper based colorimetric nanobiosensor using peroxidase-like catalytic activity of DNA-Au/Pt bimetallic nanoclusters *Microchem. J.* **159** 105475

[124] Wu M F, Wang Y, Li S, Dong X X, Yang J Y and Shen Y D *et al* 2019 Ultrasensitive immunosensor for acrylamide based on chitosan/SnO_2-SiC hollow sphere nanochains/gold nanomaterial as signal amplification *Anal. Chim. Acta* **1049** 188–95

[125] Díaz Nieto C H, Granero A M, Garcia D, Nesci A, Barros G and Zon M A *et al* 2019 Development of a third-generation biosensor to determine sterigmatocystin mycotoxin: An early warning system to detect aflatoxin B1 *Talanta* **194** 253–8

[126] Huang W, Zhang H, Lai G, Liu S, Li B and Yu A 2019 Sensitive and rapid aptasensing of chloramphenicol by colorimetric signal transduction with a DNAzyme-functionalized gold nanoprobe *Food Chem.* **270** 287–92

[127] Ma M, Sun J, Chen Y, Wen K, Wang Z and Shen J *et al* 2018 Highly sensitive SERS immunosensor for the detection of amantadine in chicken based on flower-like gold nanoparticles and magnetic bead separation *Food Chem. Toxicol.* **118** 589–94

Chapter 11

Polymeric bionanomaterials for agricultural and environmental applications

Ulucan-Karnak Fulden and Kuru Cansu İlke

The use of polymeric based bionanomaterials in several fields emerged in the last quarter of the 20th century and continues to evolve at an increasing rate. Polymeric bionanomaterials are widely used in scientific researches, medical, environmental and agricultural applications. As an emerging novel technology, many bionanomaterials have been developed to remove contaminants from environmental samples for avoiding pollution. Also, they have been developed as carrier systems for agrochemicals aimed at pesticide control and increased crop yields. In this chapter, an overview of the development of polymer-based bionanomaterials and applications for agricultural and environmental is provided.

11.1 Introduction

The term nanoscale has several definitions, and nanosized materials are commonly characterized with a diameter of 1–100 nm [1, 2]. In all definitions, it is seen that the differences between nanosize and nanomaterials can be met in a common definition (table 11.1). Nanomaterials can be generated synthetically or naturally. Nature has several nanomaterials sources that are originated from animals and plants [3].

Nanomaterials can be categorized as nanocrystalline and nano-structured materials. Nano-structured materials may be subcategorized as non-polymeric, polymer-based and lipid-based nanoparticles. In the polymer-based category, they can be interpreted as micelles, dendrimers, nanogels, nanoparticles, protein nanoparticles and drug conjugates. Non-polymeric nanoparticles can be listed as nanodiamonds, quantum dots, carbon nanotubes, metallic and silica-based nanoparticles. Liposomes and solid lipid nanoparticles are two types of lipid-based nanoparticles (figure 11.1) [4].

doi:10.1088/978-0-7503-3863-9ch11

Table 11.1. Nanoscale unit definitions.

Definition	According to
10^{-9} meter	International System of Units (Système international d'unités, SI)
Materials that have at least one dimension in the range of approximately 1 to 100 nm and exhibit dimension-dependent phenomena	The US Food and Drug Administration (USFDA)
Material with any external nanoscale dimension or having internal nanoscale surface structure	The International Organization for Standardization (ISO)
A manufactured or natural material that possesses unbound, aggregated or agglomerated particles where external dimensions are between 1–100 nm size range	EU Comission
1 to 1000 nm size range and nanomaterials with internal or external structures on the nanoscale dimension	British Standards Institution

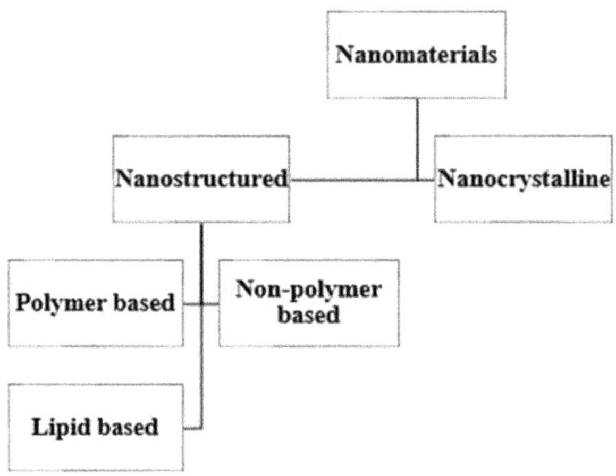

Figure 11.1. Nanomaterials main categorization. Adapted from [4], copyright (2020) with permission from MDPI, distributed under a Creative Commons Attribution License 4.0 (CC BY).

Historical developments of nanosized materials started with the discovery of liposomes. Liposomes have been used since the 1960s as nanosized materials in several applications. In 1976, controlled release with polymeric systems became popular. PEGylated forms of liposomes and polymeric nanoparticles were developed in the 1980s. Since the 2000s, several nanoparticle-based formulations have been designed for different purposes. With developing technology, nanoparticle design strategies were also developed and evolved towards mini-machines or microrobots with site specific targeting and stimuli responsive properties [5].

Bionanomaterials are described as molecular materials that contain biological molecules (for example, cells, viruses, proteins, antibodies, RNA, DNA, oligosaccharides, lipids) and result in molecular structures with nanosize(s) [6]. Organic, inorganic, and hybrid bionanomaterials are the three types of bionanomaterials and they can be produced by environmentally friendly, simple, cost-effective and more reproducible ways than with traditional methods. Bionanomaterials can be synthesized from DNA, carbohydrates, proteins, lipids and also molecular mixtures that are complex. Inorganic bionanomaterials can be classified into oxides and metallic. Hybrid bionanomaterials can be driven by molecular precursors in the presence of biological templates. All of these methods and components of the reactions can determine the properties of synthesized bionanomaterials [7].

Polymer is a kind of structure which has large molecular weight. After chemical reaction, a covalent bond between molecules is formed and creates the polymer. Polymers are light, cheap, formable, inert and suitable for different purposes. Atoms which make bonds with each other and generate the skeleton of the polymer molecule are called the main chain. Polymers are bonded to the main chain and also chemical groups which are called the side group. Polymers are classified with respect to their different features. They are classified as natural or synthetic according to their source, organic or inorganic according to main chain atoms, stepwise or addition according to their synthesis reaction mechanism and linear, branched or cross-linked according to the physical structure of the chain [8].

Solid particles or particulate dispersions with a diameter of 10–1000 nm are known as polymeric nanoparticles (PNPs). Therapeutic agents may be entrapped, encapsulated, or adsorbed to the nanoparticle surface within these PNPs. The advantages of polymeric nanoparticles can be summarized as enhancing bioavailability, targeting of higher concentration to a desired location, easy to use for drug delivery or tissue engineering, developing oral and intravenous administration with higher efficiency [9].

Polymer based nanomaterials can be prepared as nanoparticles, polymersome, micelle, polyplex, dendrimer, polymer-lipid hybrid and polymer-drug conjugate forms [10]. Designing of a nanoparticle includes some properties such as stealth, charge, targeting, stimuli responsiveness, size, morphology and stability. These properties have specified characteristic qualifications such as toxicity, specifity, biocompatibility, immunogenicity, selectivity. For instance cationic nanoparticles have high toxicity, low specificity and high cellular uptake. These can be determined by advanced experiments [11].

Nature is endowed with different types of nanomaterials which can originate from plants and animals and they can be named as bionanomaterials. For example, starch and cellulose are natural, productive, renewable and biodegradable polymers. In order to ensure environmental health, bionanomaterials have the potential to be used instead of present petrochemical-based materials in different fields such as agriculture, drug delivery, environmental process, tissue engineering and food packaging [3].

Polymeric bionanomaterials are generated from renewable polymer sources (skin, plants, exoskeletons of arthropods, spider webs, hair and silkworm cocoons) such as biopolymers. Biopolymers can be categorized into three types in terms of their monomeric units: polysaccharides, protein/polypeptides, and polynucleotides. As cellulose polysaccharides, alginate and chitin/chitosan are the most commonly used

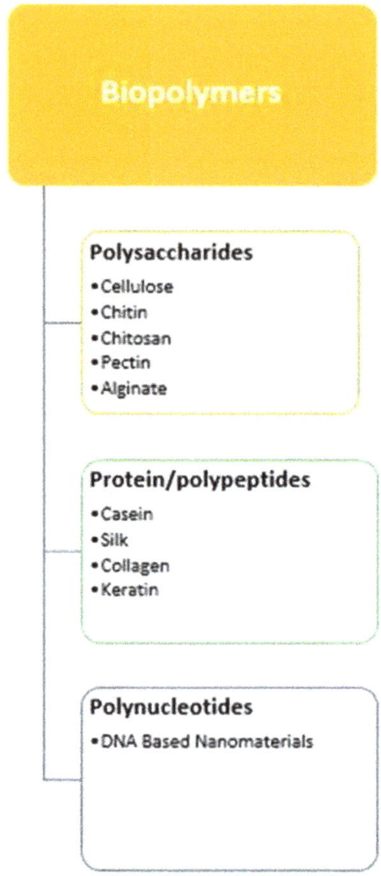

Figure 11.2. Biopolymer classifications.

ones. Silk, collagen and keratin are examples of protein-containing polymers. DNA can be regarded as a polynucleotides based polymer (figure 11.2) [12, 13].

Chitosan and pectin based polymeric bionanomaterials have been designed and commonly utilized as agrochemical carrier systems in order to control pesticides and increase crop yields. Chitosan is chosen for the ability of absorbing to plant surfaces to prolong the contact time, its antipathogen effect and its stimulation effect on plants. Pectin is chosen because it is an ingredient of the plant cell wall and it is biocompatible and biodegradable [14]. Nanobiocomposites of natural polymers with tree gum, montmorillonite clay, zinc oxide nanoparticles are included in the literature for environmental applications as catalyst, pollutant remover, substrates etc [15].

Hence, this chapter will explain the basic information of polymeric bionanomaterials such as origins, synthesis and types. Also it will emphasise especially their agricultural and environmental applications with several purposes. All of these applications are very important to understand the ability of polymeric bionanomaterials for sustainability of agriculture and the environment due to their great physicochemical properties.

11.1.1 Preparation techniques

Preparation techniques of polymeric nanoparticles can be classified as dispersion of preform polymer form or polymerization of monomers form. For synthesis as dispersion form, solvent evaporation, dialysis, super critical fluid technology and nanoprecipitation are used. For synthesis from monomers, emulsion based methods are used [16, 17]. These methods can be classified as conventional nanoparticle preparation methods. They can also be categorized as chemical, physical and biological methods, as in figure 11.3 [18].

There are also advanced methods for preparation such as controlled/living radial polymerization, electrohydrodynamic atomization and, microtube pumping, ring opening polymerization, spin penetration and dissolving of macromolecule techniques. All of these methods of nanoparticles preparation have advantages and disadvantages, therefore, selection of a method in order to synthesize the nanoparticles can be related with aim of purpose and desired properties for application [19, 20]. Natural polymers derived polymeric nanoparticles are commonly synthesized with a cross-linking method that allows free amino groups to interact with the aqueous phase by linking the polymer's chains [21].

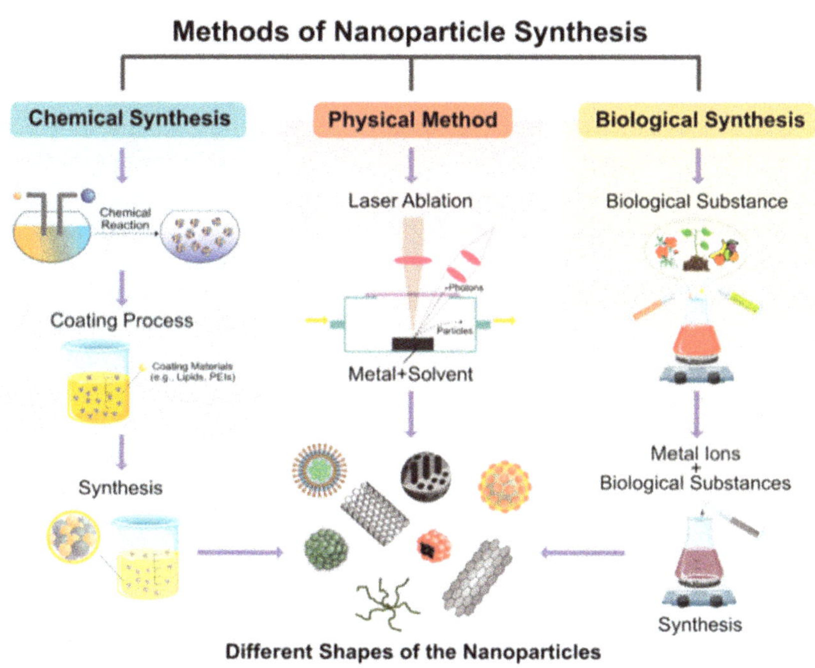

Figure 11.3. Schematization of conventional nanoparticle synthesis methods, adapted from [18], copyright (2019) with permission from MDPI, distributed under a Creative Commons Attribution License 4.0 (CC BY).

Nanosized polymeric materials are primarily generated in the literature using two methods: (i) monomer polymerization, and (ii) dispersion of pre-made polymers. Solvent evaporation, nanoprecipitation, supercritical fluid technology, salting out, polymerization (emulsion, RAFT mediated, surfactant free, condensation), solubilization, denaturation, ionic gelation techniques are methods of preparation for different polymers that are commonly utilized to synthesize nanomaterials [22, 23].

With developing technology and increased demand of various novel biocompatible and smart polymers, modern techniques were developed such as microelectromechanical, microfluidic, electrodropping and microneedle based systems. These systems provide homogeneous particle size, self assembly distribution, reproducibility and continuous flow. Researches are still progressing to investigate developing novel synthesizing techniques with the requirements of researchers and end users [24, 25]. Ultrasonication and microwave assisted synthesis process can be considered as rapid and environmental friendly techniques [26].

Rapid industrialization, urbanization, and population growth cause the Earth's atmosphere to deteriorate, exposing a vast amount of toxic and undesirable substances. Nanoparticle syntheses, as with any other scientific field, are directed towards nanoparticle synthesis methods that do not create any extra pollution and use natural materials. Furthermore, nanoparticles are commonly used in human contact areas, necessitating synthesis methods that do not use hard toxic chemicals [27].

Considering the disadvantages of physical and chemical nanoparticle synthesis methods, it was determined that a new method was needed and thus green synthesis emerged as an eco-friendly technique in the nanoparticle synthesis. The progression of efficient green synthesis, which provides natural reduction without high energy consumption, uses stabilizing agents instead of toxic, expensive chemicals and, has directed researchers to biological methods in nanoparticle synthesis [27, 28]. As reducing agents, the green synthesis technique employs biomimetic and natural materials such as algae, bacteria, enzymes and plant extracts [29, 30].

In nanoparticle synthesis, sodium citrate, transferrin and cetyl trimethyl ammonium bromide are the most commonly used stabilizing agents. Various amines, oligonucleotides, peptides, antibodies and lipids are also widely used for functionalization. Green synthesis is the bottom-up approach, in which natural product extracts are used instead of expensive and toxic reducing agents in chemical synthesis methods in metal nanoparticle synthesis [31]. Since they are readily available for mass production of nanoparticles, and these products are environmentally friendly, plants or extracts tend to be the best agents among biological entities that can be used in nanoparticle synthesis. For example, using plant phenolic content to synthesize gold nanoparticles allows for kinetic control of the metallization reaction, which influences nanoparticle development [32]. Plant components such as the root, stem, leaf, fruit, seed, shell, and flower can all be utilized to make nanoparticles [33]. Mukherjee *et al* published a green approach to gold nanoparticle synthesis using green tea derivatives [34]. Ahmed *et al* used aqueous leaf extract to synthesize silver nanoparticles [35]. Ghosh's group worked on another leaves extract to obtain copper nanoparticles [36]. Another research used various fruit juices to perform green synthesis [37]. Tea extracts' polyphenolic components have also been used to inhibit cell proliferation and synthesize nanomaterials that are being studied to trigger cancer cell apoptosis [38].

Using green synthesis, various nanoparticles can be synthesized which can further be progressed in various applications. These types of synthesis methods are affected by pH, temperature, pressure, method type and incubation time. Therefore, optimization studies must be carried out for the best results [39]. Environmental contamination such as the oxidation of organic dyes and pollutants, and heavy metal removal include metallic nanomaterials that are synthesized using green methods [40].

Preparation methods for bionanomaterials are especially important for applicability in industry. Chosen methods for synthesis can be specified in terms of deposition and instability properties. Bionanomaterials production is generally categorized as one-step (dry process and nanoprecipitation) and two-step (emulsification) methods. Figure 11.4 summarizes these methods and their submethods. In this chapter, preparation methods will not be explained detail but it is better to know some important issues about them. All methods have different advantages and disadvantages and you need to choose the best method according to your demands. For example, in terms of cost and large-scale manufacturing, the two-step approach is proclaimed to be the best approach for industrial bionanomaterials production in powder form. Also, if you want to manage the size of bionanomaterials, for monodispersed particles, you can use diffusion-controlled or reaction-controlled growth methods, with reaction-controlled growth defining the particles' final form [41].

11.1.2 Surface modifications

Prepared nanoparticles can be used with surface modification or affinity based chromatography strategies in order to enhance selectivity. Polymer surface modification is made due to physicochemical, mechanical and biological methods. In physicochemical methods, treatment with active gases, solutions and their combinations were performed. Roughening and micromanipulation strategies can be interpreted as mechanical methods. To make it clear of biological methods, physical adsorption, self-cross-linking of biomolecules, chemical conjugation and cell seeding were reported [42]. Due to different polymers and their various applications, different methods of surface modifications can be employed. For instance, PDMS

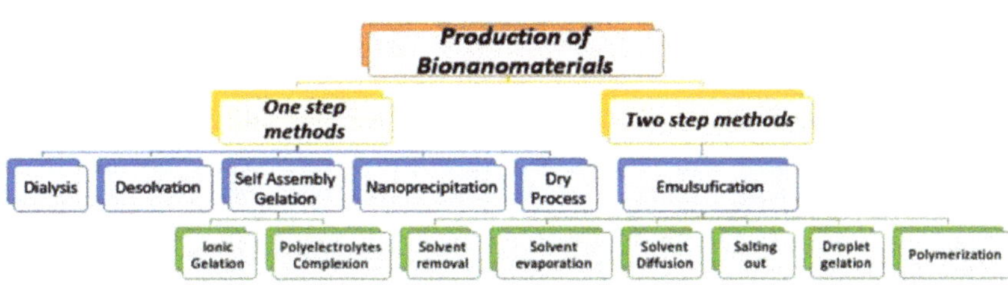

Figure 11.4. Production methods of bionanomaterials, created from information in [41].

channels for microfluidic applications can be modified via plasma treatment or using corona [43].

There are some surface modification techniques of nanoparticles such as grafting, surfactant binding and silanization [44]. But there are more methods for surface modification of polymers such as polymer coating, plasma technology, cross-linking, radiation, ion beam modification and so on [45]. Also, affinity based chromatography techniques and molecular imprint technology are used for modification of nanopolymers. We especially discourse on these approaches as follows.

11.1.2.1 Grafting

Grafting is a technique for changing the chemical topology and functionality of organic and inorganic materials' surfaces. Two grafting methods for covalently grafting polymer chains on the inorganic particles surface have been described in the literature. The first is the 'grafting to' process, which involves end-functionalized polymers reacting with a suitable surface. 'Grafting from' is the second method that involves growing polymer chains from an initiator-terminated self-assembled monolayer [46]. There are many studies with grafting process. For example, Slepicka *et al* grafted platinum nanoparticles onto PLLA foil [47], Dara *et al* grafted silver nanoparticles onto chitosan [48], Vahidhabanu *et al* grafted chitosan onto magnetic halloysite [49] etc. Researches show that the grafting from method is more anisotropic than the grafting to method. This is linked to the grafting to strategy, which results in a polymer covering that is more uniform of nanoparticle surfaces. The grafting to process results in increased miscibility and with a given grafting density, a reduced proclivity for self-assembled structures. As a result, the grafting to strategy can imply better control over the properties of polymer nanocomposites [50].

11.1.2.2 Surfactant binding

Surfactants can be grafted onto nanoparticles to modify their surfaces. Covalent assembly or noncovalent adsorption may be used in this step. Attaching thiol end groups to metals, ether linkages, and terminal OH groups functionalization are all examples of covalent grafting (oxides, polymers). Ion exchange and pairing, and hydrogen bonds are examples of non-covalent adsorption mechanisms. To graft and alter surfactants, any possible reaction at the surface of nanoparticles may theoretically be used [51]. Surfactant can effect the dispersion and aggregation of the nanoparticles in a solution [52]. Surfactants are used in coatings for emulsification, emulsion polymerization, wetting and dispersion. They can be divided into nonionics, anionics, cationics and zwitterionics [53]. For example Kaboorani *et al* used cationic surfactant for modification of cellulose nanocrystals [54], Bonnaud *et al* combined food-grade cationic surfactants with food-grade biopolymers to develop a food-grade cationic surfactant (alginate, pectin, xanthan etc) [55]. Putro *et al* conducted experiments on starch nanoparticles modified with nonionic, anionic and cationic surfactants for hydrophobic drug delivery [56]. Cationic surfactants have the best affinity on nanoparticles for especially controlled drug release studies and they are eco-friendly, soluble and biocompatible. This is why most studies are related with cationic surfactants. But the other surfactants can also be chosen due to physical structure of the nanoparticle surfaces [57].

11.1.2.3 Silanization

Silanization is an effective technique for improving surface stability and control. Silane is a binding agent with a bifunctional molecule that binds to the chemical residues of the materials to be silanized. The hydroxyl groups on the material surfaces react with the silane compounds. In addition, silane molecules are hydrolyzed on the material surface or in the solvent, and then a covalent bond is formed with the surface [58]. There are several silanization agents that can be used for activation of material surfaces.

Since it is non-toxic, low cost, commercially available and includes a pseudospecific imidazole ring for precise interaction with antibodies, 3-(2-imidazoline-1-yl) propyl(triethoxysilane) (IMEO) is used extensively. These molecules have a 'sticky end,' -Si(OEt)3, that can stick to the various solid matrixes surface. IMEO 3-(2-imidazolin-1-yl) propyl (triethoxysilane) has reactive imidazole groups so it can be used as a specific ligand. It enlarges the binding surface area by covalently binding over the hydroxyl groups to polymeric system surface [59–61]. APTES is a surface functionalizer that contains three hydrolyzable groups and an amine functional group (-NH2) [62, 63]. It is used in a study of Kuru *et al* for silanization and phenyl boronic acid for developing boronate affinity interactions of p(HEMA) to detect sugars such as galactose, fructose and raffinose due to their diol groups [64] and used by a group on silica nanoparticle in order to enhance oil recovery [65]. Matinlinna *et al* used 3-methacryloxypropyltrimethoxysilane and two commercial dental forms of silane in order to modify methacrylated resins and Ti substrates [66], and worked on the effect of γ-methacryloxypropyltrimethoxysilane and vinyltriisopropoxysilane on resin-titanium composite [67].

11.1.2.4 Affinity chromatography

Affinity chromatography is a form of liquid chromatography in which the components of a sample are purified or analyzed using a biologic agent as a stationary step in a column. It can be named due to type of ligands. Ligands can be classified as biological (antibodies, lectins, nucleic acids etc) and non-biological (boronate, metal ion chelates, dyes). Retained targets can be determined according to ligand affinity properties to biological molecules [68].

Bioaffinity chromatography is a kind of adsorption chromatography which is about reversible or selective binding affinities of complementary compounds of biologically active molecules. There are some examples for bioaffinity chromatography: complexes of enzyme with their cofactors, substrates, inhibitors or antibody antigen complexes or specific complexes of lectin with carbohydrates or nucleic acids nucleotide complexes or complexes of hormones or toxins with their receptors or complexes of transporter proteins with vitamins and sugars [69].

Complex formation between biomolecules purified by ligand is performed by general molecular forces such as ionic interactions, hydrophobic forces, dipole–dipole interactions, London dispersion forces, and charge transfer interactions.

Simultaneous happening of these many forces is the basic factor that forms the selectivity and effectiveness of the bond [70]. It has different types such as lectin, dye, immobilized metal ion, boronate, protein affinity and immunoaffinity chromatography [71].

This method is useful for selective biochemical separations, particularly when the stationary phase is a biologic agent. Its benefits include a large number of binding agents and great selectivity. For this reason, this technique can be used in biotechnology, microbiology, immunology, chemistry, environmental technology, agricultural technology, medicine, pharmaceutical technology and so on. It is not possible to mention all examples of affinity chromatography in this chapter, but it must be mentioned as the technique for separation, purification and recognition of the biomolecules [72].

11.1.2.5 Molecular imprinting technology

Molecular imprinted polymers (MIPs) are usually generated by free-radical co-polymerisation process including biological ligands and in the presence of template molecules, cross-linkers are used. A template molecule was used with functional monomers in order to create a pre-complex. Then polymerization via cross- linkers was performed and the template molecule was removed [73].

The primary advantages of polymeric nanostructures created with this technology include great selectivity, high affinity for the target molecule, increased physical robustness, strength, temperature and pressure tolerance, good plasticity, and inertness to acids, bases, metal ions, and organic solvents. This technique has low cost and simple preparation steps. The polymers' storage life is potentially quite long, and their recognition capacity can be maintained at room temperature for several years [74, 75]. As in each technique, molecular imprinted polymers also have disadvantages. The main disadvantage is template bleeding which means not removing all the template molecule during the polymerization and cleaning process. Another disadvantage can be preparing polymers with non-specific adsorption due to improper monomer ratio. Reproducibility and availability of the templates are also challenging for molecular imprinting processes [76]. According to use of polymerization types, there are also drawbacks due to requirements of some chemicals or resulting from procedures [77].

Due to molecular identification of biomolecules, molecularly imprinted polymers have superior properties like ease of preparation, higher stability and potentially lower costs. These advantages can play a role in environmental and agricultural applications in terms of separation, isolation, membrane filtration, drug delivery and monitoring of contaminants [78]. There are many studies in the literature that are about developing of molecular imprinted polymers for determining pesticides, drug or toxins. Applications are discussed in section 11.2.

11.1.3 Physicochemical properties and characterizations

Determining of the physical properties of a polymeric nanomaterial can be performed with several advanced characterization techniques. With these techniques, the developed nanomaterial can be evaluated in detail in order to understand its chemical and physical nature. Physicochemical properties can be mainly listed as morphology, particle size distribution, chemical composition and crystal structure, molar mass distribution, surface area, zeta potential, pH, stability, drug association, *in vitro* release kinetics for the nanopolymers (figure 11.5). In the following section, characterization of these properties will be clarified briefly.

11.1.3.1 Morphology

In order to understand surface structures for nanostructures and evaluate size, shape and morphology of these surfaces, transmission electron microscopy (TEM), atomic force microscopy (AFM) and scanning electron microscopy (SEM) are used [79].

SEM method produces a high resolutional three-dimensional (3D) image and it is utilized for determining the particle size, morphology of the surface and shapes. This device can include an energy-dispersive (EDS) detector that can be used for semi-quantitative and semi-qualitative information on elemental analysis of the surface. The nonconductive materials surface needs to be coated with a gold or platinum thin layer Also, some surface pretreatment may be performed such as ion sputtering or chemical etching. Chemical components density, surface defects, residues and damage can be obtained and evaluated via SEM micrographs [80, 81].

TEM is an effective tool for determining the synthesis–structure–property relationship. The transmission electron microscope is used to research polymeric materials and finds certain correlations between morphology and elemental ingredients [82]. TEM method is not always the proper method for imaging, for example for imaging soft materials the resolution is often limited. Also, radiation sensitivity of polymers can differ and therefore recent developments are still ongoing in electron microscopy techniques for advanced polymer characterizations of such forms as low dose high resolution (LD-HREM), high resolution (HREM), ion beam and so on

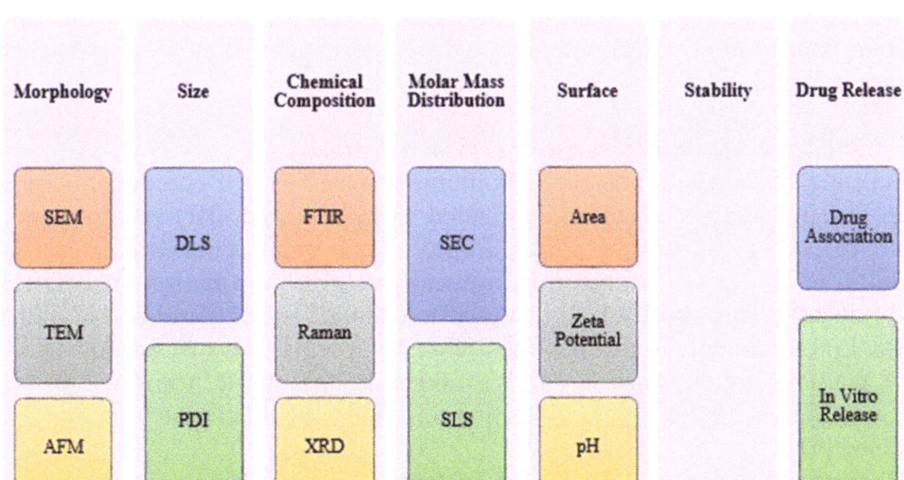

Figure 11.5. Characterization methods used for polymeric nanomaterials.

[83]. TEM images give size, shape and structural information (helix, lattice, sphere etc) [84].

For the characterization of polymeric materials, atomic force microscopy (AFM) has long been in use. It can be coupled with an infrared spectroscope and this combination can be useful for the chemical characterization of polymer-included materials such as polymer blends, biopolymers, multilayers and composites [85]. The AFM technique provides information on topography and local stiffness variations that are directly connected to material density and elastic modulus. The topographic images are created as a result of the study, using a distance between the sample and the device's tip. The other image can be generated in real time and can show a variety of surface heterogeneities. AFM images can also offer data about the polymers' surface of morphology, distribution of phases, cross-link density, compatibility, polymeric chains conformation [86]. Analysis gives Ra, Rq and Rmax values. Ra is the mean-center or arithmetic mean and Rq is the root mean square, and Rmax is the maximum surface roughness [87]. Ra is used for the roughness of the machined surfaces. It can be used to identify general variations in general profile high characteristics and to keep track of an existing production line [88].

11.1.3.2 Particle size distribution

The most important parameters for nanoparticles are morphology, particle size and distribution. These parameters can be measured by electron microscopy (EM), atomic force microscopy (AFM), dynamic light scattering (DLS), and photon-correlation spectroscopy (PCS). DLS is the most fastest, widely used and popular method to determine particle size and distribution. This approach works by shining monochromatic light onto a Brownian motion solution of spherical particles. When light strikes a moving object, it generates a Doppler shift, which alters the wavelength of incoming light due to the altering particle size [89]. The size distribution according to the density shown in the graphs is calculated with the help of advanced mathematics by assuming that the correlation function consists of the sum of different contributions from different particle sizes, rather than a force that fits an average size [90].

11.1.3.3 Chemical composition and crystal structure

The chemical composition and the bonding arrangement of components in polymeric based materials can be investigated using infrared (IR) spectroscopy. IR spectra can be transformed with using Fourier equations and numerical analysis. As a result, the IR signal is transformed optically to give information about intensity and frequency. This approach can be used to evaluate solids, gases and liquids with minimal pretreatment. It also has the ability to operate in transmission, reflection, and attenuated total reflection (ATR) modes. In this mode, transmission is best for quantitative analysis; and reflection is employed when the polymer is not dissolved [91].

IR spectroscopy is very useful for understanding that two particular types of atoms bound together with a specific bond form vibrate at a frequency that is very similar to each other. So, if certain peaks appear in an IR spectrum, we can determine the certainty of molecules included in this structure by comparing the IR spectra tables from the literature [92]. With IR spectra tables, functional groups, vibration modes and wavelength can be determined and compared with analysis results. Then, synthesis of new polymers, functionalization of surfaces and modification of the polymers can be investigated. But some metals or elements cannot be investigated with infrared spectroscopy due to the limitations of absorption, then Raman spectroscopy may be the alternative technique.

Raman scattering is also used for obtaining information on polymeric systems with their vibrational properties. The Raman effect is caused by the inelastic scattering of light that occurs when a molecule is irradiated with monochromatic light. Raman scattering is commonly known with the advantage of allowing the analysis of thick polymer samples, since the functional groups of thin-film-formed polymers absorb IR radiation [93]. High atomic numbered elements, superconducting compounds, are available Raman scatterers, whereas ionic structure, metals (because of surface plasmons), some transition metal ions from the 3d or 4f groups are difficult to analyse with Raman [94].

X-ray diffraction (XRD) is an advanced analytical method for determining the structure and chemical composition of a powdered, thick- or thin-film-formed polymeric material. It is also used for evaluation of the amount of crystalline content of the material. High energy x-ray light is applied onto the sample and it is analyzed by obtaining the diffractogram pattern. From the diffractogram, width of the peaks can be calculated and the degree of crystallinity of the material and its amorphous or lattice morphology can be determined. This advanced method represents the degree of crystallinity, the angle of diffraction and the crystalline index. So physico-chemical changes in polymeric material can be discussed with the analysis results [95, 96].

11.1.3.4 Molar mass distribution of the polymer

Size-exclusion chromatography (SEC) and static light scattering (SLS) can be used to determine the molar mass distribution of polymers [97]. SEC is a molecular characterization technique for branched polymers. It is a useful and precise instrument that uses multiple detectors to assess topology of the branching, branching degree, the branch molecular weight distribution, and the polymer chain molecular weight distribution [98]. It is also called gel permeation chromatography (GPC), and it separates analytes based on their hydrodynamic length. For size exclusion separations, both core–shell and sub-3 m particles are indicated as stationary phases [99].

SLS is based on scattering the incident light beam and measuring the light intensity at different angles. This technique is utilized for determination of the molecular mass directly [100]. SLS is ideal for measuring molar masses within the range 10^3–10^8 g mol^{-1} and the state of association of proteins in solution. When SLS is used for obtaining the geometric radius of gyration R_G, DLS can be used for obtaining the hydrodynamic Stokes radius, R_S. The combination of DLS and SLS techniques can improve measurements of the molar masses in several dimensions or association [101].

11.1.3.5 Surface area and chemistry

Bangs proposed an equation to calculate nanopolymers' surface area, based on the number of particles in a 1 ml suspension and the nanopolymers' typical mass-volume plot [102].

$$N = 6_{*}10^{10} \quad {}_{*}S\ /\pi_{*}\rho s_{*}d^{3} \tag{11.1}$$

N: the nanopolymers number,
S: % solids,
d: diameter (μm),
ρs: the polymer density (g mL^{-1}). The specific surface area of the synthesized nanopolymers is calculated in m^2 g^{-1} using the equilibrium surface area equation.

Surface area of sphere = $4 \times \pi \times r^2$ (π, 3,14; r, nanopolymer radius (m))[103] (9.2).

For analytical measurement, standard Brunauer–Emmett–Teller (BET) and Barrett–Joyner–Halenda (BJH) techniques are used for calculation of pore volume, specific surface area, and pore diameter [104].

With BET analysis, the surface areas of solids in m^2 g^{-1} unit can be measured based on the technique of nitrogen (N_2) gas adsorption in a liquid nitrogen environment at −198 °C. Especially, the differences in crystal structure between solid substances and heat treatment of the same substance at different temperatures or acid etc allows examination of changes in microstructures as a result of their treatment with chemicals. It can measure up to a minimum of 0.01 m^2 g^{-1}. The results can be taken from the computer monitor as multi-point, single-point BET surface area and multipoint Langmuir surface area values [105].

When BET is used for surface area analysis, the BJH calculation is used to determine distribution of pore size and pore volume of solid materials [106]. These two analyses can be combined in order to determine surface chemistry of nanomaterials.

11.1.3.6 Zeta potential

Determining the surface changes of the particles is commonly utilized with measuring the electrical potential of a particle. This is related with the movement of the particle in the liquid that is called shear plane. The potential of this is measured as zeta potential of the nanoparticles in suspension and this parameter is very important for the stability of suspension and adsorption research due to charge of the surfaces [107]. Suspension stability can also be affected by H+ ions in solution. Surface charge is changed when pH is increased or decreased and isoelectrical point is the criteria of the agglomeration (pH>IEP small agglomerate, pH=IEP large agglomerate, pH<IEP small agglomerate) [108].

DLS and phase analysis light scattering (PALS) can also be used to determine zeta potential [109]. Between 10 and +10 mV zeta potential are generally considered neutral. The zeta potential of strongly cationic nanoparticles is larger than +30 mV, while the zeta potential of strongly anionic nanoparticles is less than +30 mV [110].

11.1.3.7 pH

pH parameter plays an important role of dissolution and precipitation of the nanoparticles. The dissociation equilibrium, hydroxylation or protonation of the ionic groups of nanoparticles, and interfacial free energy are all linked to the pH of the solution [111]. Theorically, it is accepted that when the pH is equal to or close to the isoelectrical point (IEP), nanoparticles tend to form clusters, unstable and precipitate. Therefore, pH parameter is also considered to affect particle size distribution, zeta potential, rheology, stability and viscosity of the nanoparticle solution [112].

11.1.3.8 Stability

The stability can be affected by several parameters such as polymer molar mass distribution, particle size, zeta potential and pH. Phase separation process is unexpected for colloidal suspensions due to slow sedimentation characteristic of submicrometric particles. But agglomeration can occur due to the time. To avoid agglomeration, freeze-drying, spray-drying is recommended [113]. Dynamic mechanical analysis (DMA) and differential scanning calorimetry (DSC) were utilized to investigate the thermal and mechanical stability of nanopolymeric materials [114].

11.1.3.9 Drug association

Most of the therapeutic drugs are water insoluble and that is the main limiting factor of their clinical applications. For this purpose, drug loaded nanoformulations were developed to improve solubility, bioavailability, stability, biodistribution, pharmacokinetics, and pharmacodynamics. Also these nanoformulations are expected with low toxicity and higher therapeutic effect. To achieve all of these purposes, drug loading efficiency must be as high as possible and drug release profiles must be presented well [115].

Drug loading in nanoparticles is highly related with the fabrication process. In general, polymeric formulations achieved 1%–2% drug loading via nanoprecipitation, while emulsion/solvent evaporation achieved up to 14% drug loading. Liposomes, microemulsions, and micelles all reach a drug loading of less than 10%. Higher drug loading is preferable due to lower nanoparticle number, time, raw material and energy needed [116].

Since free drug is separated from the solution, determining the amount of drug associated with nanoparticles is a particularly difficult method. Mostly, ultracentrifugation is used for the separation but an ultrafiltration–centrifugation complex is also used. The difference between the free and total drug concentrations is used to measure the drug concentration associated with nanoparticles. And parameters related with drug association are encapsulation efficiency and drug loading capacity [97].

Encapsulation efficiency (EE) and drug loading capacity (LC) can be calculated with these equations [117]:

EE: (*total amount of drug-free drug*)/*total amount of drug**100% (9.3);

LC: (*total amount of drug-free drug*)/*nanoparticle weight**100% (9.4).

11.1.3.10 In vitro *release kinetics*

Drug release must be characterized in order to determine the performance of the developed systems. There are different methods which can be utilized to monitor drug release such as equilibrium dialysis, ultracentrifugation, ultrafiltration, and diffusion cells. Under physiological conditions, drug release measurements should be recommended optimally, taking into account aspects such as sample time and volume, medium, temperature, and pH [118].

Active agent release profile from the nanopolymer must be determined in order to understand the nanoparticle's release behaviour (controlled, slow, targeted etc). Releasing of the active agent can occur according to different diffusion, swelling and erosion mechanisms. To determine certain characteristics, diffusion and release kinetics must be calculated via empirical models. Release rate is also affected by size, shape, surface chemistry [119].

All of these characterization methods that are reviewed in this chapter are the most commonly used ones in the literature. Of course, there are more methods to characterize and evaluate polymers in terms of thermal, mechanical strength etc. Also, methods were specialized for specific purposes with different detectors, lights and other technologies. Whole approaches can be designed, modified and applied for the specific application of polymers to obtain the most intended information about developed polymeric bionanomaterials.

11.2 Applications of polymeric bionanomaterials

Polymeric bionanomaterials can be used in agricultural, environmental, food industry and biomedical fields due the their outstanding properties that we have mentioned before. In this chapter, we especially want to discourse on agricultural and environmental applications in more detail. Some of the examples from the newest and most important studies are chosen from the literature over ten years (from 2011 to today) and discussed as follows.

11.2.1 Agricultural applications

Developed nanomaterials were utilized in various fields of plant systems, such as plant growth via controlled release of agrochemicals and nutritional value enhancement, gene preservation etc [120]. Polymers and their composites in nanoscale can be used especially in encapsulation applications of agrochemicals (fertilizers, pesticides, and growth agents) due to their tailorable, low toxic and biocompatible properties (figure 11.6) [121, 122]. Different encapsulation techniques can be designed in order to reduce crop loss and enhance productivity in agriculture. Also, polymeric systems have been designed and suggested for the purpose of enhancing the production of

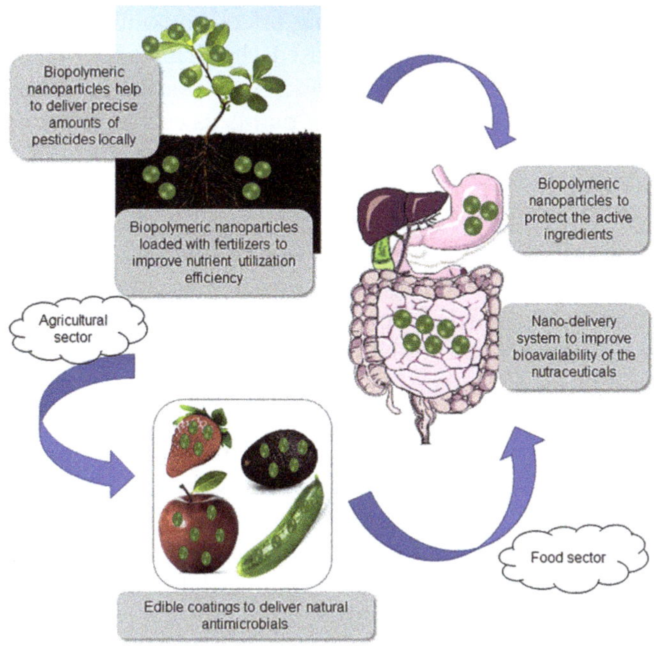

Figure 11.6. Characterization methods used for polymeric nanomaterials. Bionanomaterials applications in the agri-food sector. Adapted from [122], copyright (2020) with permission from Elsevier, Creative Commons CC-BY-NC-ND.

foods, preserving agricultural products until they reach the final consumer, for example with edible coatings, packaging etc [123].

11.2.1.1 Plant growth

Crop pests and diseases, as well as postharvest diseases, have a huge effect on agricultural yield loss and farmer earnings. Synthetic pesticides are commonly used by farmers to control pests in order to increase crop yields, posing possible threats to staff, customers, and the environment [124].

The most focus is paid to polymer-based nano-formulations for developing less harmful, biodegradable and natural origin plant growth agents [125]. Some polymeric bionanomaterials, especially glucan, alginate and chitosan based ones have been investigated for plant growth inducer and protection agent in terms of their biodegradability, biocompatibility, functionality and low toxicity [126]. Agrochemicals can be immobilized into polymeric based nanoparticles and in this way they might reduce toxic effects of active ingredients. By this methodology, the utilizing of agrochemical-loaded nanoparticles is preferred to reduce dosage for efficacy. It is important to highlight that nanotechnology will be a milestone in agriculture with the aid of increasing crop yields and minimizing chemical input [127].

Pereira *et al* developed two studies about controlled release of giberrelic acid (GA3), which is a plant growth stimulant. They developed chitosan based systems because chitosan is known for its enhancing effect on solubility and prevention of thermal degradation of GA3. They generated chitosan/tripolyphosphate and alginate/chitosan nanoparticles in the first study, which resulted in increased area of leaf and photosynthetic pigments levels on bean plants as compared to free hormone or chitosan/tripolyphosphate nanoparticles [128]. In another study of the same group GA3-loaded c-PGA/chitosan nanoparticles were reported, and this system was successful in order to obtain prolonged release of GA3. And, unlike the free hormone, this nanoencapsulation system showed a change in GA3 biological activity on germination rates, root growth, and leaf area [129].

Copper-chitosan nanoparticle (CuChNp) was developed and implemented for evaluation of its effect on finger millet plant by Sathiyabama *et al.* This study used finger millet plant as a model plant and developed a nanoparticle that was used for this plant. Both with foliar spray and as a combined application resulted with enhanced growth profile and increased yield by nearly 89%. In CuChNp treated plants, a significant increase in defense enzymes was also exhibited. The prepared CuChNp performed a role in boosting growth and protecting plants from blast fungus, according to the research [130].

Martins' group developed zinc (Zn) based and biopolymer composite nanomaterials for zinc release. Microcrystalline cellulose, chitosan, and alginate were used for creating a composite with ZnO nanoparticles. The results of the study reveal that ZnO NPs/alginate beads meet the optimum Zn needs for the crops [131].

Cu-chitosan nanoparticles (NPs) were developed to improve defense responses against Curvularia leaf spot (CLS) disease of maize. Also, plant growth promontory activity was investigated and these nanoparticles enhanced the defense response via increasing defense enzymes and antioxidant activities. These types of nanoparticle assisted treatments showed growth promontory effect in the parameters of plant height, root length, stem diameter, chlorophyll content and root number. Cu release profile was also investigated. This study is very important in terms of nanomaterial research in plant defense, disease protection and sustainable growth with biodegradable nanoparticles [132].

All of these studies are important because slow and controlled release of ingredients that have effects on plant growth significantly modulate physiological and biochemical responses of the plants. Development of novel bionanomaterials to regulate and increase the growth of plants and in this way reduce the utilization of agricultural fertilizer, helps to realize more sustainable agriculture.

11.2.1.2 Combatting plant pathogens

Plant diseases account for the majority of agricultural losses. Rapid and intricate evolution is a result of their genomic diversity. Engineered nanomaterials can be used in the treatment of a variety of plant pathogens and nanoparticles especially are able to interact with pathogens, their vectors, and host plants with unique techniques due to their physiochemical properties [133].

Sandhya *et al* developed chitosan–pectin nanoparticles containing carbendazim that is an antifungal agent. The encapsulation efficiency was calculated as 99.2%. This study found that these nanoparticles had higher antifungal activity than pure fungicide against *Fusarium oxysporum* and *Aspergillus parasiticus*, whereas non-targeted species (bacterial strain) were not toxic [134].

Sathiyabama *et al* designed a chitosan guar nanoparticle (CGNP) that has high antimicrobial activity to be utilized as a biological protectant against rice phyto-pathogens. These developed nanoparticles were evaluated for seed germination and seedling growth. They exhibited reduction on rice pathogens *Xanthomonas oryzae* and *Pyricularia grisea* under *in vitro* conditions. This study concluded that these nanoparticles can be utilized as an antimicrobial agent for controlling blast and blight disease of rice [135].

The purpose of Petroudy's group's study was producing green superabsorbent polymers (SAPs) based on cellulose derivatives and to investigate strengthening capabilities of the wheat straw cellulose nanofiber. TEMPO-mediated cellulose nanofibers (TCNFs) and sodium carboxymethyl cellulose with varying amounts of hydroxyethyl cellulose were used to make SAPs. Increasing the amount of citric acid utilized as a cross-linking agent lowered the swelling potential of SAPs, according to the findings. Swelling capacity was determined to be 200 gr gr^{-1} with the addition of 10% TCNF. These developed SAPs have biodegradable and non-toxic properties, that is why they can be used for potential horticultural applications sectors as an alternative to synthetic ones for use in applications of soil conditioning and controlled release of agrochemicals [136].

For plant disease management, a research group developed alginate–gelatin nanocomposite beads. These beads were utilized for struggle against two types of *Pseudomonas fluorescens* (VUPF5 and T17–4) that lives on potato and causes *Fusarium solani* disease. The maximum encapsulation efficiency was found to be 91.23% for VUPF5 and 87.23% for T17–4 at 1.5% gelatin concentration. These bionanomaterial-based encapsulating strategies can be used to preserve the inoculum of plant probiotic bacteria from hazardous circumstances [137].

Nanobiotechnology is a great technology for nanoscience with biogenic substances. It is known that treating disease with a single control measure is extremely difficult due to the variety of diseases of plants and mutating pathogens. In the future, nanobiotechnological approaches to plant disease control would be an alternative to conventional technologies [125].

11.2.1.3 Plant monitoring

For the plant monitoring field, nanobiosensors and nanobiomaterials especially, which can be used to inspect crops in fields, have become more popular as a result of technological advances in the agricultural world. These systems can aid precision farming by increasing agricultural crop production through better time, fertilization, and environmental management. Nanobiosensors and devices may be used to monitor the delivery of fungicide to a specific location. These nanobiosystems can accurately quantify environmental variables such as weather conditions (humidity, temperature, air velocity and air quality), as well as water level. They can also be used to make the most of the fertilizer, pesticide, and herbicide treatments that have been applied. The nutrient status of the crop can be determined, as well as the presence of any pathogens such as rodents, insects, or microorganisms. It is also possible to decide the best time to plant crops and harvest them [138].

In a study of Tang *et al*, a chitosan-based water ink was developed as a stretchable and flexible stain sensor. This developed sensor can be stretched up to 60% and exhibited measure factor of 64 within strain 1%–8%, and results exhibited that the method decreases the time and energy requirement, realizes wearable monitoring and on-site applications. Therefore, this chitosan-based ink suggests itself as an ideal candidate for the development and implementation of wearable devices [139].

Majdinasab *et al* developed alginate beads for use as a colorimetric sensor in intelligent packaging for food. The pH indicator red cabbage extract was encapsulated in alginate beads to create this sort of sensor. The CIELab method was used to calculate the color change of these beads during fish preservation. Protein breakdown, the accumulation of volatile amine, an increase of pH, and a color change in alginate beads were all observed as a result of expanding the population of bacteria during the storage and manufacture of proteolytic enzymes, according to the findings. Therefore, this developed sensors system has a great pH sensitivity and it is able to be used for deterioration of fish monitoring or other protein based foods with a low-cost colorimetric sensor [140].

Starch nanoparticles were made using a water-in-oil microemulsion process, and the starch nanoparticles were covered with poly-L-lysine in the study of Liu *et al.* The fluorescence material $Ru(bpy)_3^{2+}·6H_2O$ was mixed with the surface of the starch nanoparticles. The nanoparticles were combined with plasmid DNA to create complexes, which were subsequently treated with DNase I and ultrasonic. Plant suspension cells' cell walls, cell membranes, and nucleus membranes can be pierced by plasmid DNA-nanoparticle complexes mediated by ultrasound. The green fluorescence protein (GFP) gene is found in more than 5% of people. This nanobiomaterial will effectively solve the issue of outside genes being unable to pass through the plant cell wall [141].

There is increasing trend to develop light, flexible, and wearable electronic devices. A study of Hao *et al* aims to perform pectin and xanthan gum based artificial skin-like temperature sensor. This developed smart thermal sensor can be implemented as human body temperature measurements, electronic skin, agricultural monitoring and environmental monitoring, etc [142].

Nanosensors are used to gather vital data from the cultivated fields in real time for precision farming. For data collection, wireless nanosensors are connected to satellite networks, allowing spatial data and remote sensing devices to be distributed throughout the field. And also, phytopathologists are attempting to apply their nano plant pathology knowledge to diagnose and track symptoms of plant disease, as well as determine environmentally friendly analytical dimensions [143].

11.2.2 Environmental applications

The removal of pollutants such as antibiotics, pesticides and heavy metals from wastewater has become a major issue around the world. Nanotechnology has gained popularity as an evolving novel technology, and several nanosized and biobased

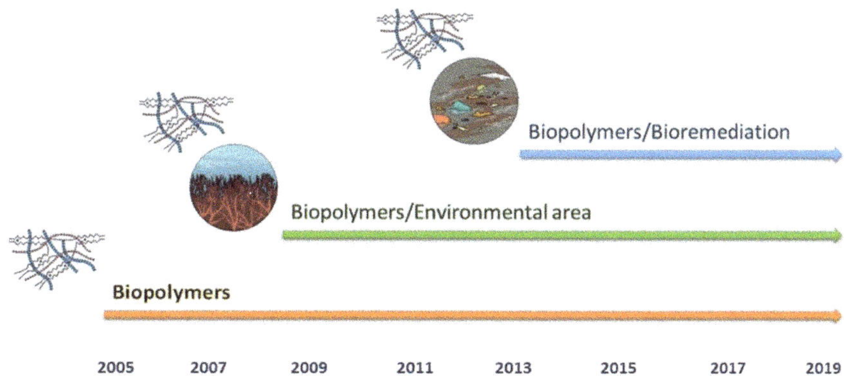

Figure 11.7. Timeline of biopolymer researches about bioremediation and environmental in the last 15 years. Adapted with permission from [146], copyright (2020), creative Commons CC-BY.

materials have been created to extract pollutants from contaminated water [144]. The chemical or physical affinity of the pollutant or size exclusion of the pollutant can be used as a separation mechanism for porous nanomaterials [145]. When a literature search is performed about biopolymers, the studies focus on their biodegradability then proposing to enhance their application, especially in bioremediation within the last 15 years (figure 11.7) [146].

Bionanopolymers also act as support material of metallic photocatalysts. Chitosan, for example, lowers the amount of intermediates formed during photocatalytic processes due to its strong adsorption and wide surface area. It can be recover trouble-free and quickly. Green polymeric bionanomaterials are also used for catalitic degradation with the properties of being low cost and environmentally friendly [147].

11.2.2.1 Heavy metal, dyes and other pollutants remediation

Bioremediation technology is known as efficient, economical and versatile, and is recognized as an environmentally friendly solution of pollution control. It is based on biodegradable pollutants via environmentally safe strategies. Toxic concentrations of pollutants are a great concern for environmental health and then special methods designed for this purpose are always necessary. The recent attitude in bioremediation field is in favour of use of bionanomaterials. Various bionanomaterials have been suggested to increase removal of pollutants as agrochemicals, pharmaceuticals, hydrocarbons, radionuclides and other heavy metals from environment [148].

By encapsulating iron oxide nanoparticles in a chitosan matrix, the Martínez-Cabanas group created a unique magnetic hybrid bionanomaterial. Metallic nanoparticles were obtained by green synthesis using eucalyptus extract. The novel hybrid materials were found to be effective as sorbents in the full elimination of arsenic and as column material for further experiments [149]. For the same purpose, Sharma *et al* developed cellulose supported zinc oxide nanoparticles in order to remove arsenic from wastewater. Microfibrillated cellulose (R-MFC) fibers were loaded with ZnO crystals and the study resulted in high efficiency at neutral pH and it is stated that it is higher than existing absorbents in the literature [150].

In a study of Kumar and colleagues, chitosan–sodium alginate based bionanomaterial scaffolds (BMS) were designed and functionalized with alumina hydroxide. These scaffolds were used for removal of chromium (Cr(VI), fluoride (F^-), and dye from water. The surface of the bionanomaterial was changed using silver nanoparticles to extend its shelf life. After surface modification of the bionanomaterial, fluoride uptakes were calculated as 168 mg g^{-1} at pH 4 and 60 mg g^{-1} at pH 7. The Kumar *et al* study exhibits a sustainable solution for contaminant removal from water [151].

Lee *et al* developed a chromatographic method for the removal of anionic dyes in their research. On a dextran gel column, chitosan-coated Fe_3O_4 nanoparticles are easily immobilized. The adsorbent surface's electrostatic interaction with the negatively charged sulfate groups on EB molecules improves dye adsorption efficiency. They conducted additional adsorption experiments with other anionic dyes of various colors to broaden the application of the current process (acid green 25, acid red, Cy5.5, acid yellow 25). Under continuous flow conditions, both of these molecules can be effectively captured, with more negatively charged dyes having a higher removal rate. These results show that the current approach is a viable method for removing anionic dye via adsorption [152].

Heavy metals, dyes, and organic contaminants are considered major pollutants around the world due to their toxic persistence in the setting. Pollutants have been found in different ways in the atmosphere and traditional methods have found it difficult to remove them due to a variety of disadvantages. The trending research for reducing these environmental contaminants with minor environmental issues is the adaptation of adsorption phenomena from conventional technologies with the adsorbents modification at nanoscale [153].

11.2.2.2 Wastewater treatment

In most cases, wastewater treatment processes are insufficient for removing toxins. For the eliminating of contaminants, photocatalysis process is a commonly used technique that is based on oxidation [154]. But researchers are still investigating novel removal methods that are environmentally friendly. Use of bionanomaterials [155] and microorganisms for catalysis [156] are the new and popular choices for wastewater treatment. The nanomaterials used wastewater treatment are attracting increasing attention, since they suggest higher adsorption capacities and increase in specific surface area. They may be regarded as cost-reductive and efficiency adjuvant in environmental pollution management [157]. Different approaches can be used for wastewater treatment such as developing polymeric films, bioplastics, 3D structures and nanosorbents from several biopolymers (figure 11.8) [158].

Wu's group developed a composite material including multi-walled carbon nanotube molecular imprinted polymers with beta-cyclodextrine for use for solid phase extraction of tartrazine from drink samples. Tartrazine is known as an azo synthetic colorant. The results of the study exhibited great levels of adsorption, good selectivity and excellent extraction of tartrazine [159].

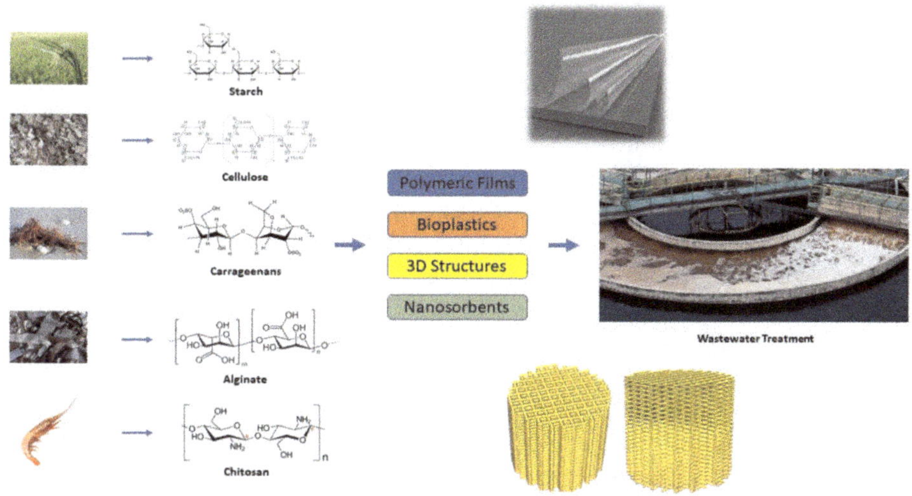

Figure 11.8. Different approaches of biopolymer-based wastewater treatments. Adapted from [158], copyright (2021) with permission from MDPI, distributed under a Creative Commons Attribution License 4.0 (CC BY).

Another research used molecular imprinted polymers to selectively remove dibutyl phthalate (DBP), a possible endocrine disrupting chemical that can be found in the setting. In this study, the α-cyclodextrin-based porous polymer was used as the molecular imprint of DBP and resulted with high selectivity against DBP. This developed polymer can also be used for purification of decontaminated water and soil [160].

Mosmeri *et al* developed alginate encapsulated CaO_2 bionanoparticles (CaO_2-NPs) for the removal of benzene from the groundwater and sandy soil. As a result of the study, it is understood that encapsulation decreased the negative effects of CaO_2-NPs on microorganisms with reduction of pH changes and of production of radicals. And 100% benzene (50 mg L^{-1}) removal was achieved and only 63% of benzene was biodegraded in the presence of encapsulated NPs within the 60 days. This study is important for the understanding of alginate encapsulated CaO_2 nanomaterials with the ability to be used for pollution bioremediation at 30 °C effectively [161].

A group of researchers devised a straightforward method for isolating carboxylic cellulose nanocrystals (CNCs) from microcrystalline cellulose (MCC). The developed nanocrystals were rod-like and they were obtained from hydrolysis of MCC. The hydrolysis time was optimized over 4 h. CNCs-4H showed coagulation-flocculation performance for kaolin, cationic dyes and textile dye wastes according to its carboxyl groups contents. These nanocrsytals have great potential for use as cationic dye adsorbents and efficient flocculants in wastewater treatment [162].

It is important to mention that the main limitations of the suggested studies are implementation of the experimental procedures in practical applications limited with laboratory scale because with low pollutant concentrations, everything is easier. But when the volume demand enhances, dose adjustment is required. More efforts are still needed to develop novel technologies in order to sustain circular economy with waste and pollution management via biodegradable polymeric nanomaterials [158].

11.2.2.3 Remediation of environmental pollutants

Environmental remediation process are based on implementing several procedures such as filtration, adsorption, absorption, photocatalysis and chemical reactions in order to remove the pollutants from different environmental media, (air, soil, water). Use of nanosized and biodegradable materials provides selective recognition of the pollutants with greener and safer alternatives for the environmental remediation of pollutants [163].

Using ultrasonic radiations, Bahal *et al* developed a more environmentally friendly method for synthesizing and analyzing new acrylic acid grafted amphoteric chitosan/TiO_2 (CAT) bionanocomposites. This developed nanocomposite was used for degradation of industrial dye (malachite green (MG)). Photocatalytic reaction of MG degradation was achieved under visible light with great potential for use in wastewater purification and treatment in a cost effective way [164].

The eco-friendly nanomaterial used by Kunztler *et al* is made from bioproducts of the microalga *Spirulina sp.* and has antibacterial properties. LEB 18 creates bioproducts like polyhydroxybutyrate (PHB), a biodegradable polymer with mechanical and thermal properties comparable to petrochemical polymers. Furthermore, microalgal phenolic compounds have antioxidant, antibacterial and antifungal properties, which is a distinct advantage for packaging growth. PHB nanofibers are hydrophobic, which means that food is protected from moisture outside the box. The material's thermal and mechanical properties have demonstrated its suitability for food packaging. This innovative nanomaterial combines a protective packaging feature that is also biodegradable with antibacterial activity that prevents microbial development and ensures food safety and preservation [165].

Sarioglu *et al* describe a report on the preparation of biocomposite materials for the removal of textile dyes. *Clavibacter michiganensis* was chosen for bacterial immobilization, setazol blue BRF-X, a commercial reactive textile dye, was chosen as the target contaminant. Electrospun polylactic acid (PLA) and polycaprolactone (PCL) nanofibrous polymeric webs were chosen for bacterial incorporation. Bacteria/PCL and bacteria/PLA webs showed effective setazol blue BRF-X dye remediation within 48 h. The bacteria-immobilized webs were then examined for five times of reuse and were found to be theoretically reusable at the end of the test, with higher bacterial immobilization and high capacity of dye removal. Overall, these results indicate that this device may be used as a starter inoculum for textile dye remediation in wastewater systems [166].

Bio-based materials have gained a lot of press recently. To take advantage of the cheap cost and ecological character of these materials, it is critical to develop creative and effective bio-techniques. CsrF, a selenite reductase enzyme, is overexpressed. In a work by Xia *et al*, *Escherichia coli* was employed as a microbial factory for producing Se(0) nanoparticles. Congo red, safranine T, and methylene blue had adsorption capacities of 1577.7, 1911.0, and 1792.2 mg g^{-1}, respectively, according to the findings. Bio-SeNPs had higher adsorption capacities for them than conventional bio-based products, by 6.8%, 25.2%, and 49.0%, respectively [167].

In recent years, researchers have focused on evaluating the potential for biofabrication of biologically sourced nanomaterials, with the goal of reducing emissions in a sustainable manner. Electrostatic interaction, photo and enzymatic catalysis, and surface active field interactions are all unique to biofabricated nanomaterials, and they all contribute to the detoxification of different contaminants [153].

11.3 Conclusions

More than 13.7 million people died globally in 2016 as a result of living in an unhealthy environment, according to the WHO. More than six million people died because of air pollution, while 1.8 million died as a result of water pollution, necessitating the use of comprehensive remediation techniques [168]. For this purpose, nanoscience and nanotechnology have been an important research focus for environmental and agricultural applications in both industry and academia in recent years. Nanoscale materials have the ability to remediate a wide variety of contaminants due to their unique surface properties. This research focus has brought the interdisciplinary field of materials science research to the forefront of scientific research through nanomaterial based studies. Polymeric nanomaterials, especially bio-based ones, are crucial in the production of new materials that can be used to influence and regulate the properties of other materials. Nanomaterials, such as nanoparticles, nanoporous materials, nanocapsules and polymer multi-layers, are becoming more widely used in industries such as polymer, environment, catalysis, energy, biomedical, pharmaceutical and automotive [169–171]. An overview of the development of polymeric bionanomaterials and applications for many purposes in various areas was emphasized and discussed in this chapter.

11.4 Challenges and future trends

Nanosized polymeric materials are still demanding on the removal, separation, recognition and drug release applications in the literature. Nanosized materials have large surface area, reactive and tunable functional-groups, mechanical strength. There are many different synthesis processes that are more effective, less expensive, environmentally friendly, and precise. The main purpose will be designing of smart and multifunctional bio-based and nanosized platforms with the highest targeting efficacy. But their safety and possible toxicity, should be investigated [172].

Green synthesis methods can be a good alternative for avoiding toxicity problems. Because they do not need the use of toxic chemicals and also they can be regarded as eco-friendly due to less energy requirement. With this synthesis method, desired shape, size or bioactivity can be obtained and therefore these nanoparticles can be utilized for several applications [173].

In the developed world, researchers and manufacturers have mainly focused on sustainable materials. Circular economy is the main popular topic for dealing with global economical challenges. Being sustainable is related with natural origin and easy decomposition. All properties must be bio-based for example: biodegradability, biocompatibility and green synthesis methods and natural biopolymer based nanoparticles provide all of these requirements.

Acknowledgments

The authors appreciate their family for their great support. Also, they would like thank each one of the editors, reviewers and publishers for their collaborations, suggestions and recommendations.

References

[1] Jeevanandam J, Barhoum A, Chan Y S, Dufresne A and Danquah M K 2018 Review on nanoparticles and nanostructured materials: history, sources, toxicity and regulations *Beilstein J. Nanotechnol.* **9** 1050–74

[2] Boverhof D R, Bramante C M, Butala J H, Clancy S F, Lafranconi M and West J *et al* 2015 Comparative assessment of nanomaterial definitions and safety evaluation considerations *Regul. Toxicol. Pharmacol.* **73** 137–50

[3] Visakh P M and Thomas S 2010 Preparation of bionanomaterials and their polymer nanocomposites from waste and biomass *Waste and Biomass Valorization* **1** 121–34

[4] Yetisgin A A, Cetinel S, Zuvin M, Kosar A and Kutlu O 2020 . Therapeutic nanoparticles and their targeted delivery applications *Molecules* **25** 2193

[5] Banik B L, Fattahi P and Brown J L 2016 Polymeric nanoparticles: the future of nanomedicine *Wiley Interdisciplinary Reviews-Nanomedicine and Nanobiotechnology* **8** 271–99

[6] Honek J F 2013 Bionanotechnology and bionanomaterials: John Honek explains the good things that can come in very small packages *BMC Biochem.* **14** 29

[7] Cauerhff A and Castro G R 2013 Bionanoparticles, a green nanochemistry approach *Electron. J. Biotechnol.* **16** 3

[8] Bereli N, Şener G, Altıntaş E B, Yavuz H and Denizli A 2010 Poly(glycidyl methacrylate) beads embedded cryogels for pseudo-specific affinity depletion of albumin and immunoglobulin G *Mater. Sci. Eng.* C **30** 323–9

[9] Rohman G and Spadavecchia J 2016 Biodegradable polymeric nanomaterials *Nanomaterials and Regenerative Medicine* (IAPC Publishing) 49–92

[10] Prabhu R, Patravale V and Joshi M D 2015 Polymeric nanoparticles for targeted treatment in oncology: current insights *Int. J. Nanomed.* **2015** 1001–18

[11] Elsabahy M and Wooley K L 2012 Design of polymeric nanoparticles for biomedical delivery applications *Chem. Soc. Rev.* **41** 2545

[12] Xiong R, Grant A M, Ma R, Zhang S and Tsukruk V V 2018 Naturally-derived biopolymer nanocomposites: interfacial design, properties and emerging applications *Mater Sci Eng R Reports* **125** 1–41

[13] Martău G A, Mihai M and Vodnar D C 2019 The use of chitosan, alginate, and pectin in the biomedical and food sector—biocompatibility, bioadhesiveness, and biodegradability *Polymers (Basel)* **11** 1837

[14] Lee Y-C and moon J-Y 2020 Bionanotechnology in agriculture, food, cosmetic and cosmeceutical *Introduction to Bionanotechnology* (Singapore: Springer) pp 199–217

[15] Kanmani P, Aravind J, Kamaraj M, Sureshbabu P and Karthikeyan S 2017 Environmental applications of chitosan and cellulosic biopolymers: a comprehensive outlook *Bioresour Technol.* **242** 295–303

[16] Rao J P and Geckeler K E 2011 Polymer nanoparticles: preparation techniques and size-control parameters *Prog Polym Sci.* **36** 887–913

[17] Katmıs A, Fide S, Karaismailoglu S and Derman S 2018 Synthesis and characterization methods of polymeric nanoparticles *Charact Appl Nanomater.* **1** 1–9

[18] Jeyaraj M, Gurunathan S, Qasim M, Kang M-H and Kim J-H 2019 A comprehensive review on the synthesis, characterization, and biomedical application of platinum nanoparticles *Nanomaterials.* **9** 1719

[19] Patel A, Khanna S, Xavier G K, Khanna K and Goel B 2017 polymeric nano-particles for tumor targeting—a review *Int J Drug Dev Res.* **9** 50–9

[20] Moreno Raja *et al* 2019 Polymeric nanomaterials *Nanocarriers for Drug Delivery* (Amsterdam: Elsevier) pp 557–653

[21] Locatelli E and Franchini M 2016 Polymeric nanoparticles: description, synthesis and applications: fundamentals and applications *Isotopes in Nanoparticles* ed L L Roig, V Gomez-Vallejo and P N Gibson (Singapore: Jenny Stanford Publishing) ch 5

[22] Mishra V, Kesharwani P and Jain N K 2014 Functionalized polymeric nanoparticles for delivery of bioactives *Drug Nano-carriers.* (April 2015) 91–123

[23] Amgoth C, Phan C, Banavoth M, Rompivalasa S and Tang G 2020 Polymer properties: functionalization and surface modified nanoparticles *Role of Novel Drug Delivery Vehicles in Nanobiomedicine* (Rijeka: IntechOpen)

[24] Bennet D and Kim S 2014 Polymer nanoparticles for smart drug delivery *Application of Nanotechnology in Drug Delivery* (Rijeka: InTechOpen)

[25] Park J I, Saffari A, Kumar S, Günther A and Kumacheva E 2010 Microfluidic synthesis of polymer and inorganic particulate materials *Annu Rev Mater Res.* **40** 415–43

[26] Mallakpour S and Behranvand V 2016 Polymeric nanoparticles: recent development in synthesis and application *Express Polym Lett.* **10** 895–913

[27] Hussain I, Singh N B, Singh A, Singh H and Singh S C 2016 Green synthesis of nanoparticles and its potential application *Biotechnol Lett.* **38** 545–60

[28] Devatha C P and Thalla A K 2018 Green synthesis of nanomaterials *Synthesis of Inorganic Nanomaterials.* (Amsterdam: Elsevier) pp 169–84

[29] Iravani S 2011 Green synthesis of metal nanoparticles using plants *Green Chem.* **13** 2638

[30] Ali M A, Ahmed T, Wu W, Hossain A, Hafeez R and Islam Masum M M *et al* 2020 Advancements in plant and microbe-based synthesis of metallic nanoparticles and their antimicrobial activity against plant pathogens *Nanomaterials.* **10** 1146

[31] Mahato K, Nagpal S, Shah M A, Srivastava A, Maurya P K and Roy S *et al* 2019 Gold nanoparticle surface engineering strategies and their applications in biomedicine and diagnostics *3 Biotech.* **9** 57

[32] Zeiri Y, Elia P, Zach R, Hazan S, Kolusheva S and Porat Z 2014 Green synthesis of gold nanoparticles using plant extracts as reducing agents *Int. J. Nanomed.* **9** 4007–21

[33] Kuppusamy P, Yusoff M M, Maniam G P and Govindan N 2016 Biosynthesis of metallic nanoparticles using plant derivatives and their new avenues in pharmacological applications—an updated report *Saudi Pharm J.* **24** 473–84

[34] Mukherjee S, Ghosh S, Das D K, Chakraborty P, Choudhury S and Gupta P *et al* 2015 Gold-conjugated green tea nanoparticles for enhanced anti-tumor activities and

hepatoprotection—synthesis, characterization and *in vitro* evaluation *J. Nutr Biochem.* **26** 1283–97
[35] Ahmed S, Saifullah, Ahmad M, Swami B L and Ikram S 2016 Green synthesis of silver nanoparticles using *Azadirachta indica* aqueous leaf extract *J. Radiat Res Appl Sci.* **9** 1–7
[36] Ghosh M K, Sahu S, Gupta I and Ghorai T K 2020 Green synthesis of copper nanoparticles from an extract of Jatropha curcas leaves: characterization, optical properties, CT-DNA binding and photocatalytic activity *RSC Adv.* **10** 22027–35
[37] Chou J, Li X, Yin Y and Indrisek N 2015 Determination of antioxidant activities in fruit juices based on rapid colorimetric measurement and characterisation of gold nanoparticles *Int J. Environ Anal Chem.* **95** 531–41
[38] Alshatwi A A, Athinarayanan J and Periasamy V S 2015 Green synthesis of bimetallic Au@Pt nanostructures and their application for proliferation inhibition and apoptosis induction in human cervical cancer cell *J. Mater Sci Mater Med.* **26** 148
[39] Pal G, Rai P and Pandey A 2019 Green synthesis of nanoparticles: a greener approach for a cleaner future *Green Synthesis, Characterization and Applications of Nanoparticles* (Amsterdam: Elsevier) pp 1–26
[40] Saif S, Tahir A and Chen Y 2016 Green synthesis of iron nanoparticles and their environmental applications and implications *Nanomaterials.* **6** 209
[41] Agi A, Junin R, Jaafar M Z, Sidek M A, Yakasai F and Gbadamosi A *et al* 2021 Formulation of bionanomaterials: a review of particle design towards oil recovery applications *J. Ind Eng Chem.* **98** 82–102
[42] Jaganathan S K, Balaji A, Vellayappan M V, Subramanian A P, John A A and Asokan M K *et al* 2015 Review: radiation-induced surface modification of polymers for biomaterial application *J. Mater Sci.* **50** 2007–18
[43] Nemani S K, Annavarapu R K, Mohammadian B, Raiyan A, Heil J and Haque M A *et al* 2018 Surface modification of polymers: methods and applications *Adv Mater Interfaces.* **5** 1801247
[44] Musino D 2018 Impact of surface modification on the structure and dynamics of silica-polymer nanocomposites *PhD Thesis* University of Montpellier
[45] Fabbri P and Messori M 2017 Surface modification of polymers *Modification of Polymer Properties* (Amsterdam: Elsevier) pp 109–30
[46] Kango S, Kalia S, Celli A, Njuguna J, Habibi Y and Kumar R 2013 Surface modification of inorganic nanoparticles for development of organic–inorganic nanocomposites—a review *Prog Polym Sci.* **38** 1232–61
[47] Slepička P, Přibyl M, Fajstavr D, Ulbrich P, Siegel J and Řezníčková A *et al* 2018 Grafting of platinum nanostructures on biopolymer at elevated temperature *Colloids Surfaces A Physicochem Eng Asp.* **546** 316–25
[48] Dara P K, Mahadevan R, Digita P A, Visnuvinayagam S, Kumar L R G and Mathew S *et al* 2020 Synthesis and biochemical characterization of silver nanoparticles grafted chitosan (Chi-Ag-NPs): *in vitro* studies on antioxidant and antibacterial applications. SN *Appl Sci.* **2** 665
[49] Vahidhabanu S, Adeogun A I and Babu B R 2019 Biopolymer-grafted, magnetically tuned halloysite nanotubes as efficient and recyclable spongelike adsorbents for anionic azo dye removal *ACS Omega.* **4** 2425–36
[50] Asai M, Zhao D and Kumar S K 2017 Role of grafting mechanism on the polymer coverage and self-assembly of hairy nanoparticles *ACS Nano.* **11** 7028–35

[51] Heinz H, Pramanik C, Heinz O, Ding Y, Mishra R K and Marchon D *et al* 2017 Nanoparticle decoration with surfactants: molecular interactions, assembly, and applications *Surf Sci Rep.* **72** 1–58

[52] Bharti B, Meissner J, Gasser U and Findenegg G H 2012 Surfactant adsorption and aggregate structure at silica nanoparticles: effects of particle size and surface modification *Soft Matter.* **8** 6573

[53] Ma X, Lee N-H, Oh H-J, Kim J-W, Rhee C-K and Park K-S *et al* 2010 Surface modification and characterization of highly dispersed silica nanoparticles by a cationic surfactant *Colloids Surfaces A Physicochem Eng Asp.* **358** 172–6

[54] Kaboorani A and Riedl B 2015 Surface modification of cellulose nanocrystals (CNC) by a cationic surfactant *Ind Crops Prod.* **65** 45–55

[55] Bonnaud M, Weiss J and McClements D J 2010 Interaction of a food-grade cationic surfactant (lauric arginate) with food-grade biopolymers (Pectin, Carrageenan, Xanthan, Alginate, Dextran, and Chitosan) *J. Agric Food Chem.* **58** 9770–7

[56] Putro J. N, Ismadji S, Gunarto C, Soetaredjo F E and Ju Y H 2020 A study of anionic, cationic, and nonionic surfactants modified starch nanoparticles for hydrophobic drug loading and release *J. Mol Liq.* **298** 112034

[57] Zakharova P, Doktorovova F and Sanchez-Lopez S *et al* 2019 Cationic surfactants: self-assembly, structure-activity correlation and their biological applications *Int J. Mol Sci.* **20** 5534

[58] Mittal K L 2009 *Silanes and Other Coupling Agents* vol 5 (Boca Raton, FL: CRC Press)

[59] Karakoc V, Yılmaz E, Türkmen D, Öztürk N, Akgöl S and Denizli A 2009 Selective separation of human serum albumin with copper(II) chelated poly(hydroxyethyl methacrylate) based nanoparticles *Int J. Biol Macromol.* **45** 188–93

[60] Öztürk N, Bereli N, Akgöl S and Denizli A 2008 High capacity binding of antibodies by poly(hydroxyethyl methacrylate) nanoparticles *Colloids Surfaces B Biointerfaces.* **67** 14–9

[61] Arkles B 2014 Silane coupling agents: connecting across boundaries (Version 3.0) *J. Organomet Chem.* **3** 1–76 Available from: www.gelest.com

[62] Landoulsi J, M J, El K, Richard C, Pulvin S and P G 2011 Silanization with APTES for controlling the interactions between stainless steel and biocomponents: reality vs expectation *Biomaterials—Physics and Chemistry* (Rijeka: InTech)

[63] Kim D, Dhand V, Rhee K and Park S-J 2015 Study on the effect of silanization and improvement in the tensile behavior of graphene-chitosan-composite *Polymers (Basel).* **7** 527–51

[64] Kuru C, Türkcan C, Uygun M, Okutucu B and Akgöl S 2014 Preparation and characterization of silanized poly(HEMA) nanoparticles for recognition of sugars *Artif Cells, Nanomedicine, Biotechnol.* **44** 835–41

[65] Ngouangna E N, Manan M A, Oseh J O, Norddin M N A M, Agi A and Gbadamosi A O 2020 Influence of (3–aminopropyl) triethoxysilane on silica nanoparticle for enhanced oil recovery *J. Mol Liq.* **315** 113740

[66] Matinlinna J P, Lassila L V J, Kangasniemi I, Yli-Urpo A and Vallittu P K 2005 Shear bond strength of Bis-GMA resin and methacrylated dendrimer resins on silanized titanium substrate *Dent Mater.* **21** 287–96

[67] Matinlinna J P, Özcan M, Lassila L V J and Vallittu P K 2004 The effect of a 3-methacryloxypropyltrimethoxysilane and vinyltriisopropoxysilane blend and tris(3-trimethoxysilylpropyl)isocyanurate on the shear bond strength of composite resin to titanium metal *Dent Mater.* **20** 804–13

[68] Hage D S, Anguizola J A, Li R, Matsuda R, Papastavros E and Pfaunmiller E *et al* 2013 Affinity chromatography *Liquid Chromatography.* (Amsterdam: Elsevier) 1–23 Available from: https://linkinghub.elsevier.com/retrieve/pii/B9780124158061000012

[69] Urh M, Simpson D and Zhao K 2009 Affinity chromatography: general methods *Methods in Enzymology* (Amsterdam: Elsevier) pp 417–38

[70] Mohamad N R, Marzuki N H C, Buang N A, Huyop F and Wahab R A 2015 An overview of technologies for immobilization of enzymes and surface analysis techniques for immobilized enzymes *Biotechnol Biotechnol Equip.* **29** 205–20

[71] Bahadir O, Saltan G, Ozbilgin S and Ergene B 2013 Affinity chromatography and importance in drug discovery *Column Chromatography* (Rijeka: InTech)

[72] Rodriguez E L, Poddar S, Iftekhar S, Suh K, Woolfork A G and Ovbude S *et al* 2020 Affinity chromatography: a review of trends and developments over the past 50 years *J. Chromatogr* **B 1157** 122332

[73] Curk T, Dobnikar J and Frenkel D 2016 Rational design of molecularly imprinted polymers *Soft Matter.* **12** 35–44

[74] Vasapollo G, Sole R D, Mergola L, Lazzoi M R, Scardino A and Scorrano S *et al* 2011 Molecularly imprinted polymers: present and future prospective *Int J. Mol Sci.* **12** 5908–45

[75] Liu G, Huang X, Li L, Xu X, Zhang Y and Lv J *et al* 2019 Recent advances and perspectives of molecularly imprinted polymer-based fluorescent sensors in food and environment analysis *Nanomaterials.* **9** 1030

[76] Murphy S 2009 *Development of Molecularly Imprinted Polymers for ACE Inhibitors* **2016** 2004–19 Available from: http://repository.wit.i.e./1449/

[77] Adumitrăchioaie A 2018 Electrochemical methods based on molecularly imprinted polymers for drug detection. a review *Int J. Electrochem Sci.* **13** 2556–76

[78] Ahmad R and Muhammad M 2019 Molecularly imprinted polymeric nanomaterials for environmental analysis *Environmental Nanotechnology* (Cham: Springer) pp 143–68

[79] Venkateshaiah A, Padil V V T, Nagalakshmaiah M, Waclawek S, Černík M and Varma R S 2020 Microscopic techniques for the analysis of micro and nanostructures of biopolymers and their derivatives *Polymers (Basel).* **12** 512 Available from: https://mdpi.com/2073–4360/12/3/512

[80] Rydz J, Šišková A and Andicsová Eckstein A 2019 Scanning electron microscopy and atomic force microscopy: topographic and dynamical surface studies of blends, composites, and hybrid functional materials for sustainable future *Adv Mater Sci Eng.* **2019** 6871785

[81] Girão A V, Caputo G and Ferro M C 2017 Application of Scanning Electron Microscopy–Energy Dispersive X-Ray Spectroscopy (SEM-EDS) *Comprehensive Analytical Chemistry* (Cham: Springer) pp 153–68

[82] Libera M R and Egerton R F 2010 Advances in the transmission electron microscopy of polymers *Polym Rev.* **50** 321–39

[83] Kuei B, Aplan M P, Litofsky J H and Gomez E D 2020 New opportunities in transmission electron microscopy of polymers *Mater Sci Eng R Reports* **139** 100516

[84] Kübel C and Martin D C 2001 Influence of structural variations on high-resolution electron microscopy images of poly[1,6-di(N -carbazolyl)2,4-hexadiyne] nanocrystals *Philos Mag* A **81** 1651–73

[85] Nguyen-Tri P, Ghassemi P, Carriere P, Nanda S, Assadi A A and Nguyen D D 2020 Recent applications of advanced atomic force microscopy in polymer science: a review *Polymers (Basel).* 12 1142 Available from: https://mdpi.com/2073–4360/12/5/1142

[86] Sousa F D B and de, Scuracchio C H 2014 The use of atomic force microscopy as an important technique to analyze the dispersion of nanometric fillers and morphology in nanocomposites and polymer blends based on elastomers *Polímeros.* **24** 661–72

[87] Bhushan B 2000 Surface roughness analysis and measurement techniques *Modern Tribology Handbook* (Boca Raton, FL: CRC Press) pp 79–150

[88] Rajesh Kumar B and Subba Rao T 2012 AFM studies on surface morphology, topography and texture of nanostructured zinc aluminum oxide thin films *Dig J. Nanomater Biostructures* **7** 1881–9

[89] Pal S L, Jana U, Manna P K, Mohanta G P and Manavalan R 2011 Nanoparticle: an overview of preparation and characterization *J. Appl Pharm Sci.* **1** 228–34

[90] Bhattacharjee S 2016 DLS and zeta potential - What they are and what they are not? *J. Control Release* **235** 337–51

[91] Barrios V A E, Méndez J R R, Aguilar N V P, Espinosa G A, Rodríguez J L D and FTIR - 2012 An essential characterization technique for polymeric materials *Infrared Spectroscopy —Materials Science, Engineering and Technology* (Rijeka: InTech)

[92] Mihály J, Sterkel S, Ortner H M, Kocsis L, Hajba L and Furdyga É *et al* 2006 FTIR and FT-Raman spectroscopic study on polymer based high pressure digestion vessels *Croat. Chem. Acta* **79** 497–501

[93] Bokobza L 2017 Spectroscopic techniques for the characterization of polymer nanocomposites: a review *Polymers (Basel)* **10** 7

[94] Gouadec G, Colomban P, Raman Spectroscopy P C and Gouadec G 2007 How spectra relate to disorder, particle size and mechanical properties *Prog Cryst Growth Charact Mater.* **53** 1–56

[95] Chauhan A and Chauhan P 2017 XRD: a pioneer technique for characterizing the polymer and fiber *J. Text Sci Eng.* **07** 3

[96] Widjonarko N 2016 Introduction to advanced X-ray diffraction techniques for polymeric thin films *Coatings.* **6** 54

[97] Zielińska A, Carreiró F, Oliveira A M, Neves A, Pires B and Venkatesh D N *et al* 2020 Polymeric nanoparticles: production, characterization, toxicology and ecotoxicology *Molecules.* **25** 3731

[98] Gaborieau M and Castignolles P 2011 Size-exclusion chromatography (SEC) of branched polymers and polysaccharides *Anal Bioanal Chem.* **399** 1413–23

[99] Lubomirsky E, Khodabandeh A, Preis J, Susewind M, Hofe T and Hilder E F *et al* 2021 Polymeric stationary phases for size exclusion chromatography: a review *Anal Chim Acta.* **1151** 338244

[100] Ahrer K, Buchacher A, Iberer G, Josic D and Jungbauer A 2003 Analysis of aggregates of human immunoglobulin G using size-exclusion chromatography, static and dynamic light scattering *J. Chromatogr* A **1009** 89–96

[101] Gast K and Fiedler C 2012 Dynamic and Static Light Scattering of Intrinsically Disordered Proteins *Intrinsically Disordered Protein Analysis* (New York: Springer) pp 137–61

[102] Bangs L 1984 *Uniform Latex Particles* (Indianapolis IN: Seragen Diagnostics Inc.)

[103] Kuru C I, Ulucan F, Kuşat K and Akgöl S 2020 A model study by using polymeric molecular imprinting nanomaterials for removal of penicillin G *Environ Monit Assess* **192** 367

[104] Bardestani R, Patience G S and Kaliaguine S 2019 Experimental methods in chemical engineering: specific surface area and pore size distribution measurements—BET, BJH, and DFT *Can J. Chem Eng.* **97** 2781–91

[105] Tan Y H, Davis J A, Fujikawa K, Ganesh N V, Demchenko A V and Stine K J 2012 Surface area and pore size characteristics of nanoporous gold subjected to thermal, mechanical, or surface modification studied using gas adsorption isotherms, cyclic voltammetry, thermogravimetric analysis, and scanning electron microscopy *J. Mater Chem.* **22** 6733

[106] De Lange M F, Vlugt T J H, Gascon J and Kapteijn F 2014 Adsorptive characterization of porous solids: error analysis guides the way *Microporous Mesoporous Mater.* **200** 199–215 Available from: https://linkinghub.elsevier.com/retrieve/pii/S138718111400482X

[107] Xu R 2008 Progress in nanoparticles characterization: sizing and zeta potential measurement *Particuology.* **6** 112–5 Available from: https://linkinghub.elsevier.com/retrieve/pii/S1674200108000357

[108] Berg J M, Romoser A, Banerjee N, Zebda R and Sayes C M 2009 The relationship between pH and zeta potential of ~ 30 nm metal oxide nanoparticle suspensions relevant to *in vitro* toxicological evaluations *Nanotoxicology.* **3** 276–83

[109] Smith M C, Crist R M, Clogston J D and McNeil S E 2017 Zeta potential: a case study of cationic, anionic, and neutral liposomes *Anal Bioanal Chem.* **409** 5779–87

[110] Clogston J D and Patri A K 2011 Zeta potential measurement *Characterization of Nanoparticles Intended for Drug Delivery Methods in Molecular Biology* vol 697 (New York: Humana Press) 63–70

[111] Zhang H, Chen B and Banfield J F 2010 Particle size and pH effects on nanoparticle dissolution *J. Phys Chem* C **114** 14876–84

[112] Wamkam C T, Opoku M K, Hong H and Smith P 2011 Effects of pH on heat transfer nanofluids containing ZrO_2 and TiO_2 nanoparticles *J. Appl Phys.* **109** 024305

[113] Bhattacharjee S and Bhattacharjee S 2019 Polymeric nanoparticles *Principles Nanomedicine* (Singapore: Jenny Stanford) pp 195–240

[114] Chrysafi I, Kontonasaki E, Anastasiou A D, Patsiaoura D, Papadopoulou L and Vourlias G *et al* 2020 Mechanical and thermal properties of PMMA resin composites for interim fixed prostheses reinforced with calcium β-pyrophosphate *J. Mech Behav Biomed Mater.* **112** 104094

[115] Liu Y, Yang G, Jin S, Xu L and Zhao C 2020 Development of high-drug-loading nanoparticles *Chempluschem.* **85** 2143–57

[116] Chu K S, Schorzman A N, Finniss M C, Bowerman C J, Peng L and Luft J C *et al* 2013 Nanoparticle drug loading as a design parameter to improve docetaxel pharmacokinetics and efficacy *Biomaterials.* **34** 8424–9

[117] Shi Y, Wan A, Shi Y, Zhang Y and Chen Y 2014 Experimental and mathematical studies on the drug release properties of aspirin loaded chitosan nanoparticles *BioMed Research Int.* **2014** 1–8

[118] Clogston J D, Crist R M and Mcneil S E 2016 Physicochemical characterization of polymer nanoparticles: challenges and present limitations *Polymer Nanoparticles for Nanomedicines* (Cham: Springer) pp 187–203

[119] Shakiba S, Astete C E, Paudel S, Sabliov C M, Rodrigues D F and Louie S M 2020 Emerging investigator series: polymeric nanocarriers for agricultural applications: synthesis, characterization, and environmental and biological interactions *Environ Sci Nano.* **7** 37–67 Available from: http://xlink.rsc.org/?DOI=C9EN01127G

[120] Verma S K, Das A K, Patel M K, Shah A, Kumar V and Gantait S 2018 Engineered nanomaterials for plant growth and development: a perspective analysis *Sci Total Environ.* **630** 1413–35

[121] Sinha T, Bhagwatwar P, Krishnamoorthy C and Chidambaram R 2019 Polymer based micro- and nanoencapsulation of agrochemicals *Polymers for Agri-Food Applications* (Cham: Springer) pp 5–28

[122] Sampathkumar K, Tan K X and Loo S C J 2020 *Developing Nano-Delivery Systems for Agriculture and Food Applications with Nature-Derived Polymers. iScience.* **23** 101055 Available from: https://linkinghub.elsevier.com/retrieve/pii/S2589004220302406

[123] Gutiérrez T J 2019 *Polymers for Agri-Food Applications* (Cham: Springer)

[124] Djiwanti S R and Kaushik S 2019 Nanopesticide: future application of nanomaterials in plant protection *Plant Nanobionics. Nanotechnology in the Life Sciences* (Cham: Springer) pp 255–98

[125] Saharan V and Pal A 2016 Current and future prospects of chitosan-based nanomaterials *Plant Protection and Growth* (Cham: Springer) pp 43–8

[126] Sathiyabama M 2019 Biopolymeric nanoparticles as a nanocide for crop protection *Nanoscience for Sustainable Agriculture* (Cham: Springer) pp 139–52

[127] Pascoli M, Lopes-Oliveira P J, Fraceto L F, Seabra A B and Oliveira H C 2018 State of the art of polymeric nanoparticles as carrier systems with agricultural applications: a minireview *Energy, Ecol Environ.* 3 137–48

[128] Pereira A E S, Silva P M, Oliveira J L, Oliveira H C and Fraceto L F 2017 Chitosan nanoparticles as carrier systems for the plant growth hormone gibberellic acid *Colloids Surfaces B Biointerfaces.* **150** 141–52

[129] Pereira A E S, Sandoval-Herrera I E, Zavala-Betancourt S A, Oliveira H C, Ledezma-Pérez A S and Romero J *et al* 2017 γ-Polyglutamic acid/chitosan nanoparticles for the plant growth regulator gibberellic acid: characterization and evaluation of biological activity *Carbohydr Polym.* **157** 1862–73

[130] Sathiyabama M and Manikandan A 2018 Application of Copper-chitosan nanoparticles stimulate growth and induce resistance in finger millet (Eleusine coracana Gaertn.) plants against blast disease *J. Agric Food Chem.* **66** 1784–90

[131] Martins N C T, Avellan A, Rodrigues S, Salvador D, Rodrigues S M and Trindade T 2020 Composites of biopolymers and ZnO NPs for controlled release of zinc in agricultural soils and timed delivery for maize *ACS Applied Nano Materials.* **3** 2134–48

[132] Choudhary R C, Kumaraswamy R V, Kumari S, Sharma S S, Pal A and Raliya R *et al* 2017 Cu-chitosan nanoparticle boost defense responses and plant growth in maize (*Zea mays L.*) *Sci Rep.* **7** 9754

[133] Farooq T, Adeel M, He Z, Umar M, Shakoor N and da Silva W *et al* 2021 Nanotechnology and plant viruses: an emerging disease management approach for resistant pathogens *ACS Nano.* **15** 6030–7

[134] Sandhya K S, Kumar D and Dilbaghi N 2017 Preparation, characterization, and bioefficacy evaluation of controlled release carbendazim-loaded polymeric nanoparticles *Environ Sci Pollut Res.* **24** 926–37

[135] Sathiyabama M and Muthukumar S 2020 Chitosan guar nanoparticle preparation and its *in vitro* antimicrobial activity towards phytopathogens of rice *Int J. Biol Macromol.* **153** 297–304

[136] Djafari Petroudy S R, Ranjbar J and Rasooly Garmaroody E 2018 Eco-friendly superabsorbent polymers based on carboxymethyl cellulose strengthened by TEMPO-mediated oxidation wheat straw cellulose nanofiber *Carbohydr Polym.* **197** 565–75 Available from: https://linkinghub.elsevier.com/retrieve/pii/S0144861718306544

[137] Pour M M, Saberi-Riseh R, Mohammadinejad R and Hosseini A 2019 Investigating the formulation of alginate- gelatin encapsulated *Pseudomonas fluorescens* (VUPF5 and T17-4 strains) for controlling Fusarium solani on potato *Int J. Biol Macromol.* **133** 603–13

[138] Younas A, Yousaf Z, Rashid M and Riaz N 2020 *Nanoagronomy* ed S Javad (Cham: Springer) pp 101–23

[139] Tang W, Yan T, Ping J, Wu J and Ying Y 2017 Rapid fabrication of flexible and stretchable strain sensor by chitosan-based water ink for plants growth monitoring *Adv Mater Technol.* 2 1700021

[140] Majdinasab M, Hosseini S M H, Sepidname M, Negahdarifar M and Li P 2018 Development of a novel colorimetric sensor based on alginate beads for monitoring rainbow trout spoilage *J. Food Sci Technol.* 55 1695–704

[141] Liu J, Wang F, Wang L, Xiao S, Tong C and Tang D *et al* 2008 Preparation of fluorescence starch-nanoparticle and its application as plant transgenic vehicle *J. Cent South Univ Technol.* 15 768–73

[142] Hao L, Ding J, Yuan N, Xu J, Zhou X and Dai S *et al* 2018 Visual and flexible temperature sensor based on a pectin-xanthan gum blend film *Org Electron.* **59** 243–6 Available from: https://linkinghub.elsevier.com/retrieve/pii/S1566119918302416

[143] Kaushal M and Wani S P 2017 Nanosensors: frontiers in precision agriculture *Nanotechnology* (Singapore: Springer) pp 279–91

[144] Yang J, Hou B, Wang J, Tian B, Bi J and Wang N *et al* 2019 Nanomaterials for the removal of heavy metals from wastewater *Nanomaterials.* **9** 424

[145] Ghadimi M, Zangenehtabar S and Homaeigohar S 2020 An overview of the water remediation potential of nanomaterials and their ecotoxicological impacts *Water (Switzerland)* **12** 1150

[146] Bedor P B A, Caetano R M J, Souza Júnior F G and de, Leite S G F 2020 Advances and perspectives in the use of polymers in the environmental area: a specific case of PBS in bioremediation *Polímeros* **30** e2020023

[147] Sarkar S, Ponce N T, Banerjee A, Bandopadhyay R, Rajendran S and Lichtfouse E 2020 Green polymeric nanomaterials for the photocatalytic degradation of dyes: a review *Environ Chem Lett.* **18** 1569–80

[148] Kumar V, Shahi S K and Singh S 2018 Bioremediation: an eco-sustainable approach for restoration of contaminated sites *Microbial Bioprospecting for Sustainable Development* (Singapore: Springer) pp 115–36

[149] Martínez-Cabanas M, López-García M, Barriada J L, Herrero R and Sastre de Vicente M E 2016 Green synthesis of iron oxide nanoparticles. Development of magnetic hybrid materials for efficient As(V) removal *Chem Eng J.* **301** 83–91

[150] Sharma S K, Sharma P R, Chen H, Johnson K, Zhan C and Wang R *et al* 2020 Cellulose-supported nanosized zinc oxide: highly efficient bionanomaterial for removal of arsenic from water *Contaminants in Our Water: Identification and Remediation Methods* (Washington DC: ACS Publications) pp 253–67

[151] Kumar A, Paul P and Nataraj S K 2017 Bionanomaterial scaffolds for effective removal of fluoride, chromium, and dye *ACS Sustainable Chemistry & Engineering.* **5** 895–903 Available from: https://pubs.acs.org/doi/10.1021/acssuschemeng.6b02227

[152] Lee S Y, Shim H E, Yang J E, Choi Y J and Jeon J 2019 Continuous flow removal of anionic dyes in water by chitosan-functionalized iron oxide nanoparticles incorporated in a dextran gel column *Nanomaterials.* **9** 1164

[153] Bhavya G, Belorkar S A, Mythili R, Geetha N, Shetty H S and Udikeri S S *et al* 2021 Remediation of emerging environmental pollutants: a review based on advances in the uses of eco-friendly biofabricated nanomaterials *Chemosphere.* **275** 129975

[154] Akpotu S O, Oseghe E O, Ayanda O S, Skelton A A, Msagati T A M and Ofomaja A E 2019 Photocatalysis and biodegradation of pharmaceuticals in wastewater: effect of abiotic and biotic factors *Clean Technologies and Environmental Policy* **21** 1701–21

[155] Trivedi R and Bergi J 2021 Application of bionanoparticles in wastewater treatment *Advanced Oxidation Processes for Effluent Treatment Plants* (Amsterdam: Elsevier) pp 177–97

[156] Aissaoui S, Ouled-Haddar H, Sifour M, Beggah C and Benhamada F 2017 Biological removal of the mixed pharmaceuticals: diclofenac, ibuprofen, and sulfamethoxazole using a bacterial consortium *Iran J. Biotechnol.* **15** 135–42

[157] Abouzeid R E, Khiari R, El-Wakil N and Dufresne A 2019 Current state and new trends in the use of cellulose nanomaterials for wastewater treatment *Biomacromolecules.* **20** 573–97

[158] Russo T, Fucile P, Giacometti R and Sannino F 2021 Sustainable removal of contaminants by biopolymers: a novel approach for wastewater treatment *Current State and Future Perspectives. Processes.* **9** 719 Available from: https://mdpi.com/2227–9717/9/4/719

[159] Wu L, Liu F, Wang G, Guo Z and Zhao J 2016 Bifunctional monomer molecularly imprinted polymers based on the surface of multiwalled carbon nanotubes for solid-phase extraction of tartrazine from drinks *RSC Adv.* **6** 464–71 Available from: http://xlink.rsc.org/?DOI=C5RA19565A

[160] Dolai J, Ali H and Jana N R 2020 Molecular imprinted poly-cyclodextrin for selective removal of dibutyl phthalate *ACS Appl Polym Mater.* **2** 691–8

[161] Mosmeri H, Alaie E, Shavandi M, Dastgheib S M M and Tasharrofi S 2017 Bioremediation of benzene from groundwater by calcium peroxide (CaO_2) nanoparticles encapsulated in sodium alginate *J. Taiwan Inst Chem Eng.* **78** 299–306

[162] Yu H-Y, Zhang D-Z, Lu F-F and Yao J 2016 New approach for single-step extraction of carboxylated cellulose nanocrystals for their use as adsorbents and flocculants *ACS Sustain Chem Eng.* **4** 2632–43

[163] Guerra F, Attia M, Whitehead D and Alexis F 2018 Nanotechnology for environmental remediation: materials and applications *Molecules.* **23** 1760

[164] Bahal M, Kaur N, Sharotri N and Sud D 2019 Investigations on amphoteric chitosan/TiO_2 bionanocomposites for application in visible light induced photocatalytic degradation *Adv Polym Technol.* **2019** 1–9

[165] Kuntzler S G, Almeida A C A, de, Costa J A V and Morais M G 2018 Polyhydroxybutyrate and phenolic compounds microalgae electrospun nanofibers: a novel nanomaterial with antibacterial activity *Int J. Biol Macromol.* **113** 1008–14

[166] Sarioglu O F, San Keskin N O, Celebioglu A, Tekinay T and Uyar T 2017 Bacteria immobilized electrospun polycaprolactone and polylactic acid fibrous webs for remediation of textile dyes in water *Chemosphere.* **184** 393–9

[167] Xia X, Zhou Z, Wu S, Wang D, Zheng S and Wang G 2018 Adsorption removal of multiple dyes using biogenic selenium nanoparticles from an *Escherichia coli* strain overexpressed selenite reductase CsrF *Nanomaterials.* **8** 234

[168] Kumar L, Ragunathan V, Chugh M and Bharadvaja N 2021 Nanomaterials for remediation of contaminants: a review *Environ Chem Lett.* **19** 3139–63

[169] Haghi A K, Noe Aguilar Gonzalez C, Thomas S and PK M (ed) 2018 *Physical Chemistry for Engineering and Applied Sciences* (Toronto: Apple Academic Press)

[170] Bayda S, Adeel M, Tuccinardi T, Cordani M and Rizzolio F 2019 The history of nanoscience and nanotechnology: from chemical–physical applications to nanomedicine *Molecules.* **25** 112

[171] Ashraf S A, Siddiqui A J, Elkhalifa A E O, Khan M I, Patel M and Alreshidi M *et al* 2021 Innovations in nanoscience for the sustainable development of food and agriculture with implications on health and environment *Sci Total Environ.* **768** 144990

[172] Kaphle A, Navya P N, Umapathi A and Daima H K 2018 Nanomaterials for agriculture, food and environment: applications, toxicity and regulation *Environ Chem Lett.* **16** 43–58

[173] Saratale R G, Saratale G D, Shin H S, Jacob J M, Pugazhendhi A and Bhaisare M *et al* 2018 New insights on the green synthesis of metallic nanoparticles using plant and waste biomaterials: current knowledge, their agricultural and environmental applications *Environ Sci Pollut Res.* **25** 10164–83

IOP Publishing

Bionanomaterials for Environmental and Agricultural Applications

Ravindra Pratap Singh and Kshitij RB Singh

Chapter 12

Role of bionanomaterials for environmental remediation

A Manjceevan and K Velauthamurty

Modern technologies are continuously being discovered to remediate contaminants from air, water, and soil. Heavy metals, pesticides, herbicides, fertilizers, oil spills, toxic gases, and industrial effluent are fair examples of the many contaminants. Synthesis of bionanomaterials with the range of 1–100 nm dimension mediated by biological material, has attracted interest due to their potential characteristic features such as higher surface area, quantum confinement effect and photocatalytic activity. There are practical implications of bionanomaterials in environmental remediation, especially photocatalysis to deplete the pollutants, detection of toxic heavy metals and removal. This chapter summarized the usage of bionanomaterials for environmental remediation.

12.1 Introduction

Accumulation of environmental pollutants is one of the global issues due to industrialization, rising world population, and the growing lifestyle towards contemporary, sophisticated life patterns that lead to non-eco-friendly activities [1]. Industrial evolution pollutes the air, water, and soil. Further, the discharge of effluent into the environment consisting of heavy metals, pesticides, herbicides, fertilizers, oil spills, toxic gases, industrial effluents, and organic compounds gradually causes them to build up in the environment and brings many adverse effects to flora and fauna [2–5]. Therefore, environmental remediation is an essential process in the sustainable world and there is a search for effective, prolonged methods and techniques to perform environmental remediation [1]. Environmental remediation refers to the removal of pollutants from water, soil, and sediments, and it brings their concentration below the levels determined based on human health and ecological risk [6, 7].

doi:10.1088/978-0-7503-3863-9ch12

Environmental remediation is performed using various methods such as physicochemical and biological methods [8, 9]. However, physicochemical methods lack eco-friendliness and are less cost-effective. In contrast, biological methods are slower and environmentally friendly [9]. Use of bionanomaterials provides a better solution and efficient removal of environmental pollutants [10]. Nanomaterials refers to material that has at least one of its dimensions in 1–100 nm [11]. Based on the unique nanoscale, bionanomaterials have characteristic properties applicable to broad disciplines, including chemistry, physics, biology, medicine, and engineering, etc. Those characteristic properties including quantum confinement effects, optical properties, high surface area-to-volume ratio, high surface charge and catalytic activity [12].

The synthesis of nanomaterials using biological material creates what are called bionanomaterials. The synthetic methods of bionanomaterials are labeled as green synthesis method and environmentally friendly process [9]. Due to several desired properties of the bionanomaterials together with excellent optical properties and high surface area-to-volume ratio, these bionanomaterials can be used in plausible applications including as catalyst, photocatalyst, sensor, antimicrobial activity, solar energy harvesting and energy storage devices. Further, these bionanomaterials are stable and reusable in several cycles. They can be used for pollution detection, monitoring, and organic and inorganic pollution depletion. This chapter mainly focuses on bionanomaterials for environmental remediation. The detailed implication includes the summary of bionanomaterial synthesis, bionanomaterials for heavy metal removal, removal of organic, inorganic pollutants, the antimicrobial action of bionanomaterials and bionanosensors for pollution detection.

12.2 Synthesis of bionanomaterials

Several methods are used to synthesize bionanomaterials using plant extracts (figure 12.1) and microorganisms that are environmentally friendly and cost-effective. Such methods include the usage of bacteria, viruses, algae, fungi, yeasts, waste materials, terpenoids, proteins, flavonoids, vitamins, glycosides, phenolic acid, carbohydrates, polymers, alkaloids, and various antioxidants [13].

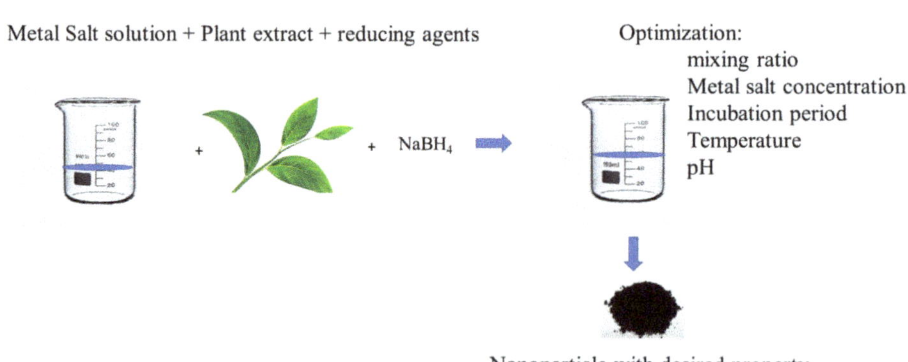

Figure 12.1. Schematic diagram of bionanomaterial synthesis.

After performing successful surface coating of respective ions by surfactant, with suitable reducing agents, metal ions reduce to a metal atom. The common reducing agents used for the synthesis of bionanometals are dimethylaminoborane, sodium citrate, ascorbate, sodium borohydride ($NaBH_4$), elemental hydrogen, Tollens reagent, N, N-dimethylformamide (DMF), and poly (ethylene glycol)-block copolymers. [14, 15]. The surfactant that exists in the surrounding nanometal prevents metal atom aggregation.

Synthesis of nanoparticles by using plant extracts is called green synthesis of bionanomaterials, in which the plant extract may act as a capping and/or reducing agent [16]. Here an extract of plant leaves, fruits, roots, and seeds can be used as a surfactant and reducing agent [17]. The plant materials consist of polyphenolic compounds. These polyphenolic compounds may act as capping agents and reducing agents. Silver nanoparticles are synthesized by using different plant extracts such as purple heart plant leaves extract [18], *Origanum vulgare* [19], plant extract of *Salvia Spinosa* [20], fresh leaves of *impatiens balsamina* and *Lantana camara* [21], etc

12.3 Environmental remediation by bionanomaterials

Bionanomaterials have many potential applications including photovoltaic devices, energy storage, water, and air treatment, monitoring and remediation, disease diagnosis, and treatment for various diseases. [22]. Due to the high surface area and higher reactivity of bionanomaterials, they have various uses in the bioremediation of environmental pollutants (table 12.1) [23]. There are numerous toxic pollutants discharged into the environment due to the non-eco-friendly activities of human beings, such as mainly discharge of industrial pollutants, higher usages of pesticide, herbicide, and fertilizers, etc [24]. Usage of bionanomaterials to eliminate such organic and inorganic pollutants became of interest due to their desired properties such as higher catalytic activity, large surface area, good charge transfer characteristic, ability to harvest a wide range of the solar spectrum, better stability, etc [25–28]. Currently, bionanomaterials are used as sensors to detect pollutants and as a catalyst in environmental remediations [29–31]. Bionanomaterials such as bionanoparticles, bionanoabsorbers and bionanomaterial embedded membrane are useful in environmental remediation, mainly to remove metal ions and degrade toxic pollutants from water sources.

Pollutants from an aqueous medium are removed by using bionanomaterial by various mechanisms such as adsorption, advanced oxidation process, chemical oxidation, flocculation, ultrafiltration, and sedimentation [32, 33]. Among them, the advanced oxidation process utilizes renewable energy for the degradation of pollutants [13].

12.3.1 Adsorption

The binding of pollutants on the surface of bionanomaterial is the crucial task of environmental clean-up by adsorption technique. The adsorption of pollutants on bionanomaterials restricts the free movement of pollutant and as a consequence it emits the thermal energy during the process. Bionanomaterials are highly preferred for use in environmental remediation as these synthesis processes are environmentally friendly

Table 12.1. Recent trends of bionanomaterial synthesis and their implications in environmental remediation.

Capping agent	Nanomaterial	Size	Application	References
Leaf extracts of *Ludwigia octovalvis*	Ag	28–50 nm	Alizarin red, congo red, methylene blue (MB), and Rhodamine b	[70]
Aqueous extract of *Bistorta amplexicaule*	Ag		4-nitrophenol	[71]
Morinda citrifolia (*M. citrifolia*) extract	Ag NPs	35 nm	MB dye	[72]
Pseudoalteromonas lipolytica	Peptide stabilized Au and Ag	Au: 10–30 nm and Ag: 5–15 nm	MB and congo red dyes	[73]
Acorus calamus seeds extract	GO-Fe_3O_4/Dop/Au	14.12–19.61 nm	MM and methyl orange (MO)	[74]
Neem leaf extract	Au core–Ag	5–35 nm	Antibacterial activity	[75]
Microalgal cultures	Ag	3–35 nm		[76]
Citrus paradisi peel	Ag	14.84 ± 5 nm	antimicrobial activity	[77]
Azadirachta indica	Ag	34 nm	antibacterial activities	[78]
Bacterium *Geobacter sulfurreducens*	nano-Pd coated Fe_3O_4	25 ± 12 nm	Cr(VI) removal	[79]
Leaf extract of *Aloe barbadensis*	zinc oxide	44 nm	Malachite green and congo red removal, powerful antimicrobial activity	[80]
Leaf extract of *Solanum Trilobatum*	Mn-Ag co-doped FeO ($Fe_{1-2x}Mn_xAg_x$)	44–51 nm	Degradation of MO	[81]
Umbrella plant	Ag NPs	below 20 nm	Degradation of methyl red, antimicrobial activity	[82]
Pod extract of *Dolichos lablab* L.	rGO/Fe_3O_4	8.86 nm	adsorptive removal of crystal violet and antifungal activities	[83]
Leaf extract of *Hibiscus rosasinensis*	Fe_2O_3 Nanoparticles	5 nm	Cytotoxic, antioxidant and bactericidal agent	[84]
Leaves of *Murraya koenigii*	Ag-CuO	60–80 nm	Antibacterial agents	[85]
Leaf extract of *Bryophyllum pinnatum*	Ag	15 nm	Photocatalytic and H_2O_2 sensing	[86]

and cost-effective. Numerous nanobioadsorbent materials have been synthesized for the enhancement of environmental remediation by adsorption. Such nanomaterials are mainly categorized as metallic nanoparticles, metal oxides bionanomaterials, metal–metal oxide bionanomaterials, and nanocomposites, etc.

Metal and metal oxide nanoparticles remove the organic and inorganic pollutants, including dye molecules, heavy metals, etc [22]. Iron oxide bionanomaterials are highly desirable for environmental remediation, applicable to large-scale wastewater treatments and perform magnetic separation at the end of the process. Iron oxide bionanomaterial can be synthesized by using different processes using biological materials including tea extract, seaweed extract, and neem leaf extract [34–36]. Iron oxide shows higher adsorption properties than other oxides. Magnetic iron oxide nanoparticles were prepared by using tea residue (MION-tea) at 450 °C and showed higher adsorption of As(III) and As(V) [35]. The adsorption capacity of arsenic (III) and arsenic (V) by MION-tea are 188.69 mg g^{-1} and 153.8 mg g^{-1}, respectively. Magnetic iron sulfide material produced by using sulfate-reducing bacteria can adsorb radioactive nucleus, mainly pertechnetate ion (TcO_4^-). With higher lifetime and high solubility, pertechnetate ions ultimately enter the food chain to pose a long-term adverse effect on the environment. The magnetic iron sulfide material provides a suitable matrix for long-term storage of radioactive materials [37]. Iron oxide bionanomaterials can be used to remove dye molecules due to their high surface area-to-volume ratio, fast kinetics, strong adsorption capacities, high reactivity, and easy magnetic separability [38]. Similarly, Fe_2O_3 and CuO synthesized by green methods by using *Aloe vera* and *Citrus limon* are used to remove As(V) and Cr(III) ions [39, 40].

Biosynthesis of ZnO nanomaterial is by using coffee husk extract(CHZnO) and sugar (SZnO) as a fuel, and the bionanomaterials synthesized by using such fuel show higher congo dye adsorption capacity such as 76 mg g^{-1} and 56 mg g^{-1} SZnO, respectively [41]. Further, zero-valent silver particles are synthesized by using *Phyllanthus emblica* fruit extract. This extract shows rich antioxidant activity and anti-inflammatory properties. Further, it consists of polyphenolic compounds and such polyphenolic compounds restrict the shape of bionanomaterials. This bionanomaterial shows a higher adsorption capacity of Hg^{2+} ions [42].

Bimetallic nanoparticles are synthesized by green methods by using microorganisms, bacteria, and plant extracts, etc. Plant extracts consist of functional groups that act as capping and reducing agents. The bimetallic Au–Ag nanoparticles are synthesized using pomegranate fruit extracts [43]. Similarly, Au–Pd [44], Ag–Se [45], etc bimetallic nanoparticles were synthesized by the green synthesis approach.

Bimetallic bionanoparticles are highly reactive species and show higher surface energy and magnetic properties. They are easily oxidized, have higher surface energy and easily attain aggregation. To prevent aggregation the stability of bimetallic bionanomaterials can be enhanced by modifying their surface with the biomolecules such as silica and polymers [22].

The composite materials consist of multifunctional bionanocomposites that may remove all heavy metal pollutants simultaneously and are capable of regeneration. Multifunctional composite material such as chitosan, silver nanoparticles (Ag NPs), copper nanoparticles (Cu NPs), and carbon nanotubes (CNTs) completely remove the metal ions in the solution [46].

12.3.1.1 Adsorption of dyes by bionanomaterials: mechanism

Dye molecules are classified based on the structures, functional groups, and ionic charge in an aqueous solution. Further, they can be divided into an ionic or non-ionic dye. Mostly, ionic dyes are categorized as a cationic or basic dye and anionic or acidic dye. Non-ionic dyes are categorized as vat dye and disperse dye. The adsorption mechanism of dye molecules depends on the functional groups of dye molecules and properties of bionanomaterials absorbent. The carbon nanomaterials act as adsorbent material, showing hydrophobic effects, preferably from π–π interaction, hydrogen bond, covalent bond, and electrostatic interactions [47]. A CNT preferentially adsorbs anionic dye rather than cationic dye [47].

12.3.1.2 Metal ion removal by zero-valent irons: mechanism

Removal of metal ion pollutants using zero-valent irons involves several proposed steps such as chemical reduction, complexation, sorption, precipitation, and co-precipitation [48]. The remediation process mainly depends on the standard reduction potentials of zero-valent iron and metal pollutants. If the metal pollutant shows more negative reduction potential than the zero-valent iron or the reduction potential near the zero-valent iron, this leads to the adsorption of the metal pollutant directly on the surface of the zero-valent iron. Conversely, if the reduction potential of the pollutant has more positive potential than the reduction potential of zero-valent iron, the metal pollutants get into a reduction on the surface and precipitation. However, if the reduction potential is slightly more positive than zero-valent iron, both reduction and adsorption may occur on the surface (figure 12.2) [48].

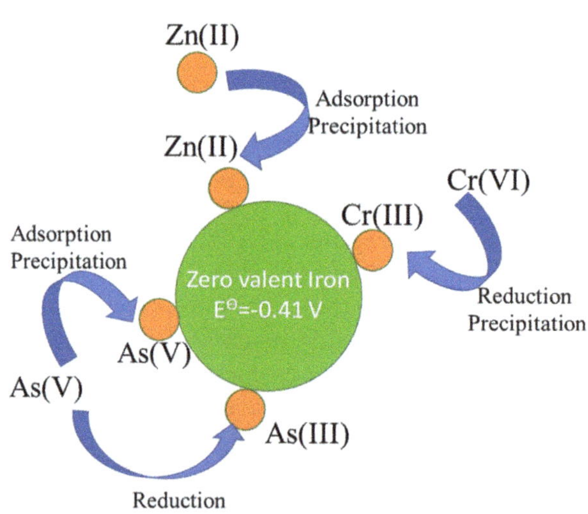

Figure 12.2. Schematic diagram of the adsorption process of metal pollutant by zero-valent iron.

12.3.2 Catalytic transformation

Due to non-environmentally friendly activities pollutants accumulate in the environment and this leads to a great challenge for the health of living beings. Transformation of organic and inorganic pollutants by oxidation and reduction reactions may convert the toxic substances into a nontoxic or less toxic form. Heavy metal pollutants such as Cd(II), Cr(II), Ni(II) and Pb(II) can be removed by using zero-valent irons. Cr(VI) can be reduced and goes into the precipitate. The heavy metals with more negative reduction potentials than zero-valent irons are removed from the solution by adsorption and precipitation. Nano-zero-valent iron reduces the many organic pollutants such as chlorinated organic pollutants, and converts them into non-toxic forms [49]. The removal of toxic substances either in organic or inorganic form by transportation and conversion into less or non-toxic forms is referred to as 'transformation' [22].

Nanosized zero-valent iron shows good adsorption and has a strong reducing ability. Under anaerobic conditions in an aqueous medium, zero-valent iron gets oxidized and converted as an iron(II) form (equation (12.1)).

$$Fe(0) + H_2O \rightarrow Fe(II) + H_2 + OH^- \qquad (12.1)$$

Under aerobic conditions, zero-valent iron gets oxidized and forms Iron(II) and H_2O_2 (equation (12.2)).

$$Fe(0) + O_2 + 2H^+ \rightarrow Fe(II) + H_2O \qquad (12.2)$$

In higher pH condition, iron (II) becomes insoluble $Fe(OH)_2$. Flocculation removes the pollutant effectively from the aqueous solution [50]. Zero-valent iron oxidizes a variety of organic pollutants in the aqueous medium. Combination of iron (II) and H_2O_2 undergoes the Fenton reaction and produces OH· radical species (equation (12.3)). Radical species are highly reactive and purified in a wide range of organic and inorganic pollutants.

$$Fe(II) + H_2O_2 \rightarrow Fe(III) + \cdot OH + OH^- \qquad (12.3)$$

Although, the nanosized zero-valent iron shows attractive properties, it has many disadvantages such as causing difficulties in recovery after treatment, aggregation and oxidation. Therefore, this zero-valent iron material undergoes various modifications such as surface coating of bionanomaterials, being prepared as a composite material, encapsulation in a matrix, and emulsification [50].

A composite material of nanosized zero-valent iron and cyanocobalamine (vitamin B_{12}) provides effective removal of a wide range of contaminants including nitrate and chlorinated organic compounds in water within minutes to hours [51]. Cyanocobalamine is an effective charge transfer mediator. The composite cyanocobalamine with zero-valent iron increases the reducing capability of nanosized zero-valent iron, prevents the agglomeration of iron nanoparticles and increases their active surface area [51].

Silver and gold nanoparticles are synthesized through biomaterial mediated process by using *Anacardium occidentale* testa [52], bark extract of *Terminalia cuneate* [53], *Pycnoporus sanguineus* [54], and so on for catalytic degradation of organic molecules with $NaBH_4$. The reaction between organic dye molecules and $NaBH_4$ is relatively low in the absence of nanomaterials. However, with the existence of bionanomaterials the degradation increases due to the high surface area on which the bionanomaterials adsorb the organic molecules, which facilitates the reaction between organic molecules and $NaBH_4$ [53].

12.3.3 Advanced oxidation process

Fenton reaction is an effective advanced treatment process and is effectively used to treat wastewater. In the presence of Fe(II) and H_2O_2, formation of radical species follows, and they degrade the organic and inorganic contaminants (equations (12.4)–(12.6)).

$$\mathrm{Fe(II)} + H_2O_2 \rightarrow \mathrm{Fe(III)} + \cdot OH + OH^- \quad (12.4)$$

$$\mathrm{Fe(III)} + H_2O_2 \rightarrow \mathrm{Fe(II)} + HOO\cdot + H^+ \quad (12.5)$$

$$H_2O_2 \rightarrow HO\cdot + \cdot OOH + H_2O \quad (12.6)$$

Biosynthesis of nanomaterials including TiO_2, ZnO and SnO_2 allows them to be used as a photocatalyst to degrade organic and inorganic pollutants in wastewater [55, 56]. In the photocatalytic degradation process, the bionanomaterials absorb the photon energy and the electron in the valence band is excited to an excited state and forms an electron–hole pair (figure 12.3 and equation (12.7)).

$$MO + h\upsilon \rightarrow MO(e)_{CB} + MO(h^+)_{VB} \quad (12.7)$$

$$MO(h^+)_{VB} + H_2O \rightarrow MO + H^+ + OH\cdot \quad (12.8)$$

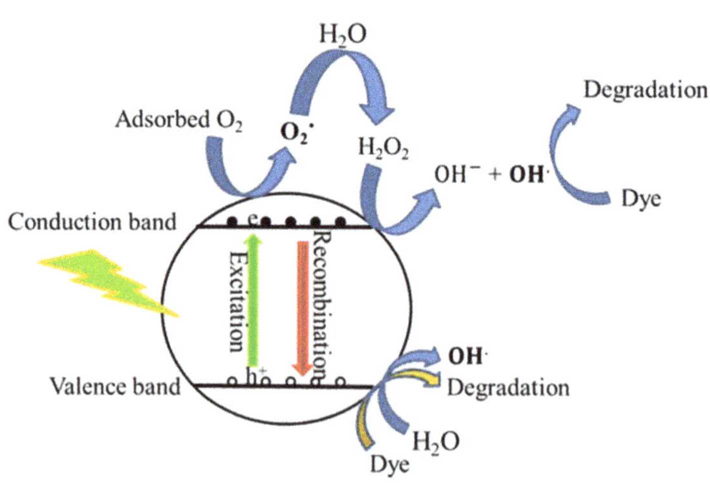

Figure 12.3. Schematic illustration of photocatalytic dye degradation mechanism of bionanomaterials.

$$MO(e)_{CB} + O_2 \rightarrow MO + O_2^{\cdot -} \quad (12.9)$$

$$O_2^{\bullet -} + H^+ \rightarrow HO_2^{\bullet} \quad (12.10)$$

$$HO_2 \cdot + HO_2 \cdot \rightarrow H_2O_2 + O_2 \quad (12.11)$$

$$MO(e)_{CB} + H_2O_2 \rightarrow OH^{\bullet} + OH^- \quad (12.12)$$

$$H_2O_2 + O_2^{\cdot -} \rightarrow \cdot OH^{\cdot} + OH^- + O_2 \quad (12.13)$$

$$H_2O_2 + h\upsilon \rightarrow 2OH^{\cdot} \quad (12.14)$$

$$\text{Organic pollutant} + OH^{\cdot} \rightarrow \text{intermediates} \rightarrow CO_2 + H_2O \quad (12.15)$$

The excited electron in the conduction band and hole in the valence band generates the radical species and facilitates the degradation of organic and inorganic molecules (equations (12.7)–(12.15)) [57]. Further, the photocatalyst's efficiency enhanced by the fabricating bionanomaterial with optimized bandgap, high recombination resistance, and defect-free catalyst [58]. Further, fabrication of bionanomaterial absorber as a composite and forming a heterojunction catalyst (figure 12.4) or z-scheme of catalyst (figure 12.5) may facilitate the effective charge collection and separation may lead to enhanced efficiency. As a result of effective charge separation, recombination attains a lower value.

There are many pieces of research performed on the removal of organic pollutants by using different mechanisms. The formation of heterojunction materials with different bandgap energy facilitates the broad absorbance of the solar spectrum and effective charge separation, enhancing the efficiency of the photocatalyst [59, 60].

Gold and silver nanoparticles were synthesized using the biosynthesis process and photocatalytic degradation studies were performed under illumination conditions [61, 62]. Under the illumination of a photon the electron in the valence band of atoms was excited to the excited state and thus the excited electron on the surface of

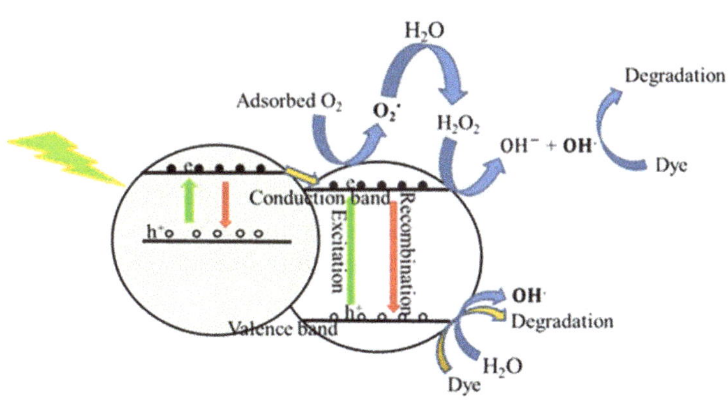

Figure 12.4. Schematic illustration of the election–hole transfer process in photocatalytic dye degradation of heterojunction nanobiomaterials with cascade band position alignment.

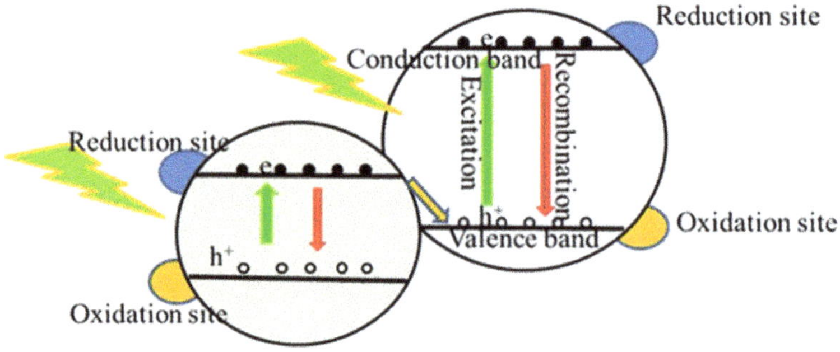

Figure 12.5. Schematic illustration of photocatalytic dye degradation mechanism of bionanomaterials with z-scheme.

the atom was given to the dye molecule and performed a subsequent dye degradation process [61].

12.3.4 Antimicrobial activities

Antimicrobial agents are critically important in plenty of sectors, such as the textile industry, water purification, medicine, and food packaging [63]. Green synthesis of nanomaterials using plant extracts shows antimicrobial properties and is reported in several studies. AgNPs show vigorous antimicrobial activity against various microorganisms such as bacteria, viruses, and fungi due to their smaller size and large surface area [64]. The particle size of AgNPs synthesized using tea leaves extract was 4.06 nm. AgNPs synthesized by the green method show antimicrobial activity against a diverse range of Gram-negative food borne pathogens. Likewise, green synthesized bionanomaterial shows a wide range of antimicrobial properties, and those are listed in table 12.2. Bionanomaterials show advantages over conventional antimicrobial agents that promote multidrug resistance. The exact mechanism behind the antimicrobial properties is unknown. However, researchers proposed that the AgNP shows antimicrobial properties due to the ability to penetrate the cell, form radicals, and inactivate the protein in the cell [64].

12.4 Nanobiosensors for environmental monitoring

Nanobiosensors are analytical devices consisting of recognition elements for binding their target, transducers, and detector. The active material produces the signal proportional to the concentration of an analyte. Nanobiosensors are a miniaturized, more sensitive, low-cost method, used to measure the parameters quickly and they facilitate on-time, on-the-spot environmental monitoring [22]. Nanobiosensors are used to measure microbial pathogens, effectively detecting environmental pollutants including heavy metals, release of nutrients from fertilizer, herbicide, pesticide, insecticide amounts, the presence of pathogens, pH and moisture content [65]. Synthesis of AgNP by using *Acacia nilotica* plant leave extract is used for detection of 4-nitrophenol. It shows a low detection limit of 15 nM and a wider linear response range from 100 nM to 350 μM. Similarly, AuNPs are synthesized from *Acacia*

Table 12.2. Bionanomaterial synthesis and its antimicrobial properties.

Capping agent	Nanomaterial	Antimicrobial properties	Reference
Leaf extract of *Cassia auriculata*	ZnO NPs	*Bacillus subtilis, Klebsiella pneumonia, Pseudomonas aeruginosa* and *Proteus mirabilis*	[87]
Leaf and root extracts of *Ricinus communis*	Ag NP	Bactericidal/fungicidal agents and urease/ xanthine oxidase enzyme inhibitors. The cytotoxicity concentration under 20 $\mu g\ ml^{-1}$	[88]
Bacillus subtilis	Ag NP	*Escherichia coli, Staphylococcus aureus, Vibrio Parahaemolyticus, and Acinetobacter baumanni*	[89]
Callus extract of the leaves	Ag NP	*Staphylococcus aureus, Bacillus subtilis, and Escherichia coli, Klebsiella pneumoniae, Pseudomonas aeruginosa, and Candida albicans.* The highest inhibition of phytopathogenic fungi *Alternaria alternata, Fusarium oxysporum, Aspergillus niger,* and *Pythium ultimum*	[90]
Grapefruit seed extract	Zinc oxide nanorods	Antimicrobial and antioxidant actions	[91]
Piper betel leaves extract	α-Fe_2O_3	Antibacterial activity towards both gram negative *(Escherichia coli and Pseudomonas aeroginosa)* and gram positive bacteria *(Staphylococcus aureus and Streptococcus mutans)*	[92]

nilotica twig bark extract and used to detect nitrobenzene. The nanobiosensor shows high sensitivity and a wide linear response range [66]. Further, Ag bionanoparticles decorated on reduced graphene oxide were used to detect nitrobenzene in a wastewater sample [67]. Heavy metals such as Cu^{2+} and Hg^{2+} were detected by using biosynthesized silver bionanomaterial [68]. Au bionanoparticles were synthesized from *Camellia sinensis* as a reducing agent and used in a nanobiosensor application to detect Ca^{2+}, Sr^{2+}, Cu^{2+}, and Zn^{2+} ions by colorimetric and plasmon resonance detection methods [69]. Likewise, several bionanomaterials are synthesized through a biologically mediated pathway, embedded into the biosensor and used for environmental pollutant monitoring.

12.5 Conclusion and prospects

Bionanomaterials have potential applications in environmental remediations, mainly detection and removal of heavy metals, degradation of organic and inorganic contaminants. Synthesis of bionanomaterials eliminates the need for toxic chemicals and high expense. In this scenario, bionanomaterials are synthesized by green synthesis eco-friendly methods, serving as an excellent alternative to chemical synthesis. However, the biosynthesis methods are still in the development stage. The properties of bionanomaterials need to be appropriately optimized for their applications. The size, shape, morphology, internal structures, and defects associated with bionanomaterials need to be optimized more. Bionanomaterial synthesis still suffers mainly based on the purification of the synthesized nanomaterials, efficiency for the desired application, stability, and their reusability. Further, more research development/studies are explicitly required, specifically bionanomaterial toxicity assessments, stability issues, pollutant removal efficiency and confirming the exact mechanism of the environmental remediation process.

Acknowledgment

The authors would like to extend their gratitude to Mr G Naveendrakumar for his technical support.

References

[1] Guerra F D, Attia M F, Whitehead D C and Alexis F 2018 Nanotechnology for environmental remediation: materials and applications *Molecules (Basel)* **23** 1760

[2] Pratush A, Kumar A and Hu Z 2018 Adverse effect of heavy metals (As, Pb, Hg, and Cr) on health and their bioremediation strategies: a review *Int. Microbiology.* **21** 97–106

[3] Ipingbemi O 2009 Socio-economic implications and environmental effects of oil spillage in some communities in the Niger delta *J. Integr. Environ. Sci.* **6** 7–23

[4] Meena R S, Kumar S, Datta R, Lal R, Vijayakumar V and Brtnicky M *et al* 2020 Impact of agrochemicals on soil microbiota and management: a review *Land Water Law Rev.* **9** 34

[5] Pereira L and Alves M 2012 Dyes—environmental impact and remediation *Environmental Protection Strategies for Sustainable Development* (Berlin: Springer) pp 111–62

[6] Ihedioha J N, Ukoha P O and Ekere N R 2017 Ecological and human health risk assessment of heavy metal contamination in soil of a municipal solid waste dump in Uyo, Nigeria *Environ. Geochem. Health* **39** 497–515

[7] Song B, Zeng G, Gong J, Liang J, Xu P and Liu Z *et al* Evaluation methods for assessing effectiveness of *in situ* remediation of soil and sediment contaminated with organic pollutants and heavy metals *Environ. Int.* **2017** 43–55 Available from https://sciencedirect.com/science/article/pii/S0160412017301435

[8] Bradl H and Xenidis A 2005 Remediation techniques *Interface Science and Technology* ed H B Bradl vol 6 (Amsterdam: Elsevier) 165–261

[9] Akhtar F Z, Archana K M, Krishnaswamy V G and Rajagopal R 2020 Remediation of heavy metals (Cr, Zn) using physical, chemical and biological methods: a novel approach *SN Applied Sciences.* **2** 267

[10] Dungani R, Abdul Khalil H P S, Aprilia N A S, Sumardi I, Aditiawati P and Darwis A *et al* 2017 Bionanomaterial from agricultural waste and its application *Cellulose-Reinforced Nanofibre Composites* ed M Jawaid, S Boufi and H P S AK (Cambridge: Woodhead Publishing) pp 45–88

[11] Boverhof D R, Bramante C M, Butala J H, Clancy S F, Lafranconi M and West J *et al* 2015 Comparative assessment of nanomaterial definitions and safety evaluation considerations *Regul. Toxicol. Pharm.* **73** 137–50
[12] Manjceevan A and Bandara J 2016 Robust surface passivation of trap sites in PbS q-dots by controlling the thickness of CdS layers in PbS/CdS quantum dot solar cells *Sol. Energy Mater. Sol. Cells* **147** 157–63
[13] Nasrollahzadeh M, Sajjadi M, Iravani S and Varma R S 2021 Green-synthesized nanocatalysts and nanomaterials for water treatment: current challenges and future perspectives *J. Hazard. Mater.* **401** 123401
[14] Iravani S, Korbekandi H, Mirmohammadi S V and Zolfaghari B 2014 Synthesis of silver nanoparticles: chemical, physical and biological methods *Res Pharm Sci.* **9** 385–406 Available from https://pubmed.ncbi.nlm.nih.gov/26339255https://ncbi.nlm.nih.gov/pmc/articles/PMC4326978/
[15] Rivero P J, Goicoechea J, Urrutia A and Arregui F J 2013 Effect of both protective and reducing agents in the synthesis of multicolor silver nanoparticles *Nanoscale Res. Lett.* **8** 101
[16] Mamat M H, Khusaimi Z, Suriani A B, Asiah M N, Soga T and Mahmood M R 2014 Plant extract as reducing agent in synthesis of metallic nanoparticles: a review *Adv. Mater. Res.* **832** 350–5
[17] Kuppusamy P, Yusoff M M, Maniam G P and Govindan N 2016 Biosynthesis of metallic nanoparticles using plant derivatives and their new avenues in pharmacological applications —an updated report *Saudi Pharm. J.* **24** 473–84 Available from https://sciencedirect.com/science/article/pii/S1319016414001492
[18] Hasnain M S, Javed M N, Alam M S, Rishishwar P, Rishishwar S and Ali S *et al* 2019 Purple heart plant leaves extract-mediated silver nanoparticle synthesis: optimization by Box-Behnken design *Mater. Sci. Eng.* C **99** 1105–14
[19] Shaik M R, Khan M, Kuniyil M, Al-Warthan A, Alkhathlan H Z and Siddiqui M R H *et al* 2018 Plant-extract-assisted green synthesis of silver nanoparticles using *Origanum vulgare L Extract and their microbicidal activities* **10** 913 Available from https://mdpi.com/2071-1050/10/4/913
[20] Pirtarighat S, Ghannadnia M and Baghshahi S 2019 Green synthesis of silver nanoparticles using the plant extract of Salvia spinosa grown *in vitro* and their antibacterial activity assessment *J. Nanost. Chem.* **9** 1–9
[21] Aritonang H F, Koleangan H and Wuntu A D 2019 Synthesis of silver nanoparticles using aqueous extract of medicinal plants' (impatiens balsamina and Lantana camara) fresh leaves and analysis of antimicrobial activity *Int. Journal of Microbiology.* **2019** 8642303
[22] Das S, Chakraborty J, Chatterjee S and Kumar H 2018 Prospects of biosynthesized nanomaterials for the remediation of organic and inorganic environmental contaminants *Environmental Science: Nano.* **5** 2784–808
[23] Rizwan M, Singh M, Mitra C K and Morve R K 2014 Ecofriendly application of nanomaterials: nanobioremediation *J. Nano.* **2014** 431787
[24] Liu L, Bilal M, Duan X and Iqbal H M J S T T E 2019 Mitigation of environmental pollution by genetically engineered bacteria—current challenges and future perspectives **667** 444–54
[25] Zhang X, Lin M, Lin X, Zhang C, Wei H and Zhang H *et al* 2014 Polypyrrole-enveloped Pd and Fe_3O_4 nanoparticle binary hollow and bowl-like superstructures as recyclable catalysts for industrial wastewater treatment *ACS Appl. Mater. Interfaces* **6** 450–8

[26] Wen M, Li G, Liu H, Chen J, An T and Yamashita H 2019 Metal–organic framework-based nanomaterials for adsorption and photocatalytic degradation of gaseous pollutants: recent progress and challenges *Environmental Science: Nano.* **6** 1006–25

[27] Alharbi N S, Hu B, Hayat T, Rabah S O, Alsaedi A and Zhuang L *et al* 2020 Efficient elimination of environmental pollutants through sorption-reduction and photocatalytic degradation using nanomaterials *Frontiers of Chemical Science and Engineering.* **14** 1124–35

[28] Manjceevan A, sulaimalebbe N and Somapala T 2021 Visible-light-harvesting hedgehog like copper bismuth oxide: optical, structural and electrochemical properties *J. Mol. Struct.* **1235** 130205

[29] Ramnani P, Saucedo N M and Mulchandani A 2016 Carbon nanomaterial-based electrochemical biosensors for label-free sensing of environmental pollutants *Chemosphere.* **143** 85–98

[30] Walekar L, Dutta T, Kumar P, Ok Y S, Pawar S and Deep A *et al* 2017 Functionalized fluorescent nanomaterials for sensing pollutants in the environment: a critical review *TrAC, Trends Anal. Chem.* **97** 458–67

[31] Khalid N R, Majid A, Tahir M B, Niaz N A and Khalid S 2017 Carbonaceous-TiO_2 nanomaterials for photocatalytic degradation of pollutants: a review *Ceram. Int.* **43** 14552–71

[32] Huang D, Wang X, Zhang C, Zeng G, Peng Z and Zhou J *et al* 2017 Sorptive removal of ionizable antibiotic sulfamethazine from aqueous solution by graphene oxide-coated biochar nanocomposites: influencing factors and mechanism *Chemosphere.* **186** 414–21

[33] Liu X, Ma R, Wang X, Ma Y, Yang Y and Zhuang L *et al* 2019 Graphene oxide-based materials for efficient removal of heavy metal ions from aqueous solution: a review *Environ. Pollut.* **252** 62–73

[34] Mahdavi M, Namvar F, Ahmad M B and Mohamad R 2013 Green biosynthesis and characterization of magnetic iron oxide (FeO) nanoparticles using seaweed (*Sargassum muticum*) aqueous extract *Molecules (Basel)* **18** 5954–64

[35] Lunge S, Singh S and Sinha A 2014 Magnetic iron oxide (Fe_3O_4) nanoparticles from tea waste for arsenic removal *J. Magn. Magn. Mater.* **356** 21–31

[36] Zambri N D S, Taib N I, Abdul Latif F and Mohamed Z 2019 *Utilization of neem leaf extract on biosynthesis of iron oxide nanoparticles* **24** 3803 Available from: https://mdpi.com/1420-3049/24/20/3803

[37] Watson J H P, Croudace I W, Warwick P E, James P A B, Charnock J M and Ellwood D C 2001 Adsorption of radioactive metals by strongly magnetic iron sulfide nanoparticles produced by sulfate-reducing bacteria *Sep. Sci. Technol.* **36** 2571–607

[38] Iravani S 2019 Bio-based synthesis of magnetic nanoparticles and their applications *Magnetic Nanostructures* (Berlin: Springer) pp 13–31

[39] Mukherjee D, Ghosh S, Majumdar S and Annapurna K 2016 Green synthesis of α-Fe_2O_3 nanoparticles for arsenic(V) remediation with a novel aspect for sludge management *J. Environ. Chem. Eng.* **4** 639–50

[40] Mohan S, Singh Y, Verma D K and Hasan S H 2015 Synthesis of CuO nanoparticles through green route using Citrus limon juice and its application as nanosorbent for Cr(VI) remediation: Process optimization with RSM and ANN-GA based model *Process Saf. Environ. Prot.* 2015 156–66

[41] Murthy T P K, Gowrishankar B S, Krishna R H, Chandraprabha M N and Rao R S 2019 Influence of fuel nature on dye adsorption efficiency of solution combustion derived zinc oxide nanoparticles: a comparative study *Mater. Res. Express* **6** 055512

[42] Sundarajan S, Sameem S M, Sankaranarayanan S and Ramaraj S J O 2013 Synthesis, characterization and application of zero-valent silver nano adsorbents.**2** Available from: https://citeseerx.ist.psu.edu/viewdoc/download?doi=10.1.1.1071.3186&rep=rep1&type=pdf
[43] Elemike E E, Onwudiwe D C, Fayemi O E and Botha T L 2019 Green synthesis and electrochemistry of Ag, Au, and Ag–Au bimetallic nanoparticles using golden rod (Solidago canadensis) leaf extract *Appl. Phys.* A **125** 42
[44] Zhan G, Huang J, Du M, Abdul-Rauf I, Ma Y and Li Q 2011 Green synthesis of Au–Pd bimetallic nanoparticles: Single-step bioreduction method with plant extract *Mater. Lett.* **65** 2989–91 Available from: https://sciencedirect.com/science/article/pii/S0167577X11007221
[45] Ibrahim A T A 2020 Toxicological impact of green synthesized silver nanoparticles and protective role of different selenium type on *Oreochromis niloticus*: hematological and biochemical response *J. Trace Elem. Med. Biol.* **61** 126507
[46] Alsabagh A M, Fathy M and Morsi R E 2015 Preparation and characterization of chitosan/silver nanoparticle/copper nanoparticle/carbon nanotube multifunctional nano-composite for water treatment: heavy metals removal; kinetics, isotherms and competitive studies *RSC Adv.* **5** 55774–83
[47] Tan K B, Vakili M, Horri B A, Poh P E, Abdullah A Z and Salamatinia B 2015 Adsorption of dyes by nanomaterials: Recent developments and adsorption mechanisms *Sep. Purif. Technol.* **150** 229–42
[48] Ullah S, Faiz P and Leng S 2020 *Synthesis, mechanism, and performance assessment of zero-valent iron for metal-contaminated water remediation: a review* **48** 2000080
[49] Dror I, Baram D and Berkowitz B 2005 Use of nanosized catalysts for transformation of chloro-organic pollutants *Environ. Sci. Technol.* **39** 1283–90
[50] Lu H, Wang J, Stoller M, Wang T, Bao Y and Hao H 2016 An overview of nanomaterials for water and wastewater treatment *Adv. Mater. Sci. Eng.* **2016** 4964828
[51] Dror I, Jacov O M, Cortis A and Berkowitz B 2012 Catalytic transformation of persistent contaminants using a new composite material based on nanosized zero-valent iron *ACS Appl. Mater. Interfaces* **4** 3416–23
[52] Edison T N J I, Atchudan R, Sethuraman M G and Lee Y R 2016 Reductive-degradation of carcinogenic azo dyes using Anacardium occidentale testa derived silver nanoparticles *J Photochem Photobiol B: Biol.* **162** 604–10
[53] Vidhu V K and Philip D 2014 Catalytic degradation of organic dyes using biosynthesized silver nanoparticles *Micron* **56** 54–62
[54] Shi C, Zhu N, Cao Y and Wu P 2015 Biosynthesis of gold nanoparticles assisted by the intracellular protein extract of Pycnoporus sanguineus and its catalysis in degradation of 4-nitroaniline *Nanoscale Res. Lett.* **10** 147
[55] Ngoepe N M, Mathipa M M and Hintsho-Mbita N C 2020 Biosynthesis of titanium dioxide nanoparticles for the photodegradation of dyes and removal of bacteria *Optik* **224** 165728
[56] Fagier M A 2021 Plant-Mediated biosynthesis and photocatalysis activities of zinc oxide nanoparticles: a prospect towards dyes mineralization *J. Nano.* **2021** 6629180
[57] Leong K H, Sim L C, Pichiah S and Ibrahim S 2016 Light driven nanomaterials for removal of agricultural toxins *Nanoscience in Food and Agriculture 3* (Berlin: Springer) pp 225–42
[58] Pan H 2016 Principles on design and fabrication of nanomaterials as photocatalysts for water-splitting *Renew. Sustain. Energy Rev.* **57** 584–601

[59] Afroz K, Moniruddin M, Bakranov N, Kudaibergenov S and Nuraje N J J M C A 2018 A heterojunction strategy to improve the visible light sensitive water splitting performance of photocatalytic materials**6** 21696–718

[60] Sahoo D P, Nayak S, Reddy K H, Martha S and Parida K 2018 Fabrication of a $Co(OH)_2$/ZnCr LDH 'p–n' heterojunction photocatalyst with enhanced separation of charge carriers for efficient visible-light-driven H_2 and O_2 evolution *Inorg. Chem.* **57** 3840–54

[61] Baruah D, Goswami M, Yadav R N S, Yadav A and Das A M 2018 Biogenic synthesis of gold nanoparticles and their application in photocatalytic degradation of toxic dyes *J Photochem Photobiol B: Biol.* **186** 51–8

[62] Tahir K, Nazir S, Li B, Khan A U, Khan Z U H and Ahmad A *et al* 2015 An efficient photo catalytic activity of green synthesized silver nanoparticles using *Salvadora persica* stem extract *Sep. Purif. Technol.* **150** 316–24

[63] Hajipour M J, Fromm K M, Akbar Ashkarran A, Jimenez de Aberasturi D, Larramendi I Rd and Rojo T *et al* 2012 Antibacterial properties of nanoparticles *Trends in Biotechnology.* **30** 499–511

[64] Loo Y Y, Rukayadi Y, Nor-Khaizura M-A-R, Kuan C H, Chieng B W and Nishibuchi M *et al* 2018 In *vitro* antimicrobial activity of green synthesized silver nanoparticles against selected gram-negative foodborne pathogens *Front. Microbiol.* **16** 1555

[65] Salouti M and Derakhshan F K 2020 Biosensors and nanobiosensors in environmental applications *Biogenic Nano-Particles and their Use in Agro-ecosystems* (Berlin: Springer) pp 515–91

[66] Emmanuel R, Karuppiah C, Chen S-M, Palanisamy S, Padmavathy S and Prakash P 2014 Green synthesis of gold nanoparticles for trace level detection of a hazardous pollutant (nitrobenzene) causing Methemoglobinaemia *J. Hazard. Mater.* **279** 117–24

[67] Karuppiah C, Muthupandi K, Chen S-M, Ali M A, Palanisamy S and Rajan A *et al* 2015 Green synthesized silver nanoparticles decorated on reduced graphene oxide for enhanced electrochemical sensing of nitrobenzene in waste water samples *RSC Adv.* **5** 31139–46

[68] Maiti S, Barman G and Konar Laha J 2016 Detection of heavy metals (Cu^{+2}, Hg^{+2}) by biosynthesized silver nanoparticles *Appl. Nanosci.* **6** 529–38

[69] Silva-De Hoyos L E, Sánchez-Mendieta V, Vilchis-Nestor A R, Camacho-López M A, Trujillo-Reyes J and Avalos-Borja M 2019 Plasmonic sensing of aqueous-divalent metal ions by biogenic gold nanoparticles *J. Nanomater.* **2019** 9846729

[70] Sarathi Kannan D, Mahboob S, Al-Ghanim K A and Venkatachalam P 2021 Antibacterial, antibiofilm and photocatalytic activities of biogenic silver nanoparticles from *Ludwigia octovalvis J. Cluster Sci.* **32** 255–64

[71] Shah L, Ali H S H M, Bello B A, Khan S A, Anwar Y and Khan M S J *et al* 2020 Catalytic performance of the biosynthesized AgNps from *Bistorta amplexicaule*: antifungal, bactericidal, and reduction of carcinogenic 4-nitrophenol % *J Green Processing and Synthesis* **9** 259–67

[72] Rajivgandhi G N, Maruthupandy M, Li J-L, Dong L, Alharbi N S and Kadaikunnan S *et al* 2020 Photocatalytic reduction and anti-bacterial activity of biosynthesized silver nanoparticles against multi drug resistant *Staphylococcus saprophyticus* BDUMS 5 (MN310601) *Mater. Sci. Eng.* C **114** 111024

[73] Kulkarni R, Harip S, Kumar A R, Deobagkar D and Zinjarde S 2018 Peptide stabilized gold and silver nanoparticles derived from the mangrove isolate *Pseudoalteromonas lipolytica* mediate dye decolorization *Colloids and Surfaces A: Physicochemical and Engineering Aspects.* **555** 180–90 Available from: https://sciencedirect.com/science/article/pii/S0927775718306009

[74] Esmaeili N, Mohammadi P, Abbaszadeh M and Sheibani H 2019 Au nanoparticles decorated on magnetic nanocomposite (GO-Fe_3O_4/Dop/Au) as a recoverable catalyst for degradation of methylene blue and methyl orange in water *Int. Journal of Hydrogen Energy* **44** 23002–9

[75] Shankar S S, Rai A, Ahmad A and Sastry M 2004 Rapid synthesis of Au, Ag, and bimetallic Au core–Ag shell nanoparticles using Neem (*Azadirachta indica*) leaf broth *J. Colloid Interface Sci.* **275** 496–502

[76] Jena J, Pradhan N, Nayak R R, Dash B P, Sukla L B and Panda P K *et al* 2014 Microalga *Scenedesmus* sp.: a potential low-cost green machine for silver nanoparticle synthesis *J. Microbiol. Biotechnol.* **24** 522–33

[77] Ayinde W B, Gitari W M and Samie A 2019 Optimization of microwave-assisted synthesis of silver nanoparticle by Citrus paradisi peel and its application against pathogenic water strain *Green Chem. Lett. Rev.* **12** 225–34

[78] Ahmed S, Saifullah, Ahmad M, Swami B L and Ikram S 2016 Green synthesis of silver nanoparticles using Azadirachta indica aqueous leaf extract *J. Radiation Res. App. Sci.* **9** 1–7

[79] Joshi N, Liu F, Watts M P, Williams H, Coker V S and Schmid D *et al* 2018 Optimising the transport properties and reactivity of microbially-synthesised magnetite for *in situ* remediation *Sci. Rep.* **8** 4246

[80] Batool M, Khurshid S, Qureshi Z and Daoush W M 2021 Adsorption, antimicrobial and wound healing activities of biosynthesised zinc oxide nanoparticles *Chemical Papers.* **75** 893–907

[81] Vandhana T and Clement Lourduraj A J 2020 Biogenic synthesis of Mn-Ag co-doped FeO ($Fe_{1-2x}Mn_xAg_x$) nanoparticles: As an effective disinfectant and anticancer agent *Inorg. Chem. Commun.* **112** 107712

[82] Elemike E E, Onwudiwe D C and Ekennia A C 2020 Eco-friendly synthesis of silver nanoparticles using Umbrella plant, and evaluation of their photocatalytic and antibacterial activities *Inorganic and Nano-Metal Chemistry.* **50** 389–99

[83] Kahsay M H, Belachew N, Tadesse A and Basavaiah K 2020 Magnetite nanoparticle decorated reduced graphene oxide for adsorptive removal of crystal violet and antifungal activities *RSC Adv.* **10** 34916–27

[84] Deepthi Reddy G, Noorjahan M, Ratnamala A, Haseena M, Manjunatha H and Chandra Babu Naidu K 2020 Meteoric synthesis of luminescent Fe_2O_3 nanoparticles: a potential cytotoxic *Antioxidant & Bactericidal Agent.* 17 2768–78

[85] Baker S, Olga P, Tatiana R, Nadezhda P, Tatyana G and Tatyana R *et al* 2020 Phyto-nano-hybrids of Ag-CuO particles for antibacterial activity against drug-resistant pathogens *J. Genetic Eng. and Bio.* **18** 53

[86] Chandraker S K, Lal M, Dhruve P, Singh R P and Shukla R 2021 Cytotoxic, antimitotic, DNA binding, photocatalytic, H_2O_2 sensing, and antioxidant properties of biofabricated silver nanoparticles using leaf extract of *Bryophyllum pinnatum* (Lam.) oken *Front Mol Biosci.* **7** 593040

[87] Ramesh P, Saravanan K, Manogar P, Johnson J, Vinoth E and Mayakannan M 2021 Green synthesis and characterization of biocompatible zinc oxide nanoparticles and evaluation of its antibacterial potential *Sensing and Bio-Sensing Research.* **31** 100399

[88] Gul A, Fozia, Shaheen A, Ahmad I, Khattak B and Ahmad M *et al* 2021 Green synthesis, characterization, enzyme inhibition, antimicrobial potential, and cytotoxic activity of plant mediated silver nanoparticle using ricinus communis leaf and root extracts *Biomolecules* **11** 206

[89] Yu X, Li J, Mu D, Zhang H, Liu Q and Chen G 2021 Green synthesis and characterizations of silver nanoparticles with enhanced antibacterial properties by secondary metabolites of *Bacillus subtilis* (SDUM301120) *Green Chem. Lett. Rev.* **14** 189–202

[90] Lashin I, Fouda A, Gobouri A A, Azab E, Mohammedsaleh Z M and Makharita R R 2021 Antimicrobial and *in vitro* cytotoxic efficacy of biogenic silver nanoparticles (Ag-NPs) fabricated by callus extract of *Solanum incanum* L *Biomolecules* **11** 341

[91] Roy S, Kim H C, Panicker P S, Rhim J-W and Kim J 2021 Cellulose nanofiber-based nanocomposite films reinforced with zinc oxide nanorods and grapefruit seed extract *Nanomaterials* **11** 877

[92] Yoonus J, Resmi R and Beena B 2021 Evaluation of antibacterial and anticancer activity of green synthesized iron oxide (α-Fe_2O_3) nanoparticles *Mater. Today Proc.* **46** 2969–74

IOP Publishing

Bionanomaterials for Environmental and Agricultural Applications

Ravindra Pratap Singh and Kshitij RB Singh

Chapter 13

Nanobiosensors for environmental risk assessment

Stephen Rathinaraj Benjamin, Eli José Miranda Ribeiro Junior, Fábio de Lima and Gilberto José de Arruda

The emergence of nanotechnology has been the subject of extensive research in recent years, and the potential of nanotechnology to reduce pollution continues to grow, promoting more revolutionary environmental changes. Nanobiosensors have gained attention over the past few decades to monitor diagnosis of pollutants in the environment and agriculture. Using nanomaterials in biosensor design enhances approach detection and reproducibility, as well as promoting quick, effective and, environmentally friendly processing techniques. This review discusses the design and implementation of the modern nanobiosensors in the fields of environmental risk monitoring and agriculture.

13.1 Introduction

Agricultural access to overall reserves of land and water is declining and major losses are being caused in agricultural production. Furthermore, increasing agriculture levels are impacted by herbicides, pesticides, and heavy metals. These problems can only be resolved successfully with the use of emerging platforms and by the increased inflow of new technology into this sector. Recent developments in nanotechnology have prompted the development of sensors of nanoscale, which are highly sensitive and versatile in agriculture and the food system. There is the potential to increase farmland by promoting organic agriculture, reducing agricultural prices, and increasing the value of commodities. The investigation of nanotechnology applications for agriculture has also received considerable attention over recent years [1, 2]. Conventional approaches for detecting pollutants are dependent on laborious and time-intensive chromatographic and immunoassay technologies. Nanobiosensors are becoming widespread for more rapid sensitive bioelectronic identification of

doi:10.1088/978-0-7503-3863-9ch13

different pollutants, such as detection of bacteria, pesticides and drug residues, heavy metals, and other toxic materials, as an important alternative to conventional analytical techniques. This chapter emphasizes various applications of nanobiosensors for agriculture, for the treatment of the soil such as nanofertilizers, nanopesticides, nanobiosensors, and the prospects in agriculture safety and research guidelines for applications of bionanotechnology.

13.2 Nanobiosensors

Nanobiosensors incorporate nano-dimensional materials functioning as transducers biorecognition elements, which signal a recognition element to detect a single or multiplex analysis. Nanobiosensors are the next generation of biosensors that are more portable and are integrated by a physicochemical transducer to detect selective analysis at ultra-low levels. Early identification and quick response improvement in crop yields can be aided by nanobiosensor technology through appropriate management of water, soil, pesticides, and fertilizers [3]. Immobilization, fictionalization, and miniaturization are the remarkable features of the nanobiosensor which integrates transduction device biocomponents into the comprehensive concept to enhance nanomaterial (NM) analytical results [4, 5]. The sensitivity and efficiency of nanobiosensors can be improved by miniaturizing such devices or using nanomaterials as sensor layers. The nanobiosensors operate to the nanoscale size and can monitor analytes accurately and rapidly. These nanobiosensors should ideally have the following characteristics: (i) excellent stability, (ii) unique function in the designed use of environmental purposes for particular analytes of interest, (iii) quick dynamics, (iv) accuracy, and reproductivity over the useful analytical scope.

13.2.1 Components of nanobiosensors

Nanobiosensors comprise a nano-sized material, which acts on a transducer as a bioreceptor, that delivers a signal to an element to recognize individual or multiplex analysis. Regarding the development and fundamental method of operation, those elements are bioreceptors, transducers, and detectors, respectively (figure 13.1). The bioreceptor (antibodies, proteins, enzymes, immunological molecules, etc) is the biosensor component that acts as a prototype to detect the material. The second part is the transducer device and its major purpose is to transform the bioanalyte

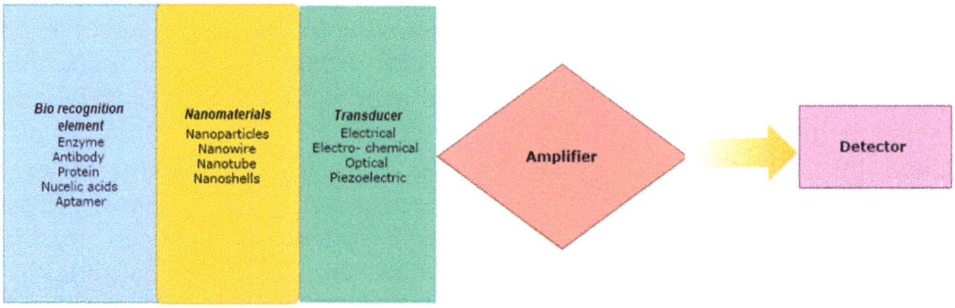

Figure 13.1. Operating principle of nanobiosensors.

interaction and the underlying bioreceptor into an electric form. The third part is the detector device that receives the electrical signal of the transducer component and amplifies it accordingly to allow correct reading and analysis of the corresponding response. Several detection methods can be used by nanobiosensors. The physical variations may be piezoelectric (measuring mass differences), electrochemical (measuring electrical distribution differences), optical (measuring light intensity differences), or calorimetric (measuring temperature differences).

13.3 Nanobiosensors in environmental monitoring

The key benefits of nanobiosensors over traditional analytical methods for environmental applications are easy portability, miniaturization, and practicability. Nanobiosensors can be used as environmental risk management techniques for biological/ecological quality assessment or chemical monitoring of both inorganic and organic priority contaminants. The important features of nanobiosensors are continuous monitoring, on-site work, and the ability to quantify contaminants in complex matrices with a limited sample. Nanobiosensors can detect a wide range of herbicides, fertilizers, insecticides, pesticides, moisture, pathogens, soil, and pH levels, and their controlled application can promote sustainable development by increasing crop productivity [3, 6] (figure 13.2). Besides that, crop production is threatened on a daily basis by pathogens and weeds, and both have an effect on the comparative farm economy; hence, plants must be controlled by appropriate action.

13.3.1 Detection of pesticides

Pesticides are one of the most important environmental pollutants owing to their widespread use in the ecosystem. Although the ingestion of pesticides is extremely

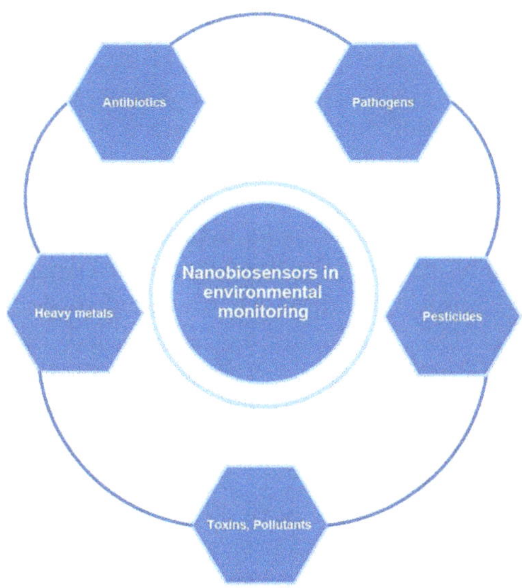

Figure 13.2. Environmental applications of nanonbiosensors.

hazardous to humans, food quality regulations promise advanced and sensitive pesticide detection methods. Pesticides are classified as organophosphates (OPs), organochlorines (OCs), synthetic pyrethroids, carbamates, and inorganic pesticides according to their particular chemical constituents. These pesticides are used in a variety of ways, such as pesticides and herbicides, pesticides, rodenticides, molluscicides, phytohormones, and nematicides [6, 7]. Several gold nanoparticle-based sensors for the detection of different pesticides have been reported [8, 9]. To monitor organophosphorus insecticides using paraoxon as a model analyte, a responsive acetylcholinesterase (AChE) amperometric biosensor was proposed that utilizes the electrocatalytic properties of gold nanorods (AuNRs) that can electrocatalytically oxidize thiocholine. The fabricated sensor demonstrated a linear range of 1 nM -5 μM and LOD (limit of detection) 0.7 nM for paraoxon and 5 nM-1 μM, LOD 3.9 nM for dimethoate. Furthermore, immobilizing enzymes in pesticide biosensor development improves sensitivity over configurations of free enzymes [10]. Due to their versatility in pesticide detection, multienzymes are often used as biosensors. Zhang *et al* [11] performed the first study by using a multienzyme/CNT (carbon nanotube) biosensor like MWCNT-OPH (Multi-walled carbon nanotube-organophosphate hydrolase) and MWCNT-AChE (Multiwalled carbon nanotube-acetylcholinesterase) composed of bilayers, such as MWCNT polyethyleneimine and MWCNT-DNA for detection of OP (paraoxon) and non-OP pesticides (carbaryl). The LOD was determined to be 0.5 mM for paraoxon and 1 mM for carbaryl. However, nanobiosensors with the ability to detect pesticides are being developed for use in agriculture. Liposome nanobiosensors were designed to monitor organophosphorus pesticides such as dichlorvos and paraoxon at extremely low concentrations using the acetyl-cholinesterase enzyme and the pH-sensitive pyranine, which effectively regulates the internal liposome environment. Vamvakaki *et al* [12] proposed a method to detect OP in very low concentrations, for example, dichlorvos and paraoxon, with a very stable liposome-based nanobiosensor. The liposome –biosensor response was regulated using a pH-sensitive fluorescence pyranine indicator, conjugated for optical transduction of enzyme activity within the liposomes. The increase in pesticide volumes produced a decrease in enzyme activity and thus a decrease of the pH signal fluorescence with a reported LOD value of 0.1 nM.

The acetylcholinesterase biosensor was established based on the assembly of MWCNT into liposome bioreactors for the detection of organophosphate pesticides A quantum dot Zn–SE immobilized AChE was used to detect OP pesticides with graphene–chitosan (Gr–CS) nanocomposite modified electrodes. This sensor detected OP pesticide methyl parathion (MPH) as a model enzyme inhibitor with a detection limit of 0.2 nmol dm^{-3} [13] The pesticides MPH and chlorpyrifos in spiked river water samples were determined using an electrochemical biosensor focused on AChE immobilized in vertically mounted CNTs and wrapped in ssDNA (Single-stranded DNA viruses). For both pesticides, the detection limit for the biosensor was found to be 1 to 10^{-12} M [14]. Chen *et al* [15] designed a specific and efficient nano-hybrid biosensor for the detection of MPH from garlic samples.

In this technique, the nanocomposite sensor film was generated by forming AuNPs into silica particles, mixing with MWCNTs, and covalent immobilization with MPH. The square wave voltammetric (SWV) responses showed well-defined peaks that are linearly proportional to MPH concentrations between 0.001 mmL^{-1} and 5.0 ml^{-1}, LOD value 0.3 ng mL^{-1}. The proposed protocol can be used as a tool for establishing high-efficiency biosensors that can quickly and efficiently identify complex enzyme electroactive organisms. A new graphene electrochemical transistor (GECT) was recently developed for the detection of MPH in cabbage samples. The developed sensor showed a sensor limit at a low of 0.1 ng ml^{-1} and ultra-large linear range from 0.4 ng ml^{-1} to 7 μg ml^{-1} due to its high carrier mobility and the excellent catalytic function of gold–zirconia (Au–ZrO_2) nanocomposites in the channel [16]. Furthermore, a novel molecularly imprinted electrochemical sensor focused on core–shell Co_3O_4@MOF-74 (cobalt oxide metal–organic framework) nanostructure was used to detect fenamiphos (FEN) in environmental samples, with quantification limit and LOD of 1.0×10^{-11} M and 3.0×10^{-12} M, respectively [17]. Dichlorvos detection was performed in environmental samples using multiple biosensors developed by the bienzyme method system consisting of choline oxidase and acetylcholinesterase enzymes [18, 19] using a platinum electrode and a composition of ionic liquid with modified AuNPs and porous carbon oxide composites. The combination of ionic liquid, porous carbon, and AuNPs enhances enzyme adsorption and action while also increasing analytical sensitivity [20]. In addition, choline oxidase inhibition enzyme based on CNTs with brilliant cresyl blue [21], AChE biosensor based on CS–TiO_2–Gr (chitosan–titanium dioxide–graphene) nanocomposites [22], graphene-nanofragments modified with chitosan nanoparticles [23], and MWCNT/Au nanocomposite [24], can be used to detect dichlorvos in environmental and food samples.

Acetamiprid is a new class of systemic broad-spectrum insecticide neonicotinoids. It is widely used to manage sucking insects, as a substitute insecticide for organophosphorus and other conventional insecticides. Nanobiosensors for the detection of neonicotinoids have been based specifically on acetamiprid residue detection with aptamers [25]. He *et al* [26] specified the required aptamer target of pesticide acetamiprid molecules in this work, based on SELEX (systematic evolution of ligands by exponential enrichment strategy). For example, Yingying *et al* [25] established a simple and rapid colorimetric test for direct acetamiprid identification in water samples with precise and efficient response to aptamer structure changes induced by acetamiprid to positive AuNPs ((+) AuNPs) with an excellent detection limit 5.6×10^{-10} M. Furthermore, Weerathunge *et al* [27] and Tian *et al* [28] designed a colorimetry method for the fast detection of acetamiprid pesticide residue in food and the environment using AuNPs.

13.3.2 Detection of pathogens

Water and food product contamination has become a worldwide epidemic. Furthermore, early identification at the lowest potential concentration of infectious microbial pathogenic agents such as microbes, fungi, or viruses is necessary to prevent outbreaks of infecting diseases, epidemics, and even pandemic conditions. Many unique nanomaterials have been proposed and demonstrated successfully by the scientific community in the field of pathogen detection, including quantum dots, CNTs, nanoribbons, nanorods, nanowires, nanoarrays, nanofibers, and NPs. Additionally, nanotechnology progress and enhanced knowledge of nanosized biomolecular structures have been accomplished due to the high sensitivity and extremely low detection limit in biosensors. In particular, electrochemical detection facilitated by nanomaterial of DNA hybrids is extremely important in biomedical fields. Many researchers documented electrochemical biosensors successfully detecting food borne pathogens. DNA-nanosensors detect various pathogenic microorganisms such as *Bacillus subtilis*, *Vibrio cholera*, *Aspergillus sp*, *Escherichia coli*, methicillin-resistant *Staphylococcus aureus* (MRSA), and *Candida sp*. AuNP-based DNA nanosensors are one of the most popular DNA nanosensors used for microorganism detection. In other research, Ariffin *et al* [29] have developed a nanowire-based sensor to identify plant virus-like cucumber mosaic viruses, papaya ringspot viruses that could be treated fatally by viruses, and the early identification of an infectious plant. In a sample, silicone-derived nanowires were used with bioreceptors to capture plant diseases like *cucumber mosaic virus* (CMV) and *papaya ringspot virus* (PRSV). Bioreceptor interactions leads to an electrical signal that can be monitored to calculate viral load while exposed to the virus. Likewise, the diagnosis of plant disease and plant health can be detected by the sensing of viruses themselves or the metabolites produced during the disease. Also in this sector, Akanbi *et al* [30] designed CNT-derived nanobiosensors to a screen-printed electrode with MWNT, which decorates gold and chitosan nanoparticles. This nano biosensor can detect quinolone, which produces an infectious plant in *Ganoderma boninense*. A cyclic voltammetry (CV) and linear sweep voltammetry (LSV) measurement has been conducted with an excellent detection ability for the nano biosensor at a linear detection range from 0.0004 to 1.0 μM at a detection limit of 3.75 nM. Sensitivity and the lower LOD are mainly attributed to the remarkable and synergistic properties of gold, chitosan nanoparticles, and MWCNTs. Currently, the pathogenic fungus *Ganoderma sp*. (*Ganoderma boninenseis*) is a soil-borne organism that passes through the roots of young plants, causing base stem rot disease in palm oil. Various electrochemical biosensors to detect *G. boninense* based on gold [31] DNA (deoxyribonucleic acid) [32], magnetic nanoparticles and DNA conjugates [33], have been employed. An electrochemical enzyme combined immunoassay was used to identify *Pantoea stewartii subsp. stewartii-NCPPB 449 (Pss)* plant bacterial pathogen in AuNPs with horseradish peroxidase-labeled antibodies [34]. *Cymbidium mosaic virus* and *Odontoglossum ringspot virus* detection methods were developed using label-free surface plasma resonance of Au nanorods and calculated in 48 and 42 pg mL^{-1} LOD, respectively [35]. Furthermore, Siddiquee *et al* [36] constructed a nanobiosensor using ZnO NPs/chitosan nanocomposite modified gold electrode to detect the soil-borne fungal pathogen *Trichoderma harzianum*.

13.3.3 Detection of antibiotics

Antibiotics are the most useful treatment for bacterial infection as one of the common pollutants.

The excessive use of antibiotics in animal farms is extremely dangerous to both humans and livestock. The identification of contaminating antibiotics is important for the animal agricultural sector as well as the antibiotic manufacturing industry. Recently, Shen *et al* [37, 38] developed a biosensor based on nanomaterials for the sensitive and rapid detection of enrofloxacin residues (ENR) by using immunomagnetic beads (IMBs) and quantum dots (QDs) in raw chicken. The established nanobiosensor exhibited a remarkable reactivity for ENR residues, which incorporates a low concentration of 0.94 ng mL^{-1} and a linear dynamic range from 1 to 100 ng mL^{-1}. Recently, Lee *et al* [39] have fabricated indirect measurement methods to deduce a powerful technique for detecting vancomycin resistance *Enterococcus* (VRE) pathogens using LSPR (localized surface plasmon resonance) based nanobiosensors. Nanoclusters are generally incorporated or manufactured in alloys with specific nanoclusters to increase sensitivity, selectivity and minimize kinetic oxygen. Furthermore, Hosseini *et al* [40] designed a fluorescent aptasensor for responsive oxytetracycline (OTC) analysis based on DNA aptamer-templated silver nanoclusters (AgNCs) quenching. The proposed nanobiosensor demonstrated a linear operating range of 0.5 nM to 100 nM and an LOD of 0.1 nm in water samples. Mohammad Razdar *et al* [41] developed the electro-chemical aptamer-based biosensor with sensitive, selective, and rapid analysis of tetracycline (TET) using a magnetic (MNP) and AuNPs composite pencil graphite electrode (PGE). In milk samples, a linear concentration level of 1.0 pM to 1.0 mM was achieved, with a detection maximum of 0.03 pM TET, which is lower than the European Union's permissible limit (225 nM).

13.3.4 Detection of metals and heavy metals

In recent decades, heavy metal pollution by air, water, and soil has significantly increased. It inhibits groundwater and other significant resources of the ecosystem and makes it poisonous and vulnerable to different conditions of disease for humans. This has become one of the main environmental issues that threaten global sustainability. Heavy metals can cause chronic exposure until that concentration in the living organism exceeds a certain amount. Heavy metal commonly includes environmental pollutants such as cadmium, mercury, chromium, metalloid arsenic, and other crucially toxic biological elements. The heavy metals are very difficult to biodegrade and can safely be stored *in vivo* by biomagnifying the food chain. Moreover, heavy metals aggregate can cause human beings harm, as they can produce prolonged intoxication and inactivation of the human body's enzymes and proteins. Heavy metals such as arsenic (As), mercury (Hg^{2+}), cadmium (Cd^{2+}), chromium (Cr^{3+}), and lead (Pb^{2+}) are considered as high priority, publicly significant metals because of their health implications even at a lower exposure of level. This causes multiple organ failure and is listed as a potential human carcinogenic agent according to the U.S. Environmental Protection Agency

(USEPA) and the International Agency for Research on Cancer (IARC). In recent years, electrochemical biosensors have become greatly beneficial to detect heavy metals. Due to their high versatility, speed, good selectivity, and specificity, electrochemical techniques can be considered one of the most promising approaches to detect heavy metal ions [42]. The electrochemical determination process for heavy metal ions includes two main components: the process of recognition and signal conversion. The identification mechanism of electrochemical sensors and nano-biosensors for heavy metal ions depends primarily on transferring a specific recognition response to a measured electrochemical signal involving recognition elements and target analytes [43]. And the signal response generated represents the concentration of the analyzed analyte, such that electrical sensors and biosensors may measure heavy metal ions quantitatively.

13.3.4.1 Arsenic

Arsenic ion is a toxic element that may cause complications with the skin, kidneys, urinary bladder, and lungs. As a result, a simple, quick, and effective method for the quantitative environmental assessment of arsenic ions must be developed. The WHO (World health organization) recommends a maximum concentration of arsenic in drinking water of 10 $\mu g\ L^{-1}$ [44]. In addition to conventional methods for assessing heavy metal toxicity, several biosensors have been created in recent years to analyze arsenic levels in diverse matrices including blood, various tissues, and microorganisms. For example, based on modified electrode ruthenium bipyridine-graphene oxide nanocomposite, Gumpu *et al* [45] have developed a favorable electrochemical sensor for measurement of arsenite (As^{3+}) and arsenate (As^{5+}). The modified nanocomposite electrode has a high oxidizing activity for As (III) and As(v) in the wide linear concentration of 0.08–15 μM with detection limits of 21 and 34 nM. Similarly, Aptamer screen-printed electrode (SPE) combined with graphene oxide (GO) and methylene blue (MB) nanocomposite as an improved signal label for the determination of total arsenic in shellfish was developed by Zhang *et al* [46]. The detection limit of this aptamer was 2×10^{-4} mg L^{-1}. Baghayeri's [47] group proposed an electrochemical sensor by functionalizing GO/metal–organic framework GO/MOF on a glassy carbon electrode (GCE) and differential pulse anodic stripping voltammetry (DPASV) measurements to detect As(III) in water samples. The GO/MOF showed sensitizing action was effective and resulted in a low detection value (0.06 ppb), higher sensitivity, and a wide linear range (0.2 to 25 ppb) for arsenic detection. The LOD value of As (III) is lower than the Environmental Protection Agency's (EPA) drinking water minimum value.

13.3.4.2 Mercury

Mercury ions (Hg^{2+}) are one of the most toxic heavy metals, with serious consequences for human health and the environment [48]. Mercury and its derivatives will spread globally across food chains and the atmosphere as significant contaminants [49]. Due to their acute toxicity, even small levels of mercury may cause an irreversible threat to animals and humans. Chronic mercuric toxicity may also disrupt the functions of the human body's internal organs and inflict neurological damage, strokes, renal failure, and other conditions [50]. The WHO recommends that Hg^{2+} ions should not exceed 1 ppb in drinking water [51]. For the quantitative analysis of mercury ions, Li and colleagues [52] have designed a high sensitivity electrochemical biosensor, which enhanced synergistic interaction between the hybridization chain on a DNA level and the core–shell nanoparticles silver@gold, with positive charges. This electrochemical biosensor of mercury ions exhibits great efficiency with a lower value of 3.6 nM. Likewise, this same group [53] has used specific base-pair thymine Hg^{2+} to fabricate the electrochemical biosensor with extremely sensitive mercury ions. Toluidine blue incorporated in the hybridization chain reaction could amplify the signal of mercury ions. The developed biosensor, which has a low detection limit (0.2 pm) and a linearity range of 1 pM to 100 nM, has been found to have a noteworthy sensitivity for mercury ions.

13.3.4.3 lead

Lead ions, a hazardous metal pollutant, can accumulate from primary manufacturers in organisms and have many adverse health consequences [54, 55]. Many nations and organizations, because of the potential risks of lead ions to public health, have strong sanitation limits on groundwater ions in drinking water [56]. The USEPA has set a maximum limit of 72 nM for lead ion contamination in drinking water [57], while the International Agency for Research on Cancer (IARC) has set a safety limit of 48.26 nM for lead ion contamination in food and drinking water [58]. Therefore, the implementation of a new analytical method is important for effective and regular environmental identification of lead ions. Zhang and co-workers [59] developed Mn_{1-x} $Zn_xFe_2O_4$-modified electrodes and used them to detect lead ions and improved the analytical efficiency of that sensor by incorporating zinc ions and a nafion layer. Moreover, several electrochemical biosensors were developed to effectively determine lead ions. For instance, Wei and colleagues [60, 61] developed an extremely responsive electrochemical aptasensor using a gold nanoparticle-modified graphene nanocomposite as a signal probe to determine lead ions with extreme sensitivity. Cantilever nanosensors are compact and versatile and are appropriate for measuring contaminants in real-time without pre-treatment and limited sample sizes [62]. Recently, Rigo *et al* [63, 64] developed a cantilever nanobiosensor for the detection of lead (Pb) in water samples. Urease and alkaline phosphatase enzymes were used as biological elements in the system to detect Pb with high sensitivity. Identifying trace amounts of Pb in water was achieved by modifying the deflection of the cantilever nanobiosensor in conjunction with liquid solution (water). Detection limits for cantilever nanobiosensors with alkaline phosphatase and urease were respectively 323 ppt, 286 ppt, and the limit of quantification was 1.12 ppb and 1.37 ppb.

13.3.4.5 Cadmium

Cadmium ion is another heavy metal contaminant that can quickly accumulate across the food chain in the ecosystem. It is extremely dangerous and threatens public health and the environment.

The WHO has defined a limit of 3 μg L^{-1} Cd^{-1} (II) water pollution [65, 66]. Therefore, fast, simple, economical, and high-efficiency techniques for the monitoring and detection of the trace volumes of cadmium ions in the atmosphere are essential for researchers. For example, Si and colleagues [52] suggested a new electrochemical sensor for cadmium ions due to reduced GO–AuNPs–tetra phenyl porphyrin nanoconjugates. Recently, Ballen *et al* [67] developed a nanobiosensor based on GO nanomaterials and urease enzymes for cadmium identification in river water samples. A high sensitivity (0.0147 Nm/ppt) and a low detection limit (18 ppt) were observed in the biosensor between the sensitive layer and the analyte.

13.4 Nanobiosensors for the detection of toxin pollutants

Cyanobacterial algal blooms induced by aquatic eutrophication contain harmful toxins, including brevetoxins and microcystins, and safe and cost-effective systems are critical for the early detection of these toxins. Nanomaterials can also help bioremediate organic pollutants from soils [68, 69]. The most widely used are nanobiosensors, for their sensor technology, and applications for soil and water analysis. Early identification of soil pollutants can contribute to preventing harmful effects. Accumulation of highly harmful metal ions in arable soils and plants is the major challenge of substantial health threats. The optical sensors are detected utilizing chemical sensors that use electromagnetic radiation to detect and attach with immobilized organic color in the sample changing optical characteristics, indicating the specific concentration of the analysis [70]. In another study, Shi *et al* [71] developed optical nanobiosensors in a recent study to detect cyanotoxin from water in which microcystin LR ovalbumin is immobilized onto a biochip surface to identify cyanotoxin residues (microcystin LR). This methodology applied an indirect competitive strategy and pre-implemented microcystin LR with a specific single-clonal antibody with a fluorescent label and exposed them to biochip surfaces and quantified microcystin LR 0.2–4 μg L^{-1}, with an LOD of 0.09 μg L^{-1}. Endogenous hormones of plants are a class of tiny organic molecular compounds formed internally through plants. These endogenous hormones control all plant growth and development processes that are widely used in agriculture [72]. Abscisic acid (ABA) plays an essential part in abiotic stress reactions, and the resistance to adverse conditions, such as droughts, low temperatures, and the high-salt stress environment, for the physiological signal transduction [73]. In addition, Wang *et al* [74] have also reported the detection of abscisic acid, a significant abiotic plant hormone by using an SPR (surface plasmon resonance)-based optical nanobiosensor in the mechanism of utilized aptamer functionalized AuNPs. The quantification of a good linear range of abscisic acid was from 5 to 10^{-7} M to 5 to 10^{-5} M with 0.33 μM of LOD. Similarly, by modifying hormonal-specific aptamer, other plant hormones can also be recognized. Recently, Bagheri *et al* [75] used a FRET (fluorescence resonance energy transfer) optical nanobiosensor in the detection of plant metabolites to detect 'scopolamine, tropane alkaloid and anticholinergic agent' extraction from the atropa plant. They used thio glycolic acid-conjugated cadmium telluride (Cd/Te) quantum dots with M_2 muscarinic receptor functionalization (Cd/Te QDs-

M_2R) and labeled rhodamine123 scopolamine, which act, respectively, as donor and ligand pairs to transfer charges. The response of nanobiosensors was linear in a range of 0.01–4 μmol L^{-1} with LODs of 0.001 μmol L^{-1} of scopolamine hydrochloride.

13.4.1 Endocrine disrupting chemicals

Endocrine disruptive chemicals (EDCs) are compounds that interfere with the body's interaction with the body's endocrine system and have adverse effects on living beings' growth, reproduction, behavioral and immune systems. EDCs are compounds that cause environmental pollution and permanent damage to living beings in addition to direct soil contamination and effluents. Bisphenol A (BPA-2,2-bis (4-hydroxyphenyl) propane) is labeled as an EDC which is used in the formation of polycarbonate polymers used in food packaging and container production. At present, various nanomaterials and modifiers with electrochemical sensor determination of BPA have been developed such as reduced graphene oxide (rGO) [76]; nanospheres [77]; CNTs [78]; AuNPs [79–81], and molecularly imprinted polymers [82]; surfactant [83]; bentonite [84]; aptamers [85]. The enzymes integrated with nanomaterial will promote electrode efficiency, providing a new and generalized method for constructing highly robust electrical biomaterials. The authors long *et al* [77] constructed Au-graphene (Au-GN), Au-Pt@SiO_2 nanocomposites (APS NPS), and tyrosinase (Tyr) for BPA detection with a wide linear concentration of 0.01 to 10 mg L^{-1} and LOD of 1.80 μg L^{-1}. Recent studies have shown that CNTs can improve the electrochemical reactivity of BPA detection [78], with a novel and sensitive electrochemical sensor based on MWCNTs, and polycrystal violet (PCV). They found good linearity 5×10^{-8}–1×10^{-4} mol L^{-1} with a LOD of 1×10^{-8} mol L^{-1}. In another study, He *et al* [86] developed a label-free, low-background signal aptasensor for the detection of BPA by using anti-BPA aptamer and Mo_2C nanotube. This method showed a linear range of 2–20 nM for the detection of BPA with a LOD of 2 nM in real water samples. Steroidal estrogens such as; 17β-estradiol, 17α-ethinylestradiol, and estrone are included in the European Commission's 2015/495 'Monitoring List' of the EU (European Union) Water Framework Directive (WFD) Substances. EDCs are very difficult to monitor since they exist in low quantities and because the limit for the identification of disposable analytical methods does not satisfy the required levels [87]. Wong *et al* [88] developed a sensitive and selective determination of 17-estradiol in synthetic urine and river water samples, using SPEs modified with copper phthalocyanine, carbon black, and nafion film. Results indicated that the linear range was 8.0×10^{-8}–7.3×10^{-6} and LOD (5.0×10^{-9}). In addition, Fan *et al* [89] designed a photoelectrochemical (PEC) sensing framework focused on anti-E2 aptamer as the biorecognition element was assembled through CdSe nanoparticle-modified TiO_2 nanotube arrays to detect 17-estradiol in lake water. The superior sensing behavior towards E2, due to the predominant tubular shape, outstanding photoelectric operation, the high packing density of aptamer on the signaling interface, and the strong capacity of the aptamer to E2 could be attributable to the effective PEC sensing interface.

The developed PEC aptasensor exhibited excellent performance in E2 measurement with a wide linear concentration from 0.05 to 15 pm with LOD 33 fM. In order to detect 17β-estradiol (17β-E2) in water employing reversible addition-fragmentation chain transfer (RAFT), a new molecularly-imprinted (MIP) electrochemical sensor was designed using Fe_3O_4 (iron oxide) graphene immobilized nanobeads developed by Li *et al* [90]. Furthermore, the Fe_3O_4-MIP@rGO-based sensor has a LOD of 0.819 nnM in a linear concentration of 0.05 to 10 mM.

13.4.2 Phenolic pollutants

The phenolic substances are prototype poisonous and listed by the USEPA as one of the 129 priority regulated contaminants that have a major hazard to public health and the ecological environment. Phenolic compounds are primarily derived as one of the main contaminants from dye, pesticides, petrochemicals, and other wastewater industries commonly spread in a natural environment. The evaluation of trace phenolic compounds in a real sample is of major importance. Phenolic pollutants can inhibit some enzymes, including tyrosinase, laccase, urease, aldehyde dehydrogenase, and alkaline phosphatase, and different approaches have been used to develop novel strategies for detecting these enzymes using biosensors [91–93]. The phenols can be monitored by all tyrosinase biosensors designed for the detection of a *p*-Cresol compound. The lowest *p*-cresol detection limit was achieved by a combination of quantum dots combined with chitosan or Fe_3O_4 nanoparticle [94–96] by combining copper oxide and mesoporous silica [97]. An inhibited tyrosinase phenolic biosensor developed on a magnetic nanoparticle allows it to detect a significantly lower concentration of the phenolic compounds. The presence of metal oxides in polymer composites based on Fe_3O_4 magnetic nanoparticles improves biosensor stability [98]. The laccase biosensor's stability for hydroquinone detection is increased considerably via the mixture of AuNPs with nafion (96% for three months) [99]. For the measurement of phenolic compounds, an optical biosensor can be constructed on a silicate film, using immobilization in laccase d enzymes by the 3-methyl-2-benzothiazolinone hydrazone (MBTH) [100]. HRP (horseradish peroxidase) biosensor is the most efficient for trace detecting several phenolic compounds, such as bisphenol A, chlorophenol derivatives [101], and nitrophenols. The analytical efficiency of HRP biosensors is because of mono phenolic compounds that are developed using conductive films of copolymers or using poly (citric acid)-block poly (ethylene glycol) to increase the electrochemical properties of HRP biosensors in combination with chitosan and glutaraldehyde combination with polyaniline nanofibers.

13.5 Plant biology and agriculture: What functions for nanobiosensors?

The exponential growth of the world's population has expanded the need for food supplies. Many farmers lose their agricultural products through pathogenic infections, poor conditions of soil, water, and environmental conditions. Researchers claim that nanobiosensors can play a pivotal role in revolutionizing farming by identifying threats to avoid agricultural losses. The shelf life for the plant, seed viability, nutritional deficiency, and biotic and abiotic stresses is determined by nanobiosensors. Nanobiosensors have the following main advantages: fast, cost-effective diagnostic device, versatile, real-time, remote monitoring. The Spanish company NT Sensors is focused on the development of nanotube-based sensors that can evaluate the concentration of ions in real-time in agricultural fields. This nanosensor (IMACIMUS Multi ION) can also calculate and detect the ionic strength of the leachate [102]. In order to ensure efficient fertilization and productivity of plants, appropriate nutrient management is required. The NT nanosensor measures macronutrients and water quality, which is particularly useful in poor agricultural conditions. The lack of adequate and effective fertilizers, nutrients, and water could lead to a major reduction in agricultural production. This sensor provides adequate nutritional and water requirements for the cultivation of crops. NanoLabs (Nanoelectronic Devices Laboratory) is a Spanish nano-technology engineering company that has developed ASAR technology to minimize the use of water in the fields of agriculture [103]. NanoLabs also helps to reduce the number of pesticides and fertilizers used in soil and water quality determinations which guarantees improved cultivation.

13.5.1 Nanobiosensors for soil-plant systems

Kalita *et al* [104] have established graphene quantum dots for soil moisture content by using interdigitated electrodes and graphene dots for moisture content sensation in two types of soils, namely white clay, and Bentonite clay. The soil moisture content is measured selectively by the gravimetric moisture content as ionic conductivity. The nanobiosensor also allows the nutrient detection of soil to assess soil content. In a further report, Mura *et al* [105] have demonstrated optical nanobiosensors to detect nitrate in water samples. In this technique, the AuNPs have cysteamine functional for nitrate detection directly using an easy and fast process (2 min for the analysis) and changes in color within the 35-ppm range and detected by naked eyes. In addition, Prasad *et al* [106] further established a phosphorous sensor to detect phosphorus in the soil with amino acid type herbicides with the use of nanofilm adapted pencil graphite electrodes, with the detection limit of 0.35 and 0.19 g ml^{-1} analyzed glyphosate and glufosinate.

13.5.2 Nanopesticides

Pesticides are widely used in cultivation to increase crop yield and production. Urea is the most extensively employed agricultural fertilizer and a source of nitrate, nitrite, and urease, which are pervasive toxins in the water that cause eutrophication and have effects on the environment [107–109]. Organophosphates, neonicotinoids, carbamates, and atrazine are among the most often used pesticides, and their residues, except at low doses, persist in soil due to the lower homogeneity. Pesticides are categorized as herbicides, insecticides, nematicides, fungicides rodenticides, or miticides. Nanopesticides provide a method of controlling pesticide distribution while still achieving greater benefits with a lower chemical dosage. The implementation and nano pesticides synthesis are enhanced to increase a pesticide's potency and durability while reducing the number of active ingredients found and minimizing or removing any possible hazards. Various nanoemulsions were used in herbicides and pesticides. Pesticide nanoemulsions have a variety of possible benefits over other processes, including good surface adhesion, high penetrability, and a wide range of applications [110, 111]. Encapsulation is a term used widely in nanotechnology to describe the mechanism of enveloping one material within another, resulting in particles with diameters ranging from a few nm to a few mm. The key explanation for encapsulation is because certain nutrients may not stay in the soil for long periods or may interact with other soil materials, resulting in undesirable results [112]. Pesticides and fertilizers are both encapsulated with nanotechnology. Nanoencapsulation is used for pesticides, notably hydrophobic pesticides (whose insolubility in water is a limiting factor). This technique allows additional stability, improves dispersion in aqueous solution, and allows a regulated release of active ingredients, increasing its efficiency in the management of pests and reducing treatment doses in crop control and human exposure [113].

13.5.3 Nanofertilizers

A nanofertilizer plays an important role in agriculture sustainability. A nanofertilizer is essentially an encapsulated substance in nanoparticles that supplies soil and plant nutrients [114]. Nitrogen deficiency in some areas is a big concern, due to leaching, long-term absorption of soil microorganisms, and pollution. In recent years, nanofertilizers have been more significant in their production and exploitation than chemical fertilizers that are heavily environmentally deteriorating by causing soil contamination. Nanofertilizers improve crop production and quality with improved productivity of nutrient usage, thus reducing the costs and contributing to sustainable agriculture [115]. Nanofertilizers reduce the rate of nutrient release in comparison with traditional fertilizers and improve their sustainability through addition or adsorption but not because of improvements in the chemical speciation due to the pH-based phenomenon, the redox potential, and soil ligand presence [116]. In addition, nanofertilizers may also be used in conjunction with microorganisms (called nanobiofertilizers) for enhancing the abiotic stress resistance for a great additional purpose. The relationship between nanomaterials and microorganisms, the shelf life of propagation of biofertilizers, and their distribution is one of the most important aspects of the development of nanobiofertilizers. Also, the association between AuNPs and plant growth-promoting rhizobacteria (PGPR) was shown to have beneficial effects [117]. Abdel-Aziz and co-workers, [118] investigated the use either alone or filled with NPK nano CS nanoparticles or CNTs as foliar application fertilizer for French bean plants and found that foliar utilization increases plant growth yield and antioxidant mechanism and biochemical quality of the yielded seed when compared to crop plants. According to Manikandan *et al* [119] nanoporous zeolite applied to nitrogen fertilizer might be used as an alternative strategy to improve the efficacy of nitrogen in sustainable agriculture.

13.5.4 Herbicides

In recent years, the emission into marine ecosystems of excessive dangerous substances from human activity, irrigation, wastewater, industry, and livestock has led to a decrease in the availability of public fresh drinking water. They have been commonly used in agriculture to increase crop yields, in particular for herbicides by eliminating weeds. The proper application of herbicides contributes to the destruction of unwanted crops, otherwise the natural growth of desirable crops will be hindered. Despite the useful aspects of the most common herbicides used by farmers, even trace amounts of herbicides consumed by humans as food or water are very dangerous and are not biodegradable; there is often a serious threat of death [120]. Atrazine is a xenobiotic compound recognized for potential endocrine disruption, which can cause significant health hazards, even in very low doses. Atrazine (2-chloro-4-ethylamino-6-isopropylamino-1, 3, 5-triazine) has become one of the herbicides most commonly used as a herb control device in broad-leaf and herbal wetlands in the world [121]. According to the WHO, the limit of atrazine in drinking water is 2 μg L^{-1} and it has been listed under IARC 2001 as a Type 3a carcinogenic agent. Recently, Shengbo Sang *et al* [122] developed a wireless ME (Magneto elastic) nanobiosensor, used for ppb atrazine detection, based on the direct competitive immunoassay protocols, as the substrate and AuNP, as the covering layer. Competitively coupled Atr–BSA (atrazine–albumin conjugate) with a heavy molecular mass and atrazine was combined on the surface of the nanobiosensor with atrazine antibodies to improve the sensitivity of the signal responses. This nanobiosensor shows excellent atrazine detection stability for 6 days with linear ranges from 1 ng ml^{-1} to 100 μg ml^{-1} and the detection limit was found to be 1 ng ml^{-1}. In addition, Tucci *et al* [123] designed a microbial biosensor for *in situ* herbicide detection based on *cyanobacteria Anabaena variabilis* in photocurrent inhibition. This sensor provides outstanding analytical sensitivity of 24.6 μA μM^{-1} cm^{-2} towards atrazine at 0.56 μM. Cristina *et al* [124] reported the determination of atrazine from tyrosinase-based biosensor from water samples. The developed biosensor exhibited atrazine 0.5–20 ppm linearity with LOD of 0.3 ppm and suitable repeatability. Furthermore, Martinazzo *et al* [125] developed a cantilever nanobiosensor for atrazine detection by using low-cost tyrosinase vegetal extract as a biological recognition element, reporting the detection limit and the quantification limit to be 7.754 ppb and 22.792 ppb, respectively. Glyphosate (N-phosphonomethyl glycine, GLY) is a non-selective, post-emergent herbicide that belongs to the category of substituted glycines. It is used to control a broad variety of weeds. Recently, a novel method for glyphosate herbicide detection was developed using a cantilever nanobiosensor functionalized with a vegetable source of peroxidase. The analysis of the nanobiosensor showed that the detection method has the ability to detect glyphosate with a detection limit of 0.028 mg L^{-1} [126]. Moreover, electrochemical biosensors are fabricated via graphite oxide and direct electrochemical detection using carbon paste electrodes for GLY detection [127, 128].

13.5.5 Fungicides

The regulation of fungal diseases in agrochemical applications is mainly affected by fungicidal tolerance in fungal pathogens. Fungicidal resistance arises through specific interactions between different elements, such as the direction the fungicide functions, the pattern of application of fungicides, the fungal pathogens' morphology, and the cropping mechanism. Carbendazim (CBM) is a systemic broad range fungicide from the benzimidazole group and is used as a pre-harvest preventive agent for the control and prevention of fruit and vegetable diseases during storage. Luo *et al* [129] studied the electrochemical determination of carbendazim on graphene oxide-multi-walled nanotubes/modified glassy carbon electrodes (GO-MWCNT/GCE). The nanostructural advantages of graphene along with better accumulation properties of the CNTs provided a striking active surface for the electroanalysis of the target compound. GO-MWCNT/GCE detected very low concentrations of the fungicide (LOD: 5 nM) and the method was validated for quantification of carbendazim in soil and tap water samples with satisfactory recoveries. In the application of the DPV method, Özcan *et al* [130] reported on the development of AgNPs on the fumed silica platform (FS@Ag) for detection of carbendazim in real samples, like tomato juice, river water, and apple juice. The authors attributed the success of this nanocomposite material to the coupling of the adsorptive capability and the increased active surface area of the nanocomposite, achieving a wide linear range 5.0×10^{-8}–3.0×10^{-6} and acceptable LOD of 9.4×10^{-10}. In recent years, electrical sensors and some functional nanomaterials have been used for the development of sensitive carbendazim (CBZ) and thiabendazole (TBZ). Dong *et al* [131] reported that a biosensor based on new sensing material ($ZnFe_2O_4$/SWCNTs nanohybrid) enhances the development of a highly sensitive platform for multiple pesticides sensing in different samples (apple, tomato, leek, paddy, and seawater). The fabrication and evaluation of a GCE modified with $ZnFe_2O_4$/SWCNTs nano-hybrid for simultaneous determination of CBZ and TBZ by DPV method. The fabricated electrode exhibited linear calibration curves ranging from 0.5 to 100 mM for CBZ and TBZ. The LOD were 0.09 mM for CBZ and 0.05 mM for TBZ. Maximiano *et al* [132] studied using a CPE modified with a thermally activated zeolite composite in the presence of a surfactant to determine the fungicide CBZ and thiram in different fruit juices by the DPV method. The electrochemical sensor exhibited a detection limit of 0.51 CBZ (MBC) and 1.01 ng mL^{-1} thiram. The Arruda group used a carbon paste electrode; CBZ was determined ultra-sensitively in water and orange juice samples [133]. The developed method achieved recovery rates of 98 to 101%, respectively, allowing it to be used routinely in laboratories for food analysis to meet the maximum residue levels rerecommended by the agencies European Commission Products (EPA) [134] and Agência Nacional De Vigilância Sanitária (ANVISA) [135].

13.6 Environmental risk assessment: what is needed?

Risk management should consider the certain potential impact of nanomaterials, however incorporating a low dose/concentration effect is a challenge. Although the environmental risk evaluation must include nanopesticides and nanofertilizers, the combined consequences of nanomaterials and pesticides are still under-studied [136]. Similarly, because nanomaterials and pesticides cause hormesis at low doses or low concentrations individually, it is not possible to preclude the preservation of hormesis inducible by low concentrations and doses of nanopesticides, this is attributed to the fact that hormesis is still shown in mixtures [137]. Although the positive effect levels/doses often are $>1 < 100$ times lower than the harmful effects that are induced [138, 139], the risk of biodiversity could be much lower at the levels/doses than generally predicted based on negative impact studies. Therefore, hormesis should be integrated into nanopesticides risk management activities by considering the effect of chemistry and mode of action for each compound [140]. Due to the high complexity of the issue, the risk assessment of nanomaterials remains a challenge. Therefore, interactions between nanomaterials in environmental conditions can contribute to their transformation and notwithstanding change their dissolution and toxicity in the lack of interactions [141]. The toxicity caused by nanomaterials can rely on physicochemical characteristics [138, 142], as well as other variables such as, contact with other pollutants, which can affect the shape and structure of nanomaterials and bioavailability in the environment and consequently the subsequent implications despite the environmental transformation of nanomaterials [143]. Nevertheless, since these characteristics of nanomaterials can influence the biphasic relationship of dosage–response [144], the evaluation of around 500 hormetic dose–responses in plants and algae demonstrates that the prevalence of nanomaterial-induced hormesis is generalized. Additionally, nanomaterial-induced hormesis usually occurs below certain concentrations/doses given the environmental variations despite the variations between experimental conditions. In order to improve the empirical evidence available for use in risk assessment systems, evaluations should encompass the entire life cycle and its consequences over generations [145]. The nanomaterial toxicity of soil animals has changed due to the aging of soil, demonstrating the need for understanding the ecological influence of aged or endured nanomaterials in environmental processes [146]. The transformation into legislation and practice of incomplete scientific information about the behavior, nature, bioavailability, and environmental consequences of nanomaterials remains an important challenge [142]. Besides that, understanding of potential regulatory translation is growing and the research community's endeavors to perform this provide a healthy perspective [138]. For example, the EU–US Research Communities, who are now seven years old, work to promote a standard interface of protocols and methodologies to address nanotechnology environmental health and safety (nanoEHS) research gaps and obstacles in nanoEHS [146]. They also provide a promising viewpoint to shorten the time needed to incorporate hormesis into the risk assessment of nanopesticides. In the future, regulation practices should take into account not only biodiversity above ground but also biodiversity below ground. This will proceed to be a concern because the destiny of below-ground species in terms of environmental issues is not investigated adequately. The considerable lack of research will continue to present the challenge of researches in the aspect of environmental threats on the fate of underground species.

13.7 Challenges and future perspectives

It is clearly evident that new micro/nano innovations, such as sensors, materials engineering, chemistry, and biochemistry, will drive future innovations and advances in sensing devices. As environmental exposure in different media increases extremely quickly and becomes a serious international risk, the design and development of new biosensor-based techniques that can accurately detect and analyze different contaminants from a wider range are essential. Since environmental contamination is rising extremely quickly in various media and is a serious global fear, the development and development of new biosensor-based techniques to detect and analyze various contaminants in a broad spectrum is quite important. However, environmental detection sensors/biosensors have certain limits, including time of response, selectivity, sensitivity, durability, and lifetime. Researchers must cooperate to minimize these restrictions for successful environmentally sustainable, accurate, and competitive analytical instruments. It is also evident that in the near future there will be an increased need for rapid detection tools including nano-biosensors. In terms of current and past major electrochemical sensor research and development activities, it is still essential to upgrade devices to prevent instrument errors and improve the performance of complex matrices. The sensors must meet all requirements to ensure the proper, stable, and sustainable monitoring of multi-analyte mixtures across interconnected networks, screening and inspection, and remote sensing in conjunction with WLANs (Wireless Local Area Networks). In this perspective, comprehensive research in the areas of biosensors, bioelectronics, and bionanotechnology would certainly have a major impact on the development of novel approaches to biosensing technology, considering the fact that they are 'smart' and user-friendly for a better future.

13.8 Conclusion

The emphasis of researchers is focused on the design and development of accurate, sustainable, and responsive analytical methods that identify, control, and remove harmful environmental pollutants. In addition to the eco-friendly performance, sensors are increasingly becoming an essential part of all environment screening and monitoring programs as they can provide more sensible, selective, stable, and low-cost performance. Changes from conventional harmful mercury-based electrodes with poor sensitivity and durability as well as lifespan complications to inert-green electrodes are among the main pathways that have been investigated. Consequently, it is also of vital importance for sustainable environmental research to choose suitable detection and material modifications, miniaturization, and green synthesis approaches. Trustworthy methods should be used in combination with method optimization to protect public health and improve environmental quality.

Acknowledgment

We are thankful for Coordination for the Improvement of Higher Education Personnel (CAPES)-Brazil.

References

[1] Parisi C, Vigani M and Rodríguez-Cerezo E 2015 Agricultural nanotechnologies: What are the current possibilities? *Nano Today* **10** 124–7

[2] Kah M, Tufenkji N and White J C 2019 Nano-enabled strategies to enhance crop nutrition and protection *Nat. Nanotechnol.* **14** 532–40

[3] Kaushal M and Wani S P 2017 Nanosensors: Frontiers in Precision Agriculture *Nanotechnology* (Singapore: Springer) pp 279–91

[4] Arduini F, Cinti S, Scognamiglio V and Moscone D 2016 Nanomaterials in electrochemical biosensors for pesticide detection: advances and challenges in food analysis *Microchim. Acta* **183** 2063–83

[5] Xiang L, Zhao C and Wang J 2011 Nanomaterials-based electrochemical sensors and biosensors for pesticide detection *Sensor Letters.* **9** 1184–9

[6] Bernardes M F F, Pazin M, Pereira L C and Dorta D J 2015 Impact of pesticides on environmental and human health *Toxicology Studies—Cells, Drugs and Environment* (Rijeka: InTech) pp 195–233

[7] Lang Q, Han L, Hou C, Wang F and Liu A 2016 A sensitive acetylcholinesterase biosensor based on gold nanorods modified electrode for detection of organophosphate pesticide *Talanta* **156–157** 34–41

[8] Yang Y, Asiri A M, Du D and Lin Y 2014 Acetylcholinesterase biosensor based on a gold nanoparticle-polypyrrole-reduced graphene oxide nanocomposite modified electrode for the amperometric detection of organophosphorus pesticides *Analyst* **139** 3055–60

[9] Wang Y, Zhang S, Du D, Shao Y, Li Z and Wang J *et al* 2011 Self assembly of acetylcholinesterase on a gold nanoparticles–graphene nanosheet hybrid for organophosphate pesticide detection using polyelectrolyte as a linker *J. Mat. Chem.* **21** 5319

[10] Lang Q, Han L, Hou C, Wang F and Liu A 2016 A sensitive acetylcholinesterase biosensor based on gold nanorods modified electrode for detection of organophosphate pesticide *Talanta* **156–157** 34–41

[11] Zhang Y, Arugula M A, Wales M, Wild J and Simonian A L 2015 A novel layer-by-layer assembled multi-enzyme/CNT biosensor for discriminative detection between organophosphorus and non-organophosphrus pesticides *Biosens. Bioelectron.* **67** 287–95

[12] Vamvakaki V and Chaniotakis N A 2007 Pesticide detection with a liposome-based nano-biosensor *Biosens. Bioelectron.* **22** 2848–53

[13] Dong J, Zhao H, Qiao F, Liu P, Wang X and Ai S 2013 Quantum dot immobilized acetylcholinesterase for the determination of organophosphate pesticides using graphene-chitosan nanocomposite modified electrode *Anal. Methods* **5** 2866–872

[14] Viswanathan S, Radecka H and Radecki J 2009 Electrochemical biosensor for pesticides based on acetylcholinesterase immobilized on polyaniline deposited on vertically assembled carbon nanotubes wrapped with ssDNA *Biosens. Bioelectron.* **24** 2772–7

[15] Ye H, Guo Z, Peng M, Cai C, Chen Y and Cao Y *et al* 2016 Methyl parathion degrading enzyme-based nano-hybrid biosensor for enhanced methyl parathion recognition *Electroanalysis* **28** 1591–6

[16] Tao T, Zhou Y, He C, He H, Ma M and Cai Z *et al* 2020 Highly sensitive methyl parathion sensor based on Au-ZrO_2 nanocomposites modified graphene electrochemical transistor *Electrochim. Acta* **357** 136836

[17] Karimi-Maleh H, Yola M L, Atar N, Orooji Y, Karimi F and Senthil Kumar P *et al* 2021 A novel detection method for organophosphorus insecticide fenamiphos: Molecularly

imprinted electrochemical sensor based on core–shell Co_3O_4@MOF-74 nanocomposite *J. Colloid Interface Sci.* **592** 174–85

[18] Meng X, Wei J, Ren X, Ren J and Tang F 2013 A simple and sensitive fluorescence biosensor for detection of organophosphorus pesticides using H_2O_2-sensitive quantum dots/bi-enzyme *Biosens. Bioelectron.* **47** 4020–407

[19] Sundarmurugasan R, Gumpu M B, Ramachandra B L, Nesakumar N, Sethuraman S and Krishnan U M *et al* 2016 Simultaneous detection of monocrotophos and dichlorvos in orange samples using acetylcholinesterase-zinc oxide modified platinum electrode with linear regression calibration *Sensors and Actuators, B: Chemical.* **230** 306–13

[20] Peng L, Dong S, Wei W, Yuan X and Huang T 2017 Synthesis of reticulated hollow spheres structure $NiCo_2S_4$ and its application in organophosphate pesticides biosensor *Biosens. Bioelectron.* **92** 563–9

[21] da Silva W, Ghica M E and Brett C M A 2019 Choline oxidase inhibition biosensor based on poly(brilliant cresyl blue)—deep eutectic solvent / carbon nanotube modified electrode for dichlorvos organophosphorus pesticide *Sensors and Actuators, B: Chemical.* **298** 126862

[22] Cui H F, Wu W W, Li M M, Song X, Lv Y and Zhang T T 2018 A highly stable acetylcholinesterase biosensor based on chitosan-TiO_2-graphene nanocomposites for detection of organophosphate pesticides *Biosens. Bioelectron.* **99** 223–9

[23] Zhang J, Hu H, Wang P, Zhang C, Wuma J and Yang L 2021 A stable biosensor based on chitosan-modified graphene for detecting organophosphorus pesticides *Biotechnology and Applied Biochemistry* (New York: Wiley)

[24] Chen Y, He K, Sun F, Wei D and Li H 2021 Efficient electrochemical sensor based on multi-Wwalled carbon nanotubes/gold nanocomposite for detection of dichlorvos pesticide in agricultural food *Int. J. Electrochem. Sci.* **16** 210321

[25] Qi Y, Chen Y, Xiu F R and Hou J 2020 An aptamer-based colorimetric sensing of acetamiprid in environmental samples: Convenience, sensitivity and practicability *Sensors and Actuators, B: Chemical.* **304** 127359

[26] He J, Liu Y, Fan M and Liu X 2011 Isolation and identification of the DNA aptamer target to acetamiprid *J. Agric. Food Chem.* **59** 1582–6

[27] Weerathunge P, Ramanathan R, Shukla R, Sharma T K and Bansal V 2014 Aptamer-controlled reversible inhibition of gold nanozyme activity for pesticide sensing *Anal. Chem.* **86** 11937–41

[28] Tian Y, Wang Y, Sheng Z, Li T and Li X 2016 A colorimetric detection method of pesticide acetamiprid by fine-tuning aptamer length *Anal. Biochem.* **513** 87–92

[29] Ariffin S A B, Adam T, Hashim U, Faridah S, Zamri I and Uda M N A 2014 Plant diseases detection using nanowire as biosensor transducer *Adv. Mater. Res.* **832** 113–7

[30] Akanbi F S, Yusof N A, Abdullah J, Sulaiman Y and Hushiarian R 2017 Detection of quinoline in g. Boninense-infected plants using functionalized multi-walled carbon nanotubes: A field study *Sensors (Switzerland)* **17** 1–14 Available from www.mdpi.com/journal/sensors

[31] Hushiarian R, Yusof N A, Abdullah A H, Ahmad S A A and Dutse S W 2015 Facilitating the indirect detection of genomic DNA in an electrochemical DNA biosensor using magnetic nanoparticles and DNA ligase *Analytical Chemistry Research.* **6** 17–25

[32] Rani E, Mohshim S A, Yusof N H, Ahmad M Z, Goodacre R and Alang Ahmad S A *et al* 2020 Sensitive and selective detection of DNA fragments associated with Ganoderma boninense by DNA-nanoparticle conjugate hybridisation *J. Mater. Sci.* **55** 14965–79

[33] Hushiarian R, Yusof N A, Abdullah A H, Ahmad S A A and Dutse S W 2015 Facilitating the indirect detection of genomic DNA in an electrochemical DNA biosensor using magnetic nanoparticles and DNA ligase *Analytical Chemistry Research.* **6** 17–25

[34] Zhao Y, Liu L, Kong D, Kuang H, Wang L and Xu C 2014 Dual amplified electrochemical immunosensor for highly sensitive detection of *Pantoea stewartii* sbusp. *stewartii ACS Appl. Mater. Interfaces* **6** 21178–83

[35] Lin H Y, Huang C H, Lu S H, Kuo I T and Chau L K 2014 Direct detection of orchid viruses using nanorod-based fiber optic particle plasmon resonance immunosensor *Biosens. Bioelectron.* **51** 371–8

[36] Siddiquee S, Rovina K, Yusof N A, Rodrigues K F and Suryani S 2014 Nanoparticle-enhanced electrochemical biosensor with DNA immobilization and hybridization of trichoderma harzianum gene *Sensing and Bio-Sensing Research.* **2** 16–22

[37] Shen Y, He Y, Fu Y, Wang J, Zhang J and Liao M *et al* 2020 A nanomaterial-based biosensor for the quantitative detection of enrofloxacin residues in raw chicken *Trans. ASABE* **63** 1869–78

[38] Shen Y, He Y, Fu Y, Wang J, Zhang J and Liao M *et al* 2018 A nanomaterials-based biosensor for rapid detection of enrofloxacin in chicken products *ASABE 2018 Annual Int. Meeting. American Society of Agricultural and Biological Engineers*

[39] Lee K, Kim Y H, Jung H and Lee H H 2017 Affinity characteristic of terminal sequence in vancomycin resistant Enterococcus (VRE) membrane peptides on nanobiosensor chip using localized surface plasmon resonance *Biochip J.* **11** 131–8

[40] Hosseini M, Mehrabi F, Ganjali M R and Norouzi P 2016 A fluorescent aptasensor for sensitive analysis oxytetracycline based on silver nanoclusters *Luminescence* **31** 1339–43

[41] Mohammad-Razdari A, Ghasemi-Varnamkhasti M, Rostami S, Izadi Z and Ensafi A A 2021 Magnetic and gold nanocomposite as a novel aptasensor for early detection of tetracycline residues *J. Food Measurement and Characterization* **15** 3387–396

[42] Du Z-Y, Zhang J, Wang C, Li L, Man Q and Lundebye A-K *et al* 2012 Risk–benefit evaluation of fish from Chinese markets: Nutrients and contaminants in 24 fish species from five big cities and related assessment for human health *Science of The Total Environment* **416** 187–99

[43] Wang N, Lin M, Dai H and Ma H 2016 Functionalized gold nanoparticles/reduced graphene oxide nanocomposites for ultrasensitive electrochemical sensing of mercury ions based on thymine–mercury–thymine structure *Biosens. Bioelectron.* **79** 320–6

[44] Chakraborti D 2016 Is WHO guideline value of arsenic in drinking water 10 ppb in the developing countries safe to drink? *BLDE University J. Health Sciences* **1** 57

[45] Gumpu M B, Veerapandian M, Krishnan U M and Rayappan J B B 2018 Electrochemical sensing platform for the determination of arsenite and arsenate using electroactive nano-composite electrode *Chem. Eng. J.* **351** 319–27

[46] Zhang W, Chen Z, Guan Y, Liu C, Zheng K and Zou X 2021 Aptamer-functionalized screen-printed electrode coupled with graphene oxide and methylene blue nanocomposite as enhanced signal label for total arsenic determination in shellfish *Sensors and Actuators, B: Chemical.* **335** 129383

[47] Baghayeri M, Ghanei-Motlagh M, Tayebee R, Fayazi M and Narenji F 2020 Application of graphene/zinc-based metal-organic framework nanocomposite for electrochemical sensing of As(III) in water resources *Anal. Chim. Acta* **1099** 60–7

[48] Mao M-X, Zheng R, Peng C-F and Wei X-L 2020 DNA–gold nanozyme-modified paper device for enhanced colorimetric detection of mercury ions *Biosensors* **10** 211

[49] Sadani K, Nag P and Mukherji S 2019 LSPR based optical fiber sensor with chitosan capped gold nanoparticles on BSA for trace detection of Hg (II) in water, soil and food samples *Biosens. Bioelectron.* **134** 90–6

[50] Yuan H, Ji W, Chu S, Liu Q, Qian S and Guang J *et al* 2019 Mercaptopyridine-functionalized gold nanoparticles for fiber-optic surface plasmon resonance Hg^{2+} sensing *ACS Sensors* **4** 704–10

[51] Benson N U, Adedapo A E, Fred-Ahmadu O H, Williams A B, Udosen E D and Ayejuyo O O *et al* 2018 New ecological risk indices for evaluating heavy metals contamination in aquatic sediment: A case study of the Gulf of Guinea *Regional Studies in Marine Science* **18** 44–56

[52] Si Y, Liu J, Chen Y, Miao X, Ye F and Liu Z *et al* 2018 rGO/AuNPs/tetraphenylporphyrin nanoconjugate-based electrochemical sensor for highly sensitive detection of cadmium ions *Anal. Methods* **10** 3631–6

[53] Yu Y, Yu C, Gao R, Chen J, Zhong H and Wen Y *et al* 2019 Dandelion-like CuO microspheres decorated with Au nanoparticle modified biosensor for Hg^{2+} detection using a T-Hg^{2+}-T triggered hybridization chain reaction amplification strategy *Biosens. Bioelectron.* **131** 207–13

[54] Wang L, le, Wen Y, Li L, Yang X, Jia N and Li W *et al* 2018 Sensitive and label-free electrochemical lead ion biosensor based on a DNAzyme triggered G-quadruplex/hemin conformation *Biosens. Bioelectron.* **115** 91–6

[55] Dolati S, Ramezani M, Abnous K and Taghdisi S M 2017 Recent nucleic acid based biosensors for Pb^{2+} detection *Sensors and Actuators,* B **246** 864–78

[56] WHO – World Health Organization 2017 Guidelines for drinking-water quality. World Health Organization

[57] Ballen S C, Steffens J and Steffens C 2021 Stability characteristics of cantilever nano-biosensors with simple and complex molecules for determination of cadmium *Sensors and Actuators A: Physical* **324** 112686

[58] Leeuwen C J and van, Hermens J L M 2007 *Risk Assessment of Chemicals* ed C J Leeuwen and T G van, Vermeire (Dordrecht: Springer) pp 1–26

[59] Zhang Y, Yang X, Zhang Y, Li W, Hao X and Liu T *et al* 2019 High-performance electrochemical sensor based on $Mn_{1-x}Zn_xFe_2O_4$ nanoparticle/nafion-modified glassy carbon electrode for Pb^{2+} detection *J. Electrochem. Soc.* **166** B341–8

[60] Wang M, Hu M, Liu J, Guo C, Peng D and Jia Q *et al* 2019 Covalent organic framework-based electrochemical aptasensors for the ultrasensitive detection of antibiotics *Biosens. Bioelectron.* **132** 8–16

[61] Wang Y, Zhao G, Zhang Q, Wang H, Zhang Y and Cao W *et al* 2019 Electrochemical aptasensor based on gold modified graphene nanocomposite with different morphologies for ultrasensitive detection of Pb^{2+} *Sensors and Actuators, B: Chemical.* **288** 325–31

[62] Lofstedt R E 2011 Risk versus hazard—how to regulate in the 21st Century *Eur. J. Risk Regulation* **2** 149–68

[63] Silina Y E, Gernaey K V, Semenova D and Iatsunskyi I 2020 Application of organic-inorganic hybrids in chemical analysis, bio- and environmental monitoring *Appl. Sci.* **10** 1458

[64] Prasad S 2014 Nanobiosensors: the future for diagnosis of disease? *Nanobiosensors in Disease Diagnosis* **3** 1–10

[65] Tavakkoli Yaraki M and Tan Y N 2020 Recent advances in metallic nanobiosensors development: Colorimetric, dynamic light scattering and fluorescence detection *Sensors Int.* **1** 100049
[66] Solanki P R, Kaushik A, Agrawal V V and Malhotra B D 2011 Nanostructured metal oxide-based biosensors *NPG Asia Materials* **3** 17–24
[67] Ballen S C, Ostrowski G M, Steffens J and Steffens C 2021 Graphene oxide/urease nanobiosensor applied for cadmium detection in river water *IEEE Sens. J.* **21** 9626–33
[68] Benjamin S R, de Lima F and Rathoure A K 2015 Genetically engineered microorganisms for bioremediation processes: GEMs for bioremediaton *Toxicity and Waste Management Using Bioremediation* (Hershey, PA: IGI Global) 1607–34
[69] Benjamin S R, de Lima F, Florean E O P T and Guedes M I F 2019 Current trends in nanotechnology for bioremediation *Int. J. Environ. Pollut.* **66** 19–40
[70] Arias-Estévez M, López-Periago E, Martínez-Carballo E, Simal-Gándara J, Mejuto J-C and García-Río L 2008 The mobility and degradation of pesticides in soils and the pollution of groundwater resources *Agriculture, Ecosystems & Environment* **123** 247–60
[71] Shi H C, Song B D, Long F, Zhou X H, He M and Lv Q *et al* 2013 Automated online optical biosensing system for continuous real-time determination of microcystin-LR with high sensitivity and specificity: Early warning for cyanotoxin risk in drinking water sources *Environ. Sci. Technol.* **47** 4434–41
[72] Miransari M and Smith D L 2014 Plant hormones and seed germination *Environ. Exp. Bot.* **99** 110–21
[73] Gomez-Cadenas A, Vives V, Zandalinas S, Manzi M, Sanchez-Perez A and Perez-Clemente R *et al* 2015 Abscisic acid: a versatile phytohormone in plant signaling and beyond *Current Protein & Peptide Science.* **16** 413–34
[74] Wang S, Li W, Chang K, Liu J, Guo Q and Sun H *et al* 2017 Localized surface plasmon resonance-based abscisic acid biosensor using aptamer-functionalized gold nanoparticles *PLoS ONE* **12** e0185530
[75] Bagheri F, Piri K, Mohsenifar A and Ghaderi S 2017 FRET-based nanobiosensor for detection of scopolamine in hairy root extraction of Atropa belladonna *Talanta* **164** 593–600
[76] Li Y, Wang H, Yan B and Zhang H 2017 An electrochemical sensor for the determination of bisphenol A using glassy carbon electrode modified with reduced graphene oxide-silver/ poly-L-lysine nanocomposites *J. Electroanal. Chem.* **805** 39–46
[77] Wu L, Yan H, Wang J, Liu G and Xie W 2019 Tyrosinase incorporated with Au-Pt@SiO_2 nanospheres for electrochemical detection of bisphenol A *J. Electrochem. Soc.* **166** 562–8
[78] Wang W, Tang J, Zheng S, Ma X, Zhu J and Li F *et al* 2017 Electrochemical determination of bisphenol A at multi-walled carbon nanotubes/poly (crystal violet) modified glassy carbon electrode *Food Analytical Methods* 10 3815–24
[79] Lee E-H, Lee S K, Kim M J and Lee S-W 2019 Simple and rapid detection of bisphenol A using a gold nanoparticle-based colorimetric aptasensor *Food Chemistry* **287** 205–13
[80] Yaman Y T and Abaci S 2016 Sensitive adsorptive voltammetric method for determination of Bisphenol A by gold nanoparticle/polyvinylpyrrolidone-modified pencil graphite electrode *Sensors (Switzerland).* **16** 1–12
[81] Wang Z, Yan R, Liao S, Miao Y, Zhang B and Wang F *et al* 2018 *In situ* reduced silver nanoparticles embedded molecularly imprinted reusable sensor for selective and sensitive SERS detection of Bisphenol A *Applied Surface Science* **457** 323–31

[82] Zhao W R, Kang T F, Lu L P, Shen F X and Cheng S Y 2017 A novel electrochemical sensor based on gold nanoparticles and molecularly imprinted polymer with binary functional monomers for sensitive detection of bisphenol A *J. Electroanal. Chem.* **786** 102–11

[83] Zhang J, Xu X and Chen L 2018 An ultrasensitive electrochemical bisphenol A sensor based on hierarchical Ce-metal-organic framework modified with cetyltrimethylammonium bromide *Sensors and Actuators, B: Chemical.* **261** 425–33

[84] Alves T S, Santos J S, Fiorucci A R and Arruda G J 2019 A new simple electrochemical method for the determination of Bisphenol A using bentonite as modifier *Mater. Sci. Eng.,* C **105** 110048

[85] Ragavan K V, Selvakumar L S and Thakur M S 2013 Functionalized aptamers as nano-bioprobes for ultrasensitive detection of bisphenol-A *Chem. Commun.* **49** 5960–2

[86] He M Q, Wang K, Wang J, Yu Y L and He R H 2017 A sensitive aptasensor based on molybdenum carbide nanotubes and label-free aptamer for detection of bisphenol A *Anal. Bioanal. Chem.* **409** 1797–803

[87] Kase R, Javurkova B, Simon E, Swart K, Buchinger S and Könemann S *et al* 2018 Screening and risk management solutions for steroidal estrogens in surface and wastewater *TrAC–Trends Anal. Chem.* **102** 343–58

[88] Wong A, Santos A M, Fava E L, Fatibello-Filho O and Sotomayor M D P T 2019 Voltammetric determination of 17β-estradiol in different matrices using a screen-printed sensor modified with CuPc, Printex 6L carbon and Nafion film *Microchemical J.* **147** 365–73

[89] Fan L, Zhao G, Shi H, Liu M, Wang Y and Ke H 2014 A femtomolar level and highly selective 17β-estradiol photoelectrochemical aptasensor applied in environmental water samples analysis *Environ. Sci. Technol.* **48** 5754–61

[90] Li Y, Zhao X, Li P, Huang Y, Wang J and Zhang J 2015 Highly sensitive Fe*3O4* nanobeads/graphene-based molecularly imprinted electrochemical sensor for 17β-estradiol in water *Anal. Chim. Acta* **884** 106–13

[91] Wang Hong-hai, Qin Y and Kai-yu Chen H X 2016 The phenol biosensor based on LDHs / SWNTs hybrid materials *Int. J. Electrochem. Sci.* **11** 777–91

[92] Oriero D A, Gyan I O, Bolshaw B W, Cheng I F and Aston D E 2015 Electrospun biocatalytic hybrid silica-PVA-tyrosinase fiber mats for electrochemical detection of phenols *Microchem. J.* **118** 166–75

[93] Li H, Hu X, Zhu H, Zang Y and Xue H 2017 Amperometric phenol biosensor based on a new immobilization matrix: Polypyrrole nanotubes derived from methyl orange as dopant *Int. J. Electrochem. Sci.* **12** 6714–28

[94] Wang S, Tan Y, Zhao D and Liu G 2008 Amperometric tyrosinase biosensor based on Fe*3O4* nanoparticles-chitosan nanocomposite *Biosens. Bioelectron.* **23** 1781–7

[95] Wang Y, Zhang X, Chen Y, Xu H, Tan Y and Wang S 2010 Detection of dopamine based on tyrosinase-Fe_3O_4 nanoparticles-chitosan nanocomposite biosensor *Am. J. Biomed. Sci* **2** 209–16

[96] Dong W, Han J, Shi J, Liang W, Zhang Y and Dong C 2017 Amperometric biosensor for detection of phenolic compounds based on tyrosinase, N-acetyl-L-cysteine-capped gold nanoparticles and chitosan nanocomposite *Chin. J. Chem.* **35** 1305–10

[97] Han E, Yang Y, He Z, Cai J, Zhang X and Dong X 2015 Development of tyrosinase biosensor based on quantum dots/chitosan nanocomposite for detection of phenolic compounds *Anal. Biochem.* **486** 102–6

[98] Yang J, Li D, Fu J, Huang F and Wei Q 2016 TiO_2-CuCNFs based laccase biosensor for enhanced electrocatalysis in hydroquinone detection *J. Electroanal. Chem.* **766** 16–23

[99] Li G, Sun K, Li D, Lv P, Wang Q and Huang F *et al* 2016 Biosensor based on bacterial cellulose-Au nanoparticles electrode modified with laccase for hydroquinone detection *Colloids Surf.*, A **509** 408–14

[100] Patel N, Shahane S, Shivam, Majumdar R and Mishra U 2018 Mode of action, properties, production, and application of laccase: a rview *Recent Patents on Biotechnology* **13** 19–32

[101] Wu C, Liu Z, Sun H, Wang X and Xu P 2016 Selective determination of phenols and aromatic amines based on horseradish peroxidase-nanoporous gold co-catalytic strategy *Biosens. Bioelectron.* **79** 843–9

[102] IMACIMUS Nutrient Analyzer | NT Sensors https://ntsensors.com/imacimus-series/nutrient-analyzer/

[103] Carbon nanotube manufacturer & product development services https://nano-lab.com/home.html

[104] Kalita H, Palaparthy V S, Baghini M S and Aslam M 2016 Graphene quantum dot soil moisture sensor *Sensors and Actuators B: Chemical* **233** 582–90

[105] Mura S, Greppi G, Roggero P P, Musu E, Pittalis D and Carletti A *et al* 2015 Functionalized gold nanoparticles for the detection of nitrates in water *Int. J. Environ. Sci. Technol.* 12 1021–8

[106] Prasad B B, Jauhari D and Tiwari M P 2014 Doubly imprinted polymer nanofilm-modified electrochemical sensor for ultra-trace simultaneous analysis of glyphosate and glufosinate *Biosens. Bioelectron.* **59** 81–8

[107] de Oliveira J L, Campos E V R, Gonçalves da Silva C M, Pasquoto T, Lima R and Fraceto L F 2015 Solid lipid nanoparticles co-loaded with simazine and atrazine: preparation, characterization, and evaluation of herbicidal activity *J. Agric. Food Chem.* **63** 422–32

[108] Grillo R, Abhilash P C and Fraceto L F 2016 Nanotechnology applied to bio-encapsulation of pesticides *J. Nanosci. Nanotechnol.* **16** 1231–4

[109] Nuruzzaman M, Rahman M M, Liu Y and Naidu R 2016 Nanoencapsulation, nano-guard for pesticides: a new window for safe application *J. Agric. Food Chem.* **64** 1447–83

[110] García M, Forbe T and Gonzalez E 2010 Potential applications of nanotechnology in the agro-food sector *Ciência e Tecnologia de Alimentos* **30** 573–81

[111] Kah M, Beulke S, Tiede K and Hofmann T 2013 Nanopesticides: state of knowledge, environmental fate, and exposure modeling *Crit. Rev. Environ. Sci. Technol.* **43** 1823–67

[112] Villena de Francisco E and García-Estepa R M 2018 Nanotechnology in the agrofood industry *J. Food Eng.* **238** 1–11

[113] Gao Y, Li D, Li D, Xu P, Mao K and Zhang Y *et al* 2020 Efficacy of an adhesive nanopesticide on insect pests of rice in field trials *J. Asia-Pacific Entomology* **23** 1222–7

[114] Zulfiqar F, Navarro M, Ashraf M, Akram N A and Munné-Bosch S 2019 Nanofertilizer use for sustainable agriculture: advantages and limitations *Plant Science* **289** 110270

[115] Usman M, Farooq M, Wakeel A, Nawaz A, Cheema S A and ur Rehman H *et al* 2020 Nanotechnology in agriculture: Current status, challenges and future opportunities *Science of The Total Environment* **721** 137778

[116] Raliya R, Saharan V, Dimkpa C and Biswas P 2018 Nanofertilizer for precision and sustainable agriculture: current state and future perspectives *J. Agric. Food Chem.* **66** 6487–503

[117] Shukla S K, Kumar R, Mishra R K, Pandey A, Pathak A and Zaidi M *et al* 2015 Prediction and validation of gold nanoparticles (GNPs) on plant growth promoting rhizobacteria (PGPR): a step toward development of nano-biofertilizers *Nanotechnol. Rev.* **4** 439–448

[118] Abdel-Aziz H M M, Hasaneen M N A and Omer A M 2019 Impact of engineered nanomaterials either alone or loaded with NPK on growth and productivity of French bean plants: Seed priming vs foliar application *South African J. Bot.* **125** 102–8

[119] Manikandan A and Subramanian K S 2014 Fabrication and characterisation of nanoporous zeolite based N fertilizer *Afr. J. Agr. Res.* **9** 276–84

[120] Saidur M R, Aziz A R A and Basirun W J 2017 Recent advances in DNA-based electrochemical biosensors for heavy metal ion detection: A review *Biosens. Bioelectron.* **90** 125–39

[121] Kobal A B, Horvat M, Prezelj M, Briški A S, Krsnik M and Dizdarevič T *et al* 2004 The impact of long-term past exposure to elemental mercury on antioxidative capacity and lipid peroxidation in mercury miners *J. Trace Elements in Med. and Bio.* **17** 261–74

[122] Safarpour H, Safarnejad M R, Tabatabaei M, Mohsenifar A, Rad F and Basirat M *et al* 2012 Development of a quantum dots FRET-based biosensor for efficient detection of Polymyxa betae. Canadian *J. Plant Pathology* **34** 507–15

[123] Tucci M, Grattieri M, Schievano A, Cristiani P and Minteer S D 2019 Microbial amperometric biosensor for online herbicide detection: Photocurrent inhibition of Anabaena variabilis *Electrochim. Acta* **302** 102–8

[124] Tortolini C, Bollella P, Antiochia R, Favero G and Mazzei F 2015 Inhibition-based biosensor for atrazine detection *Sensors and Actuators, B: Chemical.* **224** 552–8

[125] Martinazzo J, Muenchen D K, Brezolin A N, Cezaro A M, Rigo A A and Manzoli A *et al* 2018 Cantilever nanobiosensor using tyrosinase to detect atrazine in liquid medium *J. Env. Sci. and Health – Part B Pesticides, Food Contaminants, and Agricultural Wastes.* **53** 229–36

[126] Muenchen D K, Martinazzo J, Brezolin A N, de Cezaro A M, Rigo A A and Mezarroba M N *et al* 2018 Cantilever functionalization using peroxidase extract of low cost for glyphosate detection *Appl. Biochem. Biotechnol.* **186** 1061–73

[127] Santos J S, Pontes M S, Santiago E F, Fiorucci A R and Arruda G J 2020 An efficient and simple method using a graphite oxide electrochemical sensor for the determination of glyphosate in environmental samples *Sci. Total Environ.* **749** 142385

[128] Oliveira P C, Maximiano E M, Oliveira P A, Camargo J S, Fiorucci A R and Arruda G J 2018 Direct electrochemical detection of glyphosate at carbon paste electrode and its determination in samples of milk, orange juice, and agricultural formulation *J. Environmental Science and Health – Part B Pesticides, Food Contaminants, and Agricultural Wastes.* **53** 817–23

[129] Luo S, Wu Y and Gou H 2013 A voltammetric sensor based on GO-MWNTs hybrid nanomaterial-modified electrode for determination of carbendazim in soil and water samples *Ionics* **19** 673–80

[130] Özcan A, Hamid F and Özcan A A 2021 Synthesizing of a nanocomposite based on the formation of silver nanoparticles on fumed silica to develop an electrochemical sensor for carbendazim detection *Talanta* **222** 121591

[131] Dong Y, Yang L and Zhang L 2017 Simultaneous electrochemical detection of benzimidazole fungicides carbendazim and thiabendazole using a novel nanohybrid material-modified electrode *J. Agric. Food Chem.* **65** 727–36

[132] Maximiano E M, Cardoso C A L and Arruda G J 2020 Simultaneous electroanalytical determination of thiram and carbendazim in samples of fresh fruit juices in the presence of surfactants *Food Anal. Methods* **13** 119–30

[133] Arruda G J, de Lima F and Cardoso C A L 2016 Ultrasensitive determination of carbendazim in water and orange juice using a carbon paste electrode *J. Env. Sci. and Health – Part B Pesticides, Food Contaminants, and Agricultural Wastes.* **51** 534–9

[134] Comisión Europea. EU Pesticides database. Regulation (EC) No 1107/2009. 2009

[135] BRASIL. Ministério da Saúde Agência Nacional de Vigilância Sanitária. Agência Nacional de Vigilância Sanitária – ANVISA 2015

[136] Sun Y, Liang J, Tang L, Li H, Zhu Y and Jiang D *et al* 2019 Nano-pesticides: A great challenge for biodiversity? *Nano Today* **28** 100757

[137] Belz R G and Piepho H-P 2017 Predicting biphasic responses in binary mixtures: Pelargonic acid versus glyphosate *Chemosphere* **178** 88–98

[138] Agathokleous E, Feng Z, Iavicoli I and Calabrese E J 2019 The two faces of nanomaterials: A quantification of hormesis in algae and plants *Environment Int.* **131** 105044

[139] Calabrese E J and Blain R B 2011 The hormesis database: The occurrence of hormetic dose responses in the toxicological literature *Regulatory Toxicology and Pharmacology* **61** 73–81

[140] Chamsi O, Pinelli E, Faucon B, Perrault A, Lacroix L and Sánchez-Pérez J-M *et al* 2019 Effects of herbicide mixtures on freshwater microalgae with the potential effect of a safener *Annales de Limnologie – Int. Journal of Limnology* **55** 3

[141] Sharma V K, Sayes C M, Guo B, Pillai S, Parsons J G and Wang C *et al* 2019 Interactions between silver nanoparticles and other metal nanoparticles under environmentally relevant conditions: A review *Science of The Total Environment* **653** 1042–51

[142] Lead J R, Batley G E, Alvarez P J J, Croteau M-N, Handy R D and McLaughlin M J *et al* 2018 Nanomaterials in the environment: behavior, fate, bioavailability, and effects-An updated review *Environmental Toxicology and Chemistry* 37 2029–63

[143] Freixa A, Acuña V, Sanchís J, Farré M, Barceló D and Sabater S 2018 Ecotoxicological effects of carbon based nanomaterials in aquatic organisms *Science of The Total Environment* **619–620** 328–37

[144] Iavicoli I, Leso V, Fontana L and Calabrese E 2018 Nanoparticle exposure and hormetic dose–responses: an update *Int. Journal of Molecular Sciences* **19** 805

[145] Iavicoli I, Leso V, Beezhold D H and Shvedova A A 2017 Nanotechnology in agriculture: opportunities, toxicological implications, and occupational risks *Toxicology and Applied Pharmacology* **329** 96–111

[146] Selck H, Handy R D, Fernandes T F, Klaine S J and Petersen E J 2016 Nanomaterials in the aquatic environment: a European Union–United States perspective on the status of ecotoxicity testing, research priorities, and challenges ahead *Environmental Toxicology and Chemistry* 35 1055–67

IOP Publishing

Bionanomaterials for Environmental and Agricultural Applications

Ravindra Pratap Singh and Kshitij RB Singh

Chapter 14

Developments, utilization and applications of nanobiosensors for environmental sustainability and safety

Kingsley Eghonghon Ukhurebor, Uyiosa Osagie Aigbe, Robert Birundu Onyancha, Monday Onoyivwe Ama, Heri Septya Kusuma, Israel Uzuazor Siloko, Joseph Onyeka Emegha, Samuel Ogochukwu Azi, Ese Lawrence Esiekpe, Abel Inobeme and Ayodotun Oluwafemi Bobadoye

Presently, the developments, utilization and applications of nanobiosensors (NNBs) are recognised as one of the evolving vital ingredients in proffering remedies for most of the biomedical and environmental challenges especially as they relate to sustainability and safety issues. Nanomaterials (NMs)-based biosensors (BSs) mechanisms as well as their advancements are now being employed globally for solving some of the challenges faced in the environmental sector, specifically sustenance and safety issues. Hence, this chapter provides a precise overview of the developments, utilization and applications of NNBs for environmental sustainability and safety (ESS).

14.1 Introduction

At this time, one of the foremost problems confronting the larger society is the contamination of the environment with heavy metals, petroleum spills, particulate matter, noxious gases, industrial, agricultural and domestic chemicals/composites, etc [1–6]. These contaminations are normally found in the air (atmospheric environment), soil (terrestrial environment) and water (aquatic environment), and if not properly monitored and managed, the resultant effects are mostly detrimental [1, 7].

The rising amount of possibly detrimental pollutants in the environment requires quick, eco-friendly, cost-effective and optimal systematic procedures to be employed and utilized in the extensive developments, utilization and applications of programs

doi:10.1088/978-0-7503-3863-9ch14

for ESS. Moreover, for the past few decades, an increasing amount of inventiveness and jurisdictive actions for the control and management of environmental issues vis-à-vis ESS have been implemented together with growing systematic and public concern [8–10].

The necessities for developments, utilization and applications of most traditional systematic procedures for the control and management of environmental issues, habitually constitute a critical impediment for their consistent development, utilization and applications on a consistent basis. The requirement for disposable arrangements or mechanisms for the development, utilization and applications in the environmental domain, has stimulated the evolution of innovative machinery and additional suitable practices. In this situation, BSs or NNBs seem like a suitable complementary systematic mechanism. Hence, it is essential to explore innovative machinery such as nanotechnology (NTech) or bionanotechnology (BNTech), specifically NNBs, which can be utilized for the monitoring, detecting, and remediating of these contaminants for the sole purpose of ESS [11, 12]. The NTech or NNBs, to be specific, offer enormous capabilities and machinery to enhance the present quality of the entire ecosystem. Similarly, they have the potential of offering feasible solutions to the present global environment challenges [4, 7].

Innovative machinery is presently and recurrently being established and explored in utilizing the diverse constituents' types as well as a wide range of measures for the contaminants' or impurities' remediation from the atmospheric environment (air), terrestrial environment (soil) and aquatic environment (water). Albeit, environmental contaminants dilapidation is occasionally stimulating due to the impediment of various composites mixture, their improved volatility and low-level reactivity [4]. The current emphasis has been on the application, utilization and development of NMs such as quantum dots (QDs) NMs, metal/metallic oxide NMs, and carbon NMs and/or BNMs for environmental remediation (ER) machinery enhancement owing to the restricted physical possessions of NMs/BNMs, enhanced reactivity, and their excellent surface interaction [4]. For effective ER of impurities, they can also be embedded (functionalized) with functional groups (FGs) that can target precisely targeted impurities (contaminants). The envisioned modification of NMs/BNMs physical possessions such as dimensions, morphology, penetrability, and chemical arrangement could give additional advantageous features that can directly influence their performance for ER purposes [13]. In proffering a solution to the raising requirement to monitor and manage the contamination of the environment (environmental pollution problems) confronting the larger society, some remediation machinery established utilizes the *in situ* and *ex situ* procedures in treating the contaminated atmospheric environment (air), terrestrial environment (soil) and aquatic environment (water). ER measures have been continually established and refined; the embracing of this state-of-the-art machinery, as well as the evolving of the processes of ER, is an essential ingredient for confronting and mitigating these environmental menaces that are now a threat to the whole world [1, 14].

These environmental menaces as they relate to the contamination or pollution of the environment could be allied to the action of anthropogenic activities (human caused events) and natural sources which might be linked to the accessibility and tenacity of several impurities that are noxious in nature. These noxious impurities are generally from agricultural processes, domestic activities, and industrial processes [1]. Studies have shown that most of the hostile and detrimental effects ensuing from these agricultural processes, domestic activities, and industrial processes are contributing not just to the climatic and environmental complications presently confronting the entire planet but also contribute significantly to some of the stimulating problems confronting the telecommunication domain, radio waves propagation to be specific [6, 15–22]. There are several reported decontamination processes and machineries that have been recognized previously all over the world [23–25]. According to Gavrilescu [26], most of these previous processes and machineries employed for ESS (which are mainly conventional) hardly hold the ability to persist with a lesser long-term environmental effect which is determined to place an impact objectively by employing an advanced priority on social and economic influences. However, to attain an advanced level of ESS, innovative processes and machineries such as BSs/NNBs mechanisms for effective environmental monitoring and management need to be developed and evolved [4].

The term sustainability according to the report from the World Commission on Environment and Development (WCED) is the ability to meet the request of the present requirements, without surrendering the competence of upcoming associates to come in contact with these requirements [4]. Ukhurebor and Adetunji [4] also, adopted their definition of the term sustainability from the World Bank (WB) definition as the condition or a procedure that could conserve unspecified requirements without advanced decreasing of the values of the external or internal qualities of the system in which several procedures that could occur as the requirements are stunned. The procedure involved in the lessening of the footprint of the environment involves the emissions of air, environmental influences, constituents, greenhouse gases (GHGs), terrestrial (land), waste, water (aquatic) as well as the utilization, developments and applications of the environmental and biodegradable ingredients that are sustained by the ‘green’ features [4]. Besides, it has been known that green and sustainable remediation (GSR) involves an all-inclusive economic influence, social and environmental remediation phases [27]. This implies that GSR addresses the protection of the wellbeing of living organisms together with the conservation of an eco-friendly (cleaner) environment, as well as the lessening of several side impacts that could affect the entire environment [4, 27, 28].

NTech/BNTech machineries such as NNBs mechanisms have been recognised as a process that involves bioscientific proceedures that exploit the application of modern technology for the manipulation of the effective generation of innovative and additional procedures for the successful attainment of more sustainable products that could improve ESS as well as having other scientific benefits. The recent utilization, development and applications of NNBs have been reported as one of the dynamic constituents in proffering remedies for most of the environmental challenges, particularly with regard to sustainability and safety problems; BSs founded on NMs as well as their progressions are the utmost employed mechanisms for solving some of the environmental challenges faced globally. Hence, this chapter presents an outline on the utilization, developments and applications of NNBs in the

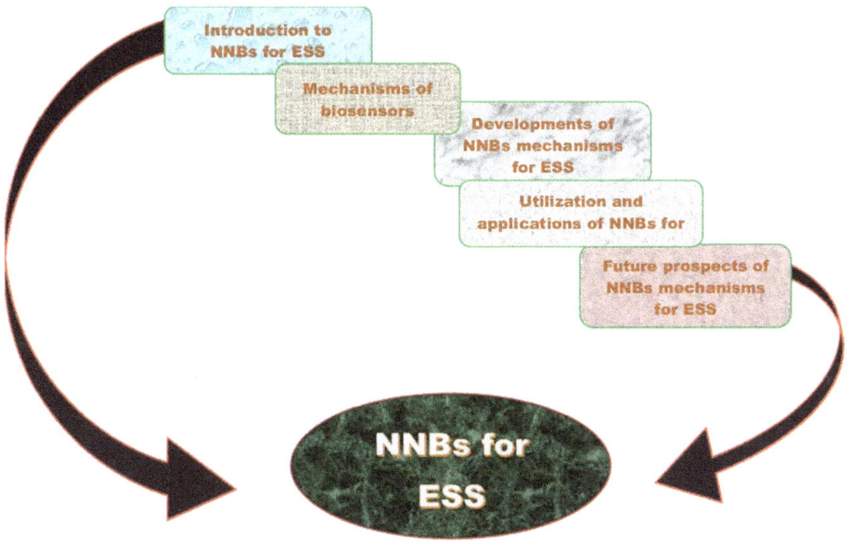

Figure 14.1. The diagrammatic illustration of major constituent parts of the chapter (ESS: environmental sustainability and safety, NNBs: nanobiosensors).

environmental science domain drawn from existing literature. The remaining segments of the chapter are designed as follows: section 14.2 discusses briefly the mechanisms of BSs. Some of the utmost developments of NNBs mechanisms for ESS are discussed in section 14.3, while section 14.4 deliberates on the utilization and applications of NNBs mechanisms for ESS. The last section (section 14.5) is the concluding section which comprises some discussions on the future prospects of NNBs mechanisms for ESS tendencies. Figure 14.1 shows the diagrammatic illustration of major constituent parts of the chapter in the form of a graphical abstract.

14.2 Overview on the mechanisms of biosensors

14.2.1 Components and mechanism of action of biosensors

BSs have a bounty of valuable benefits or applications in the several scientific and technological domains such as fields of agricultural science, environmental science, biomedical sciences, etc [29–34]. Owing to these benefits or applications in the various fields, scientists/researchers are presently exploiting, establishing, utilizing, exploring and advancing several categories of BSs in various areas of human activities [35, 36]. According to Turner *et al* [37], ever since the very first notable discovering of the BS by Leland Clark Jnr in the early 1960s, that was known as chemical/glucose ($C_6H_{12}O_6$)-based (commonly recognized as Clark electrode/BS), which was principally employed for the detection of the amount of oxygen (O_2) circulating in human blood [38], there have been several developments, advancements and evolutions in the mechanisms of BSs [29]. However, it was in the mid-1970s that BSs become somewhat popular and afterwards came the beginning of great advancement accomplishments, together with the boundless competence involved and the utilization and application of BS technologies with advanced propensities encompassing bioelectronics, BBNs, NTech, electrochemistry, etc [39–42], and now there are several BBNs established for numerous worthwhile initiatives.

The components that make up BSs are influential for their operational processes [29, 30]. Supposedly, all BSs are intended to accomplish or execute certain functions specifically. The technique of operation of a BS depends on the bioreceptor (antibody, DNA, enzyme, etc) as well as the sensing technique (the mechanisms of the transducer) [43, 44]. In most cases, the transducer's electrical signal is minor and intersects on a justly high baseline. On a normal basis, the operation of the BS's signal entails the presumption of a baseline position signal, collected from a comparable transducer in the absence of any coated biocatalyst. The BS's reaction is relatively slow, and this eases the problem of electrical sound percolation. However, the signal could be digitalized and transmitted to the microprocessor component where the information is handled, directed by ideal units and transferred to where the data are stored [45]. BSs' mechanisms depend on the sensing transducers or rudiments [29]. Invariably, they can operate with a living creature (inside or outside) but the target distinctive components to which the device is utilized to detect, monitor or measure frequently originate from the living creature or the environment. Hence, BSs' features may be characterized based on the origin of their actions of mechanism as a targeted analyte, bioreceptor, sensing/transduction approach (contemporary approaches are; optical, electrical, nano, photoelectric, piezoelectric, thermometric, photeoelectrochemical and electrochemical sensing approaches) and the nature of organic material utilized [35, 46–51]. The contemporary BS mechanisms have predominant advantages and benefits, which offer additional sensitivity, precision, swift, distinguishable and multiplicative results when compared to earlier established BSs [39]. Presently, with the evolution of NTech/BNTech, advanced NMs/BNMs are currently considered and their resourceful features vis-à-vis their advantages in ESS [29, 52]. BSs such as NNBs, polymers' BSs, and microbes' BSs encompass the alliance of material science, NTech/BNTech, molecular engineering, biotechnology (BioTech), applied and natural sciences (such as physics, mathematics, chemistry, biomedical sciences, agricultural science and environmental science) [29].

A BS can function inside or/and outside of a living system; but, the target distinctive parameter to which the mechanism is utilized for the detection or measurement regularly originates from the living system or the environment [29, 30]. As earlier stated, BSs are characterised based on the targeted analyte, the bioreceptor, the sensing or transduction procedures and the nature of the utilized biological parameters. Apparently, all these characteristics depend on the source of the mechanistic action of the BSs [29, 30].

14.2.2 Categories of biosensors

Typically, BSs are categorized into two main groups which are dependent or instituted on the nature of mechanisms of the sensor and transduction procedures.

The mechanism of the sensor comprises antibodies (immunosensors), microorganisms, (microbes), enzymes, biological (organic or living) organelles and tissues BSs. However, the transduction procedures are pinpointed on the physico-chemical distinction resulting from mechanisms of the BSs. However, the various mechanisms of the transducer BSs can possibly be calorimetric-based, electrochemical-based, optical-based and piezoelectric-based [53], and according to several research reports there are various categories of these mechanisms of the transducer BSs and some of the main categories of the piezoelectric transducer-based BSs are the ultrasonic-based and acoustic-based, those of the electrochemical transducer BSs are conductometric-based, amperometric-based and potentiometric-based, while those of the optical transducer BSs are chemiluminense-based, absorbance-based and fluorescence-based [29, 54, 55]. The schematic illustration of some of the main categories of the various mechanisms of the transducer BSs is shown in figure 14.2.

Another means of categorizing BSs is the period, time or order (which in this context is known as generation or period during which such BSs were released) of discovering, releasing or establishment of such BSs' mechanisms; in this context we have the first-generation BSs, second-generation BSs, and third-generation BSs [29, 30].

BSs that are produced and made available over a given duration of time with comparable features fit into a particular generation of BSs; like other mechanisms such as computers as well as other mechanical and electronic mechanisms, the advanced generations often display more complexities than those of the previous generations. These complexities according to Ukhurebor [29], could possibly be defined with respect to the rate of consumption of energy, precision, the integration scale, multi-functionalities, magnitudes, modification, etc.

The development and utilization of BSs (NNBs to be specific) have attained some considerable benefits and applications in several endeavours such as biomedical sciences, ESS, agricultural science (food production, processing and safety), social/management sciences, etc. BSs' discovery has occasioned great evolution of detailed and leading analytical and problem-solving mechanisms via biological sensing mechanisms known as NNBs [32, 33]. However, the various categories of BSs' mechanisms have their development, utilization and applications as well as machinery modifications.

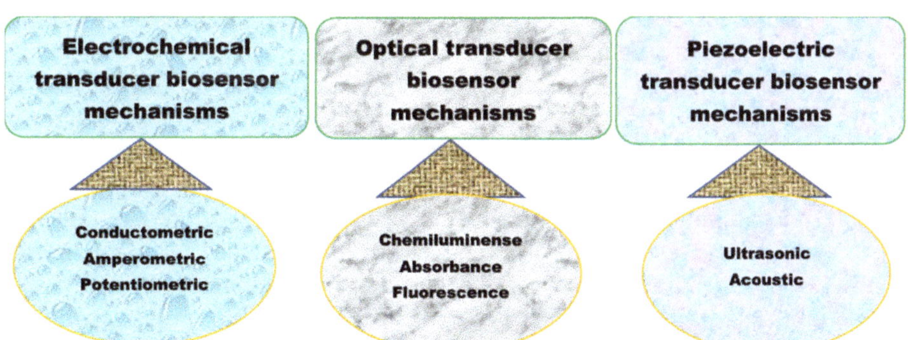

Figure 14.2. Some of the main categories of the various mechanisms of the transducer biosensor.

Presently, there are several established types of BSs such as the electrochemical-based BSs (ECBSs), mass-based BSs (MBBSs), thermal-based BSs (TBSs), calorimetric-based BSs (CBSs), photoelectrochemical-based BSs (PECBSs), microbial-based BSs (MBSs), and optical-based BSs (OBSs), etc and these various types of BSs have their subtypes [29, 30]. However, the recent trend in the NTech/BNTech field is the NNBs that use NMs such as carbon nanotubes (CNTs), dendrimers, hydrogels, liposomes, micelles and polymers for handling biomedical, environmental and other challenges confronting humanity.

14.3 Developments of NNBs mechanisms for ESS

The necessities for the application of most traditional investigative procedures in the environmental domain for monitoring and analysis of environmental pollutants, frequently institute a significant impairment for their application on a consistent and systematic basis. The necessity for non-refundable systems or mechanisms for environmental monitoring and applications, has reinvigorated the development and evolution of innovative technologies and more appropriate procedures.

In the ESS backgrounds, NNBs appear as an appropriate alternative or better still a complementary investigative mechanism. NNBs could be considered as a sub-set of chemical-based BSs in which an organic mechanism is utilized for the detection of analytes basically for ESS tendencies [29, 30, 56–58]. Biosensing mechanism systems and procedures are being established, developed, advanced and have evolved as appropriate mechanisms for diverse applications, such as controlling and monitoring of bioprocess, controlling and monitoring food quality, agricultural processing, military operations, as well as for biomedical applications [29, 30, 58].

Presently, the use of some recent NMs such as CNTs, dendrimers, hydrogels, liposomes, micelles and polymers to nano-encapsulation particles that control or determine the features of the environment are now possible. Nano-encapsulation procedures improve the features of the environment. NMs in the environmental sector are utilized in this circumstance in two different ways to preserve and conserve the environment for ESS:

- NNBs are used as additives in the polymeric media from which the developments, utilization and applications materials are produced. The addition of NMs advances the mechanical, permeability, and antimicrobial characteristics of the environment.
- Also, the NNBs are included in the core parameters for the development, utilization and applications for ESS that describe the quality of the environment. Generally, some metal and metal oxide NMs are utilized for improving ESS.

In the environmental domain, the foremost advantages offered by BSc/NNBs over conventional investigative procedures are the prospect of transferability, miniaturization, work on-location, and the capability to measure, quantify and monitor contaminants in multifaceted matrices with negligible sample groundwork. NNBs are utilized as tools for environmental quality monitoring, analysis and management.

There are presently several establishments and institutions that are working in the aspect of BSs/NNBs and bioelectronics globally; some of these establishments and institutions are involved in the fabrication and marketing of BSs/NNBs, while others only offer the relevant ingredients and instruments for the manufacture of BSs/NNBs. Some of these establishments and institutions are working on existing BBNs machineries and only a limited number of them are evolving and developing innovative technologies [59].

The application of innovative biodevices to real-time environmental models is an essential aspect in the concluding phases of development. However, notwithstanding the huge number of recently developed BSs/NNBs, some literature references overlook the actual phase and only concentrate on reporting the applications of the BSs/BBNs. There is still some work in most of the revised systems, before their applications to real models could be properly established, and the study of the matrix influences, the stability problems as well as the cautious comparison with recognized procedures are critical steps in this method.

Even though large-scale returns from NNBs used in the environmental domain are considerably less than those from the biomedical domain, public attention and funding from governments and other relevant agencies have produced great research determination aimed at the development, utilization and applications of NNBs in the monitoring, measurement and analysis of contaminants as well as other environmental hazards for ESS [59]. Most BSs/NNBs mechanisms have been verified only on purified water or buffered solutions, but additional NNBs that could be utilized in real models have recently appeared. In this setting, BBNs for possible environmental utilization and applications continue to demonstrate developments in several research areas such as genetic (hereditary) modification of enzymes and microbes, the enhancement of recognition components, immobilization as well as BS interfaces.

14.4 Utilization and applications of NNBs mechanisms for ESS

The recognition of the advantages or benefits of NNBs for ESS is growing relatively in the scientific world (research) and has stimulated great attention from those in the industrial segment as well. However, the commercial (large-scale) utilization and application of NNBs for ESS are at present somewhat new and scarce. The rapid developments, utilization, applications as well as the understanding of the mechanisms of NNBs for ESS that designate the features of NMs make the NTech/BNTech field of study auspicious and will undoubtedly soon be connected to the larger market. The success of the developments, utilization and applications of NTech/BNTech for ESS will depend largely on the level of acceptance of these evolving technologies (NTech/BNTech). This creates to some extent some challenges for researchers (scientific community) that are involved in this field, governments as well as other relevant stakeholders.

Some existing examples of BSs/NNBs established and utilized for the detection and remediation of heavy metal and phenolic composites in the environment for ESS tendencies are summarized in table 14.1, while table 14.2 encompass some

Table 14.1. Some established BSs/NNBs that are developed and utilized for the detection, monitoring and remediation of heavy metal and phenolic composites in the environment for ESS tendencies.

Analytes	Recognition biological components (biocomponents)	Transduction mechanisms	Background	References
Copper (Cu), cadmium (Cd) nickel (Ni) and Zinc (Zn)	*Pseudomonas fluorescens* 10586s pUCD607 with the lux insertion on a plasmid	Optical-based BSs/NNBs	Soil	[60]
Cd	DNA	Electrochemical-based BSs/NNBs	normal (standard) solutions	[61]
Cd	Phytochelatins	Optical-based BSs/NNBs	Normal (standard) solutions	[62]
Cd, mercury (Hg), and arsenic (As)	Urease enzyme	Electrochemical-based BSs/ NNBs	Normal (standard) solutions	[63]
Nickel (Ni) ions	*Bacillus sphaericus* strain MTCC 5100	Electrochemical-based BSs/ NNBs	Effluents from industry and foods	[64]
Cd, Cu, Ni, Zn, aluminum (Al), lead (Pb) and iron (Fe)	*Chlorella vulgaris* strain CCAP 211/12	Electrochemical-based BSs/NNBs	Municipal waters	[65]
Cd	*Escherichia coli* RBE23–17	Electrochemical-based BSs/ NNBs	Wastewater	[66]
Zn, Cu and cobalt (Co)	*Pseudomonas sp.* B4251, *bacillus cereus* B4368 and *E.coli* 1257	Electrochemical-based BSs/NNBs	Water	[67]
Mercury (II) ion (Hg^{+2}) and lead (II) ion (Pb^{+2})	DNA	Optical BSs/NNBs	Water	[68]
Copper (I) ion (Cu^{+}) and copper (II) ion (Cu^{+2})	DsRed (red fluorescent) protein	Optical-based BSs/NNBs	Normal (standard) solutions	[69]
Cd, Cu and Pb	Sol–gel-immobilizedurease	Electrochemical-based BSs/NNBs	Synthetic (artificial) effluents	[70]
Binary mixtures: phenol (C_6H_6O)/ clorophenol (C_6H_5ClO), catechol ($C_6H_6O_2$)/C_6H_6O), cresol (C_7H_8O)/clorocresol (C_7H_7ClO) and C_6H_6O/C_7H_8O	Laccase and tyrosinase	Amperometric multicanal-based BSs/ NNBs	Wastewater	[71]
M-cresol (C_7H_8O) or catechol ($C_6H_6O_2$)	DNA	Amperometric-based BSs/NNBs	Wastewater	[72]
C_6H_6O	Mushroom tissue (tyrosinase)	Amperometric-based BSs/ NNBs	Wastewater	[73]
C_6H_6O, p-cresol (C_7H_8O), C_7H_8O and $C_6H_6O_2$	Polyphenol oxidase	Amperometric-based BSs/ NNBs	Wastewater	[74]

Table 14.2. Some established BSs/NNBs developed and utilized for the detection and monitoring of pesticides and herbicides in the environment for ESS tendencies.

Analytes	Form of interaction	Recognition biocatalyzer	Transduction mechanisms	References
Simazine (C_7H_{12} ClN_5)	Biocatalytic	Peroxidase	Potentiometric-based BSs/NNBs	[75]
Isoproturon ($C_{12}H_{18}N_2O$)	Biocatalytic	Condensed antibody	Immunosensor immunoreaction-based BSs/NNBs	[76]
Parathion ($C_{10}H_{14}NO_5PS$)	Biocatalytic	Parathion hydrolase	Amperometric-based BSs/NNBs	[77]
Paraoxon ($C_{10}H_{14}NO_6P$)	Biocatalytic	Alkaline phosphatase	Optical-based BSs/NNBs	[77, 78]
Carbaril ($C_{12}H_{11}NO_2$)	Biocatalytic	Acetilcolinesterase	Amperometric-based BSs/NNBs	[79]
2,4- diclorofenoxiacetic ($C_8H_6C_{12}O_3$)	Immunoanalysis	Acetilcolinesterase	Amperometric-based BSs/NNBs	[80]
Diuron ($C_9H_{10}C_{12}N_2O$), paraquat (Methyl viologen; $C_{12}H_{14}C_{12}N_2$)	Biocatalytic	Cyanobacterial	Bioluminescence-based BSs/NNBs	[81]

BSs/NNBs that are established and utilized for the detection and monitoring of pesticides and herbicides, as modified from Silva *et al* [59].

Several BSs or NNBs reported, established and utilized in the environmental sector demonstrate the potential to be advanced for either sole-sample setups for field screening tendencies or incessant setups for field monitoring and analysis applications. A discussion regarding the clean-up of a Superfund location may offer examples of the possibility and categories of environmental screening, monitoring and analysis issues for which BSs/NNBs could offer exceptional solutions.

Most of the studies on metal-specific microbial BSs/NNBs have been carried out by employing a liquid suspension of feasible biosensor microbes. However, an additional advanced method is to utilize these microbes in the BSc/NNBs mechanisms, such as by putting out of action the cells (in what is known as immobilization) onto optical fibres associated with a photo discovery device. This category of fibre-optical-based BSs has been earlier built for Cu, geno-toxicants, or wide-ranging noxiousness of industrial waste. These fibre-optical-based BSs could simply be carried to the field and utilized for online measurement and monitoring. There are presently several developed fibre-optical-based BSs or NNBs for the monitoring and analysis of environmental models, such as soils and residues *in situ* [59]. In all, the modification in inhibition of enzymes urease and acetylcholine esterase via the heavy metal ions developed the prospect of designing and evolving BSs or NNBs for discrimination between diverse categories of water impurities and the mechanism is simple, transferrable and robust.

Presently, there are some developed and well-established BSs/NNBs such as the CNTs-based BSs/NNBs, quantum dots (QDs) NMs-based BSs/NNBs, metal/metallic oxide NMs-based BSs/NNBs, and microbial fuel cell-based BSs/NNBs which function or operate on the principle of optical and other tendencies; they are basically designed, utilized and developed for the detention, control, monitoring, and management of biochemical oxygen in the environment as well as for the detection, control, monitoring and management of environmental contaminations ensuing from heavy metals, phenolic composites, pesticides, herbicides and other chemicals in the environment for ESS tendencies [82–93]. According to the review study by Neethirajan *et al* [94], BSs/NNBs and biosensing machineries with their wide-ranging applications, are being extensively utilized in tackling some of the utmost challenges in the agricultural sector (especially as it relates to food production and sustainability) as well as the ESS issues.

14.5 Conclusion and future prospects of NNBs mechanisms for ESS

The present environmental challenges that are confronting the entire globe have become a cause of great concern. Consequently, the proper utilization, exploration and exploitation of the various components of the ecosystem play an essential role in ESS and other dynamic progressions within the entire ecosystem. It is expected that by implementing and utilizing suitable procedures, industrial aids will increase and the adverse effects ensuing from most industrial events particularly as they relate to environmental menaces will considerably be mitigated. Undoubtedly, ESS procedures offer a dynamic role in enhancing economic, environmental, social as well as other benefits. It is exceptionally critical to secure and sustain environmental processes and maintain eco-friendly development of industrial innovations. State-of-the-art processes such as the utilization of NTech/BNTech like NNBs machinery for the monitoring and management of the environmental processes are crucial and require further enhancement and development.

The present utilization and application of NTech/BNTech promises a revolution in the enhancement and development beyond our presently appalling environmental effectiveness via NNBs machineries. Evolving our understanding of the hypothetical basics of the environmental systems coupled with the utilization of progressive concepts, models, and/or theories such as the chaos and string theories may expose novel avenues. NTech/BNTech requires a systematic understanding of materials science, as well as its fabrication and machineries, in combination with knowledge of the environmental systems. The precision of this challenge could possibly attract brilliant minds to work in the environmental sectors to the mitigation of environmental challenging issues. NTech/BNTech in the environmental domains could still take a couple of years to move to a more advanced level, especially since it has to evade the drawbacks experienced with NTech/BNTech generally. For this to materialise, continuous funding and consideration by researchers and policy developers, together with practical expectations, would be critical for this promising and emerging field to blossom.

The development of NNBs for ESS is rapidly becoming prominent and as described in this chapter, NMs and recognition agents are incessantly being combined in creative and innovative means. The topical development in BNNs mechanism is aimed at overcoming the deficiencies of initial developed sets of BSs such as the non-specific binding, the variation in the magnitude of the particles, NMs aggregation, and NMs stability. Questions about the assessment selectivity and sensitivity in multifaceted environmental matrices remains; nevertheless, a rising number of studies and reports are employing representative matrices to determine the stability and selectivity of their BSs. The robustness of BSs that are employed in fields is a necessity if individuals are going to be allowed to examine their respective environments.

For an improved, accessible and friendly environment for all and sundry, the study of smart systems is in pressing demand for ESS. To accomplish this task, BSs or NNBs are evolving as a critical analytical device for active monitoring and targeted biomarker discovery for quality environment assessment. Advancements in sensing constituent development together with enhanced integration made biosensing mechanisms smart, which meets the appropriate requirements for ESS. Also, the interfacing of biosensing models with microfluidic classifications and the internet of environmental things made biosensing based investigations accessible. Additionally, the introduction of artificial intelligence (AI) and the internet of things (IoT) are similarly becoming an imperative instrument in analysing biosensing data to properly understand and predict tendencies in correlating various environmental procedures and implications. In developing biosensor mechanisms of anticipated performance, it is critical to enhance and develop a biosensing model which tolerates and offers for the anticipated micro-environment to a bio-active required for extreme binding without compromise in the functionality for executing selective detections for ESS.

In summary, this present chapter has revealed some thorough and precise information about the recent developments, utilization and applications of NNBs for ESS. Conclusively, it is recommended that future research studies on environmental procedures and implications should lay more prominence on developing eco-friendly/less harmful, more operative, cost-effective, and progressively ecological techniques such as the ones involving NTech/BNTech, especially the NNBs mechanisms together with the use or introduction of AI and IoT. The government and relevant stakeholders as well as researchers should put in place evidence-based and action-oriented policy frameworks for the management and mitigation of the dilapidation of the environment. Correspondingly, incessant prospects for imminent studies and establishment of the theoretical route as well as the systematic foundation for the application of eco-friendly techniques and measures in environmental procedures and implications should be reinvigorated. The scientific community, in addition to producing solutions via nanotechnological means for ESS, should produce the systematic basis for assessing risks of the use of NNBs, especially as it relates to migration, toxicity as well as the procedures or means of disposing of final waste products for domestic and industrial sources. The government and all relevant stakeholders should consolidate on the rules, regulations and standards for the development, utilization and applications of NNBs for ESS, as well as pursue the incorporation of these nanotechnological studies (NTech/BNTech) in the educational sector, since the fact that the knowledge of the form of NTech/BNTech would further push for its overall acceptance as well as its appropriate development, utilization and applications.

Acknowledgement

The authors express their gratitude to their respective institutions and to authors whose publications were used in writing this chapter.

References

[1] Ukhurebor K E, Aigbe U O, Onyancha R B, Nwankwo W, Osibote O A, Paumo H K, Ama O M, Adetunji C O and Siloko I U 2021 Effect of hexavalent chromium on the environment and removal techniques: a review *J. Env. Manage.* **280** 111809

[2] Nwankwo W, Adetunji C O, Ukhurebor K E, Panpatte D G, Makinde A S and Hefft D I 2021 Recent advances in application of microbial enzymes for biodegradation of waste and hazardous waste material *Microbial Rejuvenation of Polluted Environment. Microorganisms for Sustainability* ed C O Adetunji, D G Panpatte and Y K Jhala vol 27 (Cham: Springer Nature) pp 35–56

[3] Ukhurebor K E, Aigbe U O, Onyancha R B and Adetunji C O 2021 Climate change and pesticides: their consequence on microorganisms *Microbial Rejuvenation of Polluted Environment. Microorganisms for Sustainability* ed C O Adetunji, D G Panpatte and Y K Jhala vol 27 (Cham: Springer Nature) pp 83–113

[4] Adetunji C O and Ukhurebor K E 2021 Recent trends in utilization of biotechnological tools for environmental sustainability *Microbial Rejuvenation of Polluted Environment. Microorganisms for Sustainability* ed C O Adetunji, D G Panpatte and Y K Jhala vol 27 (Cham: Springer Nature) pp 239–63

[5] Ukhurebor K E, Nwankwo W, Adetunji C O and Makinde A S 2021 Artificial intelligence and internet of things in instrumentation and control in waste biodegradation plants: recent Developments *Microbial Rejuvenation of Polluted Environment. Microorganisms for Sustainability* ed C O Adetunji, D G Panpatte and Y K Jhala vol 27 (Cham: Springer Nature) pp 265–79

[6] Nwankwo W and Ukhurebor K E 2019 An x-ray of connectivity between climate change and particulate pollutions *J. Adv. Res. Dyn. Control Syst. (discontin.)* **11** 3002–11

[7] Yunus I S, Harwin, Kurniawan A, Adityawarman D and Indarto A 2012 Nanotechnologies in water and air pollution treatment *Rev. Environ. Sci. Biotechnol.* **1** 136–48

[8] Arora N 2013 Recent advances in biosensors technology: a review *Octa Journal of Biosciences* **1** 147–50

[9] Silva C L M, Melo A F and Salgado A M 2011 Biosensor for environmental applications *Environmental Biosensors* ed V Somerset (Rijeka: IntechOpen) pp 3–16

[10] Salouti M and Khadivi Derakhshan F 2020 Biosensors and nanobiosensors in environmental applications *Biogenic Nano-Particles and their Use in Agro-ecosystems* ed M Ghorbanpour, P Bhargava, A Varma and D Choudhary (Singapore: Springer)

[11] Ukhurebor K E and Adetunji C O 2021 Relevance of biosensor in climate smart organic agriculture and their role in environmental sustainability: what has been done and what we need to do *Biosensors in Agriculture: Recent Trends and Future Perspectives. Concepts and Strategies in Plant Sciences* ed R N Pudake, U Jain and C Kole (Cham: Springer) pp 115–36

[12] Adetunji C O, Nwankwo W, Ukhurebor K E, Olayinka A S and Makinde A S 2021 Application of biosensor for the identification of pests and diseases mitigating against increase in agricultural production: recent advances *Biosensors in Agriculture: Recent Trends and Future Perspectives. Concepts and Strategies in Plant Sciences* ed R N Pudake, U Jain and C Kole (Cham: Springer) pp 169–89

[13] Guerra F D, Attia M F, Whitehead D C and Alexis F 2018 Nanotechnology for environmental remediation: materials and applications *Molecules.* **23** 1760

[14] Yadav K K, Singh J K, Gupta N and Kumar V J 2017 A review of nanobioremediation technologies for environmental cleanup: a novel biological approach *Journal of Materials and Environmental Science.* **8** 740–57

[15] Nwankwo W and Ukhurebor K E 2021 Big data analytics: a single window IoT-enabled climate variability system for all-year-round vegetable cultivation *IOP Conf. Series: Earth and Environmental Science* 655 012030

[16] Chinedu P U, Nwankwo W, Ukhurebor K E and Benson B U 2020 Throughput in wireless LAN: Implications of some meteorological variables in undulating landscape *Nigerian Research Journal of Engineering and Environmental Sciences.* **5** 633–43

[17] Ukhurebor K E and Nwankwo W 2020 Estimation of the refractivity gradient from measured essential climate variables in Iyamho-Auchi, Edo State, South-South region of Nigeria *Indonesian Journal of Electrical Engineering and Computer Science.* **19** 276–84

[18] Nwankwo W, Olayinka S A and Ukhurebor K E 2020 Green computing policies and regulations: A necessity? *Int. Journal of Scientific & Technology Research.* **9** 4378–383

[19] Ukhurebor K E and Azi S O 2019 Review of methodology to obtain parameters for radio wave propagation at low altitudes from meteorological data: New results for Auchi area in Edo State, Nigeria *Journal of King Saud University – Science.* **31** 1445–451

[20] Ukhurebor K E, Olayinka S A, Nwankwo W and Alhasan C 2019 Evaluation of the effects of some weather variables on UHF and VHF receivers within Benin City, South-South region of Nigeria *Journal of Physics: IOP Conf. Series* 1299 012052

[21] Ukhurebor K E and Odesanya I 2019 Relationship between meteorological variables and effective earth radius factor over Auchi, Edo State, South-South, Nigeria *Covenant Journal of Physical & Life Sciences.* **7** 1–10

[22] Ukhurebor K E and Umukoro O J 2018 Influence of meteorological variables on UHF radio signal: recent findings for EBS, Benin City, South-South, Nigeria *IOP Conf. Series: Earth & Environment Science* 173 012017

[23] Gavrilescu M and Macoveanu M 2000 Attached growth process engineering in wastewater treatment *Bioprocess Engineering.* **23** 95–106

[24] Khan M A, Hsain T and Hejazi R 2014 An overview and analysis of site remediation technologies *J. Env. Manage.* **71** 95–122

[25] Kumar A, Bisht B S, Joshi V D and Dhewa T 2011 Review on bioremediation of polluted environment: a management tool *Int. Journal of Environmental Sciences.* **1** 1079–93

[26] Gavrilescu M 2011 Sustainability *Comprehensive Biotechnology* ed M Moo Young and C Webb vol IV 2nd edn (New York: Wiley) pp 95–923

[27] Reddy K R and Adams J A 2010 Towards green and sustainable remediation of contaminated sites *6th Int. Congress of Environmental Geotechnics (New Delhi)*

[28] Ellis D E and Handley P W 2009 *Sustainable Remediation White Paper–Integrating Sustainable, Principles, Practices and Metrics into Remediation Projects, Remediation.* (New York: John Wiley)

[29] Ukhurebor K E 2020 The role of biosensor in climate smart organic agriculture towards agricultural and environmental sustainability *Agrometeorology* ed R S Meena (London: IntechOpen)
[30] Kerry R G *et al* 2021 A comprehensive review on the applications of nano-biosensor based approaches for non-communicable and communicable disease detection *Biomaterials Science* **9** 3576–602
[31] El-Said W A, Abdelshakour M, Choi J H and Choi J W 2020 Application of conducting polymer nanostructures to electrochemical biosensors *Molecules.* **25** 1–11
[32] Salek-Maghsoudi A, Vakhshiteh F, Torabi R, Hassani S, Ganjali M R, Norouzi P, Hosseini M and Abdollahi M 2018 Recent advances in biosensor technology in assessment of early diabetes biomarkers *Biosensors and Bioelectronics* **99** 122–35
[33] Vigneshvar S, Sudhakumari C C, Senthilkumaran B and Prakash H 2016 Recent advances in biosensor technology for potential applications—an overview *Frontiers in Bioengineering and Biotechnology* **4** 1–9
[34] Antonacci A, Arduini F, Moscone D, Palleschi G and Scognamiglio V 2016 Biosensors for sustainable food—new opportunities and technical challenges *Comprehensive Analytical Chemistry* ed V Scognamiglio, G Rea, F Arduini and G Palleschi vol 74 (Amsterdam: Elsevier) pp 315–40
[35] Mehrotra P 2016 Biosensors and their applicationsa review *Journal of Oral Biology and Craniofacial Research.* **6** 153–59
[36] Farzin L, Shamsipur M, Samandari L, Sadjadi S and Sheibani S 2020 Biosensing strategies based on organic-scaffolded metal nanoclusters for ultrasensitive detection of tumor markers *Talanta* **214**
[37] Turner A P F, Karubeand I and Wilson G S 1987 *Biosensors Fundamentals and Applications* (Oxford: Oxford University Press)
[38] Clark L C Jr and Lyons C 1962 Electrode systems for continuous monitoring in cardiovascular surgery *Annals of the New York Academy of Sciences.* **102** 29–45
[39] El-Said W A, Yea C H, Kwon I-K and Choi J W 2009 Fabrication of electrical cell chip for the detection of anticancer drugs and environmental toxicants effect *BioChip Journal* **3** 105–12
[40] Nikoleli G-P, Nikolelis D P, Siontorou C, Karapetis S, Bratakou S and Tzamtzis N 2018 Nanobiosensors based on graphene electrodes: recent trends and future applications *Applications of Nanomaterials* (Amsterdam: Elsevier)
[41] Fracchiolla N S, Artuso S and Cortelezzi A 2013 Biosensors in clinical practice: focus on oncohematology *Sensors (Basel).* **13** 6423–447
[42] Turner A P 2013 Biosensors: sense and sensibility *Chemical Society Reviews* **42** 3184–196
[43] Khan M Z H, Hasan M R, Hossain S I, Ahommed M S and Daizy M 2020 Ultrasensitive detection of pathogenic viruses with electrochemical biosensor: state of the art *Biosensors and Bioelectronics.* **166** 112431
[44] Khan M Z H 2020 Recent biosensors for detection of antibiotics in animal derived food *Critical Reviews in Analytical Chemistry.*
[45] Zhang L, Guo W and Lu Y 2020 Advances in cell-free biosensors: principle, mechanism, and applications *Biotechnology Journal* **15** 2000187
[46] Kahn K and Plaxco K W 2010 Principles of biomolecular recognition *Recognition Receptors in Biosensors* ed M Zourob vol 9 (New York: Springer) pp 3–46
[47] Bhushan B 2009 Biomimetics: lessons from nature—an overview *Philosophical Trans. of the Royal Society A: Mathematical, Physical and Engineering Sciences* **67** 1445–86

[48] Monošík R, Stred'anský M and Šturdík E 2012 Biosensors—classification, characterization and new trends *Acta Chimica Slovaca* **5** 109–20

[49] Niu X, Zhong Y, Chen R, Wang F, Liu Y and Luo D A 2018 'Turn-on' fluorescence sensor for Pb^{2+} detection based on graphene quantum dots and gold nanoparticles *Sensors Actuators* B **255** 1577–581

[50] Long R, Guo Y, Xie L, Shi S, Xu J and Tong C *et al* 2020 White pepper-derived ratiometric carbon dots for highly selective detection and imaging of coenzyme A *Food Chemistry* **315** 126171

[51] Farzin L, Shamsipur M, Samandari L, Sadjadi S and Sheibani S 2020 Biosensing strategies based on organic-scaffolded metal nanoclusters for ultrasensitive detection of tumor markers *Talanta* **214** 120886

[52] Rodriguez-Mozaz S, Alda M J and Marco M P 2005 Biosensors for environmental monitoring: a global perspective *Talanta* **65** 291–97

[53] Reyes De Corcuera J I and Cavalieri J 2003 Prototype instruments for laboratory and on-line measurement of lipoxygenase activity *Food Science and Technology Int.* **9** 5–9

[54] Kerry R G, Sahoo S, Das G and Patra J K 2019 *Theranostic Application of Nanoparticulated System: Present and Future Prospects.* ed T A Inamuddin, M I Rangreez, Ahamed and A M Asiri vol 47 (Millersville, TN: Materials Research Foundation) pp 241–88

[55] Singh S, Kumar V, Dhanjal D S, Datta S, Prasad R and Singh J 2020 Biological biosensors for monitoring and diagnosis *Microbial Biotechnology: Basic Research and Applications* (Cham: Springer) pp 317–35

[56] Rogers K R 2006 Recent advances in biosensor techniques for environmental monitoring *Analytica Chimica Acta* **568** 222–31

[57] Senveli S U and Tigli O 2013 Biosensors in the small scale: methods and technology trends *IET Nanobiotechnology* **7** 7–21

[58] Salouti M and Derakhshan F K 2020 Biosensors and nanobiosensors in environmental applications *Biogenic Nano-Particles and their Use in Agro-ecosystems* ed M Ghorbanpour, P Bhargava, A Varma and D Choudhary (Singapore: Springer)

[59] Sharpe M 2003 It's a bug's life: biosensors for environmental monitoring *J Environ Monit* **5** 109–13

[60] Mcgrath S P, Knight B, Killham K, Preston S and Paton G I 1999 Assessment of the toxicity of metals in soils amended with sewage sludge using a chemical speciation technique and a lux-based biosensor *Env. Toxicol. and Chem.* **18** 659–63

[61] Wong E L S, Chow E and Gooding J J 2007 The electrochemical detection of cadmium using surfaceimmobilized DNA *Electrochem. Commun.* **9** 845–49

[62] Lin T-J and Chung M-F 2009 Detection of cadmium by a fiber-optic biosensor based on localized surface plasmon resonance *Biosens. Bioelectron.* **24** 1213–218

[63] Pal P, Bhattacharyay D, Mukhopadhyay A and Sarkar P 2009 The detection of mercury, cadium, and arsenic by the deactivation of urease on rhodinized carbon *Environ. Eng. Sci.* **26** 25–32

[64] Verma N and Singh M A 2006 Bacillus sphaericus based biosensor for monitoring nickel ions in industrial effluents and foods *J. Autom. Methods and Management in Chem.* **2006** 083427

[65] Claude D, Houssemeddine G, Andriy B and Jean-Marc C 2007 Whole cell algal biosensors for urban waters monitoring *Novatech.* **7** 1507–514

[66] Biran R, Babai K, Levcov J and Rishpon R E Z 2000 Online and *in situ* monitoring of environmental pollutants: electrochemical biosensing of cadmium *Environ. Micro.* **2** 285–90

[67] Gruzina T G, Zadorozhnyaya A M, Gutnik G A, Vember V V, Ulberg Z R, Kanyuk N I and Starodub N F 2007 A bacterial multisensor for determination of the contents of heavy metals in water *J. Water Chem. and Technol.* **29** 50–3

[68] Knecht M R and Sethi M 2009 Bio-inspired colorimetric detection of Hg^{2+} and Pb^{2+} heavy metal ions using Au nanoparticles *Analytical and Bioanalytical Chem.* **394** 33–46

[69] Sumner J P, Westerberg N M, Stoddard A K, Hurst T K, Cramer M, Thompson R B, Fierke C A and Kopelman R 2006 DsRed as a highly sensitive, selective, and reversible fluorescence-based biosensor for both Cu^{+} and Cu^{2+} ions *Biosens. Bioelectron.* **21** 1302–08

[70] Ilangovan R, Daniel D, Krastanov A, Zachariah C and Elizabeth R 2006 Enzyme based biosensor for heavy metal ions determination *Biosens. Bioelectron. Equipment.* **20** 184–89

[71] Kwok N-Y, Dongb S and Loa W 2005 An optical biosensor for multi-sample determination of biochemical oxygen demand (BOD) *Sensors & Actuators, B: Chemical.* **110** 289–98

[72] Sode K, Yamazaki T, Lee I, Hanashi T and Tsugawa W 2016 Biocapacitor: a novel principle for biosensors *Biosens. Bioelectron.* **76** 20–8

[73] Moorcroft M J, Davis J and Compton R G 2001 Detection and determination of nitrate and nitrite: a review *Talanta* **54** 785–803

[74] Chen H, Mousty C, Cosnier S, Silveira C, Moura J J G and Almeida M G 2007 Highly sensitive nitrite biosensor based on the electrical wiring of nitrite reductase by [ZnCr-AQS] LDH *Electroanal. Commu.* **9** 2240–45

[75] Rodriguez-Mozaz S, Alda M J L and Barceló D 2006 Biosensors as useful tools for environmental analysis and monitoring *Analytical and Bioanalytical Chem.* **386** 1025–41

[76] Su-Hsia L and Ruey-Shin J 2009 Adsorption of phenol and its derivatives from water using synthetic resins and low-cost natural adsorbents: a review *J. Env. Manage.* **90** 1336–49

[77] Wezel A P and Opperhuizen A 1995 Narcosis due to environmental pollutants in aquatic organisms: residue-based toxicity, mechanisms, and membrane burdens *Crit. Rev. in Toxicology* **25** 255–79

[78] Luong J H T, Male K B and Glennon J D 2008 Biosensor technology: technology push versus market pull *Biotechnol. Advances* **26** 492–500

[79] Zheng Y, Yang C, Pu W and Zhang J 2009 Carbon nanotube-based DNA biosensor for monitoring phenolic pollutants *Microchimica Acta* **166** 21–6

[80] Bagni G, Osella D, Sturchio E and Mascini M 2006 Deoxyribonucleic acid (DNA) biosensors for environmental risk assessment and drug studies *Analytica Chimica Acta* **573–74** 81–9

[81] Parellada J, Narvaez A, Lopez M A, Dominguez E, Fernandez J J, Pavlov V and Katakis I 1998 Amperometric immunosensors and enzyme electrodes for environmental applications *Analytica Chimica Acta* **362** 47–57

[82] Guerra F D, Attia M F, Whitehead D C and Alexis F 2018 Nanotechnology for environmental remediation: materials and applications *Molecules* **23** 1760

[83] Sun J Z, Peter K G, Si R W, Zhai D D, Liao Z H and Sun D Z *et al* 2015 Microbial fuel cell-based biosensors for environmental monitoring: a review *Water Science and Technology* **71** 801–09

[84] Du Z, Li H and Gu T 2007 A state-of-the-art review on microbial fuel cells: a promising technology for wastewater treatment and bioenergy *Biotechnology Advances* **25** 464–82

[85] Gutierrez J C, Amaro F and Martin-Gonzalez A 2015 Heavy metal whole- cell biosensors using eukaryotic microorganisms: an updated critical review *Frontiers in Microbiology* **6** 48

[86] Bănică F-G 2012 *Chemical Sensors and Biosensors: Fundamentals and Applications* (Chichester: Wiley) p 576

[87] Ukhurebor K E, Adetunji C O, Bobadoye A O, Aigbe U O, Onyancha R B, Siloko I U, Emegha J O, Okocha G O and Abiodun I C 2021 Bionanomaterials for biosensor technology *Bionanomaterials: Fundamentals and Biomedical Applications* ed R P Singh and K R B Singh (Bristol: IOP Publishing)

[88] Onyancha R B, Aigbe U O, Ukhurebor K E and Muchiri P W 2021 Facile synthesis and applications of carbon nanotubes in heavy-metal remediation and biomedical fields: a comprehensive review *J. Mol. Struct.* **1238** 130462

[89] Willner M R and Vikesland P J 2018 Nanomaterial enabled sensors for environmental contaminants *J. Nanobiotechnology* **16** 1–16

[90] Duhan J S, Kumar R, Kumar N, Kaur P, Nehra K and Duhan S 2017 Nanotechnology: the new perspective in precision agriculture *Biotechnology Reports* **15** 11–23

[91] Kumar V and Guleria P 2020 Application of DNA-nanosensor for environmental monitoring: recent advances and perspectives *Current Pollution Reports*

[92] Ukhurebor K E, Singh K R B, Nayak V and UK-Eghonghon G 2021 Influence of SARS-CoV-2 pandemic: a review from the climate change perspective *Environmental Science: Processes & Impacts* **23** 1060–78

[93] Ukhurebor K E, Athar H, Adetunji C O, Aigbe U O, Onyancha R B and Abifarin O 2021 Environmental implications of petroleum spillages in the Niger Delta region of Nigeria: a review *J. Env. Manage.* **293** 112872

[94] Neethirajan S, Ragavan V, Weng X and Chand R 2018 Biosensors for sustainable food engineering: challenges and perspectives *Biosensors.* **8** 1–34

www.ingramcontent.com/pod-product-compliance
Ingram Content Group UK Ltd.
Pitfield, Milton Keynes, MK11 3LW, UK
UKHW051903100926
8556IPUK00005B/751

9 780750 338615